international®
AIR POWER
REVIEW

AIRtime Publishing
United States of America • United Kingdom

international® AIR POWER REVIEW

Published quarterly by AIRtime Publishing Inc.
US office: 120 East Avenue, Norwalk, CT 06851
UK office: CAB International Centre, Nosworthy Way,
Wallingford, Oxfordshire, OX10 8DE

© 2002 AIRtime Publishing Inc.
Draken cutaway © Aerospace Publishing Ltd
F-22 cutaway © Mike Badrocke/Aviagraphica
Photos and other illustrations are the copyright
of their respective owners

Softbound Edition ISSN 1473-9917 / ISBN 1-880588-44-7
Hardcover Deluxe Casebound Edition ISBN 1-880588-45-5

Publisher
Mel Williams

Editor
David Donald
e-mail: airpower@btinternet.com

Assistant Editors
John Heathcott, Daniel J. March

Sub Editor
Karen Leverington

US Desk
Tom Kaminski

Russia/CIS Desk
Piotr Butowski, Zaur Eylanbekov
e-mail: zaur@airtimepublishing.com

Europe and Rest of World Desk
John Fricker, Jon Lake

Correspondents
Argentina: Jorge Felix Nuñez Padin
Australia: Nigel Pittaway
Belgium: Dirk Lamarque
Brazil: Claudio Lucchesi
Bulgaria: Alexander Mladenov
Canada: Jeff Rankin-Lowe
France: Henri-Pierre Grolleau
India: Pushpindar Singh
Israel: Shlomo Aloni
Italy: Luigino Caliaro
Japan: Yoshitomo Aoki
Netherlands: Tieme Festner
Romania: Danut Vlad
Spain: Salvador Mafé Huertas
USA: Rick Burgess, Brad Elward, Peter Mersky, Bill Sweetman

Artists
Mike Badrocke, Chris Davey, Zaur Eylanbekov, Keith Fretwell, Malcolm Laird/
Ventura Publishing, Aleksey Mikheyev, Mark Rolfe, Mark Styling, Iain Wyllie

Designer
Zaur Eylanbekov

Controller
Linda DeAngelis

Origination by Universal Graphics and Chroma Graphics, Singapore
Printed in Singapore by KHL Printing

International Air Power Review is published quarterly in two editions
(Softbound and Deluxe Casebound) and is available by subscription or as
single volumes. Please see details opposite.

Acknowledgments
We wish to thank the following for their kind help with the
preparation of this issue:

Paul H. Earnshaw/BAE Systems
Chick Ramey/Boeing
Eric W. Hehs/Lockheed Martin
Jan Jørgensen
Dave W. Russell/Lockheed Martin
Laurie Tardif/Pratt & Whitney
Terry Panopalis
Simon Watson

The author of the HC-85 report would like to express his sincere thanks
to the following for their help and support: Cdr James Iannone, Capt
Danny Bell and Lt Cdr Bill Pevey at HELWINGRES and to Lt Cdr Brian
Fitzsimmons and all at HC-85.

The author of the CACOM-3 feature would like to thank Coronel Romero
of the Relaciones Publicas and Teniente Mauricio Tejedor Medina of
CACOM-3 of the FAC for their generous help and Foto Rudolf, Bogotá,
for the air-to-air photograph.

The editors welcome photographs for possible publication but can
accept no responsibility for loss or damage to unsolicited material.

Subscriptions & Back Volumes

**Readers in the USA, Canada, Central/South America and the rest
of the world (except UK and Europe) please write to:**
  AIRtime Publishing, P.O. Box 5074, Westport, CT 06881, USA
  Tel (203) 838-7979 • Fax (203) 838-7344
  Toll free 1 800 359-3003
  e-mail: airpower@airtimepublishing.com

Readers in the UK & Europe please write to:
  AIRtime Publishing, RAFBFE, P.O. Box 1940,
  RAF Fairford, Gloucestershire GL7 4NA, England
  Tel +44 (0)1285 713456 • Fax +44 (0)1285 713999

**One-year subscription rates (4 quarterly volumes),
inclusive of shipping & handling/postage and packing:**
  Softbound Edition
  USA $59.95, UK £48, Europe EUR 89, Canada Cdn $99,
  Rest of World US $79 (surface) or US $99 (air)
  Deluxe Casebound Edition
  USA $79.95, UK £68, Europe EUR 123, Canada Cdn $132,
  Rest of World US $99 (surface) or US $119 (air)

**Two-year subscription rates (8 quarterly volumes),
inclusive of shipping & handling/postage and packing:**
  Softbound Edition
  USA $112, UK £92, Europe EUR 169, Canada Cdn $187,
  Rest of World US $148 (surface) or US $188 (air)
  Deluxe Casebound Edition
  USA $149, UK £130, Europe EUR 236, Canada Cdn $246,
  Rest of World US $187 (surface) or US $227 (air)

Single-volume/Back Volume Rates by Mail:
  Softbound Edition
  US $16, UK £12, Europe EUR 19.50, Cdn $26 (plus s&h/p&p)
  Deluxe Casebound Edition
  US $20, UK £17, Europe EUR 27.50, Cdn $32 (plus s&h/p&p)

All prices are subject to change without notice.
Canadian residents please add GST. Connecticut residents please add sales tax.

**Shipping and handling (postage and packing) rates
for back volume/non-subscription orders are as follows:**

	USA	UK	Europe	Canada	ROW (surface)	ROW (air)
1 item	$4.50	£4	EUR 8	Cdn $7.50	US $8	US $16
2 items	$6.50	£6	EUR 11.50	Cdn $11	US $12	US $27
3 items	$8.50	£8	EUR 14.50	Cdn $14	US $16	US $36
4 items	$10	£10	EUR 17.50	Cdn $16.50	US $19	US $46
5 items	$11.50	£12	EUR 20.50	Cdn $19	US $23	US $52
6 or more	$13	£13	EUR 23.50	Cdn $21.50	US $25	US $59

Volume Five
Summer 2002

CONTENTS

MAJOR FEATURES PLANNED FOR VOLUME SIX
Focus Aircraft: Tupolev Tu-95/142 'Bear', **Warplane Classic:** Republic F-105 Thunderchief, **Technical Briefing:** General Dynamics F-111 in Australia, **Photo Feature:** Uruguay, **Air Power Analysis:** Greece, **Special Feature:** Brazilian navy, **Air Combat:** P-51 Mustang in Korea, **Variant File:** Boeing B-47 Stratojet

PROGRAMME UPDATE

Lockheed Martin F-35

Following the historic contract award in October 2001 to Lockheed Martin to develop its Model 235 design for the Joint Strike Fighter requirement, work has progressed to the System Development and Demonstration (SDD) phase. The first milestone, ASRR, was passed on 6 February 2002, on schedule. In late 2002 the overall shape of the aircraft should be frozen, followed by a Preliminary Design Review in early 2003. A year later the first of three Critical Design Reviews will be undertaken, the second following in late 2004 and the final CDR in mid-2005.

By this time the first of 14 flying SDD aircraft should be virtually complete, ready for a scheduled October 2005 first flight. It will be an F-35A CTOL aircraft. The first STOVL F-35B is due to fly in early 2006, and the first CV F-35C before the end of the year. The 14 SDD aircraft (five F-35As, five F-35Bs and four F-35Cs) will be supported by eight non-flying airframes.

Details have also been given of the Low-Rate Initial Production (LRIP) aircraft which, under current plans, will cover 465 aircraft in six batches, after which full-rate production will take over. The LRIP batches are broken down as follows:

LRIP 1: from late spring 2006 to early 2009. 10 aircraft (6 CTOL, 4 STOVL)
LRIP 2: from beginning 2007 to end 2009. 22 aircraft (14 CTOL, 8 STOVL)
LRIP 3: from beginning 2008 to end 2010. 54 aircraft (20 CTOL, 20 STOVL, 9 CV, 5 UK)
LRIP 4: from beginning 2009 to end 2011. 91 aircraft (30 CTOL, 32 STOVL, 20 CV, 9 UK)
LRIP 5: from beginning 2010 to end 2012. 120 aircraft (44 CTOL, 32 STOVL, 32 CV, 12 UK)
LRIP 6: from beginning 2011 to end 2013. 168 aircraft (72 CTOL, 36 STOVL, 48 CV, 12 UK)

According to current plans, the F-35A CTOL variant for the US Air Force will achieve IOC a year after the USMC's STOVL F-35B. This highlights the urgency of the Marine Corps to field a replacement for its stretched AV-8B Harrier II fleet. As both variants share greater commonality than with the Navy's F-35C, test work performed by the F-35A will speed the F-35B programme.

PROJECT DEVELOPMENT

Europe

German A400M decision

Although the Bundestag confirmed German plans on 24 January to acquire 73 Airbus A400M military transport aircraft for the national armed forces, as its share of the seven-nation programme, it proved politically impossible to gain approval for little more than half the required funding. Full Bundestag financial backing of Germany's national A400M commitments before the 31 January deadline was regarded as essential for programme continuation. Initial contracts for all but 73 of the required 196 aircraft had been signed by Germany's other partners, through OCCAR, the European armaments procurement agency, with the Airbus Military Company (AMC) last December.

On 29 January, Defence Minister Rudolf Scharping admitted that, despite parliamentary agreement, Germany was still unable to provide legal guarantees for its full $8.3 billion commitment to buy 73 A400Ms. This followed opposition party legal action, claiming government contravention of constitutional budget procedures in committing the next parliament to paying the balance of the full German A400 programme in 2003. The opposition parties also challenged AMC consortium demands for penalty payments acceptance, equivalent to the cost of 15 aircraft, if 4.4 billion euros in funding for the Luftwaffe's remaining 33 A400Ms is not forthcoming from the next German government by 2003.

German budget committee A400 funding approval announced on 20 March therefore remained unchanged at only 5.1 billion euros ($4.45 billion), for procurement of 40 aircraft. AMC maintains that without firm guarantees from Germany for the remainder of its requirement, originally due by 31 January, agreed programme costings for planned purchases of a minimum 196 A400Ms are no longer valid. Scharping insisted, however, that Germany had made clear its intention to buy 73 A400s, giving its partners firm grounds for a programme go-ahead.

Meanwhile, original turboprop powerplant plans for the A400M, based on the proposed Aero Propulsion Alliance (APA) TP400, incorporating the SNECMA M88 high-pressure core with Rolls-Royce's three-shaft technology, were abandoned because of excessive design weight and fuel consumption. Formerly known as Turboprop International, the APA consortium has also lost FiatAvio, following Italy's withdrawal from the A400M programme, as well as Techspace Aero in Belgium. Its surviving members comprise Rolls-Royce, ITP, MTU and SNECMA, who are working on a new three-shaft core for the required 7457-kW (10,000-shp) powerplant.

India

LCA production launched

A major milestone in India's two-decade indigenous Light Combat Aircraft (LCA) development programme was reached in February, with Hindustan Aeronautics Ltd (HAL) chairman N.R. Mohanty's announcement of its first production order. This

The first Eurofighter Instrumented Production Aircraft (IPA2) took to the air at Alenia's Caselle plant on 5 April 2002 (above), piloted on its 25-minute inaugural flight by Maurizio Cheli. Two more flights were mounted later in the day. EADS-Deutschland's IPA3 followed from Manching on 8 April, flown for 31 minutes by Chris Worning and Lt Col Robert Hierl, and the UK-assembled IPA1 flew its 26-minute first flight on 15 April from Warton, with Keith Hartley and Paul Hopkins aboard (right). Meanwhile, BAE two-seater DA4 accomplished Eurofighter's first guided AMRAAM launch, with a successful hit against a Mirach drone target over the Benbecula range off the Western Isles of Scotland.

As can be seen, initial aircraft will be USAF or USMC aircraft, the first US Navy and UK machines being included in LRIP 3. The first delivery of an LRIP aircraft is expected in 2008, and the first for the UK should be handed over in 2010. The early LRIP machines will join the flight test aircraft in clearing the F-35 for service, including operational test and evaluation work (OT&E) which will continue after IOC (Initial Operating Capability) has been achieved. IOC is set for the three variants as: late 2010 for the USMC/STOVL F-35B, late 2011 for the USAF/CTOL F-35A and late 2012 for the USN/CV F-35C.

Meanwhile, on 7 February 2002 Canada joined the JSF team as the second international partner. Its stake of $150 million is for Level 3 participation, providing significant work in the future for Canadian industry. On 9 February the Netherlands announced it would join the programme, although a parliamentary ratification may be required (subsequently deferred due to the Dutch cabinet resignation in April), and in March Turkey agreed to join SDD as a Level 3 partner. In an attempt to repeat the F-16 'Sale of the Century', Lockheed Martin is pressing Belgium, Denmark and Norway to join the Netherlands in the F-35 programme at an early stage. Denmark and Norway both have outstanding fighter requirements under study, while Belgium is looking for an F-16 replacement from around 2015. Australia and Italy are also studying F-35 involvement, with decisions expected in the first half of 2002.

Bell-Boeing V-22 Osprey

Bell-Boeing Osprey has received a $63.5 million contract modification for Phase I of the MV-22 Block Upgrade Program. The Osprey, which has been grounded since December 2000, is scheduled to resume developmental flight tests in late April 2002 when the MV-22B enters an 18-month developmental flight test period at the Naval Air Warfare Center Aircraft Division at NAS Patuxent River, Maryland. Two engineering, manufacturing and development (EMD), and three low-rate initial production (LRIP) Ospreys will participate, and each will be equipped with improved engine nacelles featuring rerouted hydraulic/electric lines and updated software. The developmental flight tests, which will validate the engineering and software changes, will include dynamic shipboard compatibility, formation flying, combat manoeuvrability, and low-speed hovering and landing conditions. The first redesigned production aircraft will be delivered to the Marines in late 2003 and the second phase of flight testing, including a second operational evaluation (OPEVAL), will be conducted between late 2004 and spring of 2005. The USAF is expected to resume flight tests of the CV-22 at Edwards AFB, California, in July 2002, using two EMD aircraft that are configured with special operations equipment. The USAF expects that the CV-22B OPEVAL will begin in 2006.

Despite the Osprey's grounding, the contractor team has continued to build new aircraft, which have been placed in short-term storage at the Bell Helicopter facility in Amarillo, Texas. In fact, the contractor team recently received a $770 million contract to build 11 additional MV-22Bs. Bell-Boeing Joint Program Office recently awarded a $5.7 million contract covering the long-term preservation and storage of 19 MV-22Bs. The aircraft will remain in storage until the Block 'A' configuration changes are developed and certified for service.

Boeing Sikorsky RAH-66

The initial prototype of the Boeing Sikorsky RAH-66 Comanche completed its final scheduled test flight at Sikorsky's test facility in West Palm Beach, Florida, on 30 January 2001. Rolled out in May 1995 at Sikorsky's Bridgeport, Connecticut facility, the prototype made its first flight on 4 January 1996 in West Palm Beach. Primarily tasked with evaluating the Comanche's flight control software and handling qualities, the aircraft conducted 318 sorties and accumulated 387.1 flight hours over a six-year period. The aircraft will now serve as a backup for the second prototype, which is scheduled to resume flight-testing during the spring of 2002. This testing will focus on validating the Comanche mission equipment package (MEP). The aircraft has already logged 93 flights and accumulated 103.5 flight hours. Delivery to combat units is expected in 2009.

involves only eight LCAs, for IAF delivery from 2006. Following agreement on security, military information and greater technology co-operation exchanges, signed on 18 January in Washington by Indian Defence Minister George Fernandes, production LCAs will now be powered by F404-GE-F2J3 turbofans.

These FADEC-equipped engines were originally intended to power only the first two LCA prototypes, and replaced in the next four development and subsequent aircraft by indigenous GTRE Kaveri turbofans. Bench-running and flight development of this engine below the fuselage of a Tupolev Tu-16 are now in progress, but it remains a long-term project, with no LCA installations currently planned.

Initial flight trials of the technology demonstrator prototype LCA (TD1), which features an unstable configuration with a digital fly-by-wire flight-control system, almost completely composite airframe, glass cockpit and advanced sensors, totalled a dozen subsonic sorties from January 2001.

Transonic exploration is beginning with TD2, expected to fly in April, shortly followed by prototype vehicle PV3, originally scheduled as the first Kaveri-engined version. Work is also continuing on two-seat operational training and naval versions of the LCA.

Joint Indo-Russian programme
India's June 2001 military and industrial agreements for the Su-30MKI programme includes planned joint development and production with IAPO and Ilyushin of a new multi-role twin PS90- or Rolls-Royce BR 710-powered rear-loading high-wing tactical transport aircraft (MTA), for both military and civil use. India has a 47 per cent investment in the $350-million 10-year MTA programme, which is derived from Ilyushin's Il-214T project.

Design and development work is now beginning in both countries on the 15/20-tonne payload MTA, for a planned start of flight development in 2005, international marketing with a proposed fly-away unit cost of only

Mitsubishi's F-2 fighter programme is progressing swiftly. Here F-2B no. 8 lands at Nagoya (Komaki AB) on 15 February 2002 after the final inspection flight by JASDF pilots.

$12-15 million from 2007, and production status by 2010. With his Indian partners, Ilyushin Aviation Complex General Director Viktor Livanov is estimating the receipt of orders for 100-200 aircraft, mainly for their armed forces.

Joint development with India is also reportedly being planned by Russia of a new-generation combat aircraft, which is currently scheduled to make its first flight by early 2006.

Italy

C-27J receives certification
The Alenia/Lockheed Martin C-27J Spartan recently received its military type certification from the Direzione Generale Armamenti Aeronautici (DGAA), the Italian Ministry of

Defence military certification authority. As part of the certification testing, paradrop capabilities and the ability to take off and land on short and unprepared runways were demonstrated. The certification is required for any aircraft that operates in a military environment. The C-27J flight test programme began in September 1999 and three test aircraft conducted 445 flights, accumulating 793 flight test hours and demonstrated 4,600 flight test points.

United Kingdom

Bombardier delivers ASTOR
On 31 January 2002 Bombardier delivered the first of five Global Express aircraft to Raytheon Systems' facility in Greenville, Texas, where the aircraft will receive structural and systems modifications associated with the Royal Air Force Airborne Stand-Off Radar (ASTOR) programme. The initial aircraft, which carries the serial ZJ690, is scheduled for completion during the

Having for many years operated rather anonymously, the Turkish Transall fleet began to adopt large squadron markings in 2001, as demonstrated by this 221 Filo aircraft.

During December 2001/January 2002 NASA tested its new Propulsion Flight Test Fixture (PFTF) beneath this F-15B.

third quarter of 2003, and delivery to the Ministry of Defence will take place in 2004. Raytheon Systems Limited (RSL) will modify four further airframes in the United Kingdom. Raytheon had previously equipped a Global Express development aircraft with aerodynamic modifications associated with the ASTOR configuration. This aerodynamic prototype has been flying since August 2002 and has accumulated nearly 100 hours.

United States

USAF tankers

Although EADS and its Airbus subsidiary would like to supply the USAF with as many as 100 tanker/cargo aircraft, the service rejected the offer recently. The US Congress recently authorised the service to negotiate the possible lease of modified Boeing 767 airliners as replacements for the elderly KC-135 tanker, and Airbus countered with an offer to provide a modified version of the A330 airliner. The USAF's rejection of the A330 cited a higher-risk technical approach, a less preferred financial arrangement and specifically indicated the A330 provided insufficient volume for fuel, and would not provide the versatility required for worldwide missions. The estimated cost to lease 100 modified 767s is $20 billion and Airbus believed it could save the service as much as $8 billion over the life of the contract. According to Boeing, the 767 airframes will be built in Everett, Washington, while military modifications will be carried out at the company's plant in Wichita, Kansas.

F-15E production continues

The first of 10 new F-15Es for the USAF recently entered the final assembly stage at Boeing's St Louis facility. In this production stage the major assemblies of the aircraft, which include the forward, centre and aft sections, as well as the wings, are spliced or assembled in preparation for the installation of system equipment. Known as E227, the aircraft will be the 227th F-15E for the USAF when it is delivered in June 2002. The USAF will accept the aircraft mid-2004.

UPGRADES AND MODIFICATIONS

Australia

AP-3C deliveries begin

Landmark progress with the RAAF's Project Air 5276 programme for mission systems upgrades of 17 RAAF Lockheed Orion Update II maritime patrol aircraft was achieved late last year from delivery of the first enhanced AP-3C from Raytheon Australia at Avalon. This followed arrival in Australia in late 2000 of the prototype AP-3C from Raytheon's Greenville, Texas, facility.

New digital systems in the initial AP-3C upgrade centre on an Elta EL/M 2022A(V)3 imaging multi-mode surveillance radar, dual Honeywell H-423 RLG INS and an MAGR 300 GPS, Unisys DDC-060 data management sub-system, and Computing Devices of Canada UYS-503 acoustic processor. Also fitted are an AQS-504 magnetic anomaly detector, Elta EL/L-8300 ELINT/electronic support measures, and Magnavox UHF/VHF secure radios. Phase 5 electro-optic suite replacement is also planned in 2004-05.

Hornet update progress

The Royal Australian Air Force has begun the second phase of its Hornet Upgrade (HUG) program. This phase of the HUG will replace the fighter's AN/APG-65 radar with the AN/APG-73 developed for the F/A-18E/F. The first production aircraft to be equipped with the new radar was inducted into a modification facility operated by Boeing Australia at RAAF Williamtown in New South Wales on 5 March 2002. Two aircraft have already been fitted with the new radar for validation and verification purposes, and the last of 71 aircraft will be completed in mid-2003. The first phase of the modification programme, which provided improved communication and navigation capabilities, will be completed in June 2002. These modifications included the installation of a sixth avionics multiplex bus, new mission computers, AN/ARC-210 secure radios, a global positioning system (GPS), enhanced Identification Friend or Foe (IFF) system and armament updates.

Bulgaria

RSK MiG 'Fulcrum' upgrade

Having abandoned surplus or new fighter procurement because of budget limitations, the Bulgarian government selected Russia's RSK MiG group in January to upgrade its existing MiG-29s. This will help Bulgaria meet its NATO and EU defence commitments. At least six of Bulgaria's 17 MiG-29s and four two-seat MiG-29UB combat trainers are expected to be refurbished and upgraded by RSK MiG from a $67 million contract by the year-end, in conjunction with the Benkovski aircraft MRO facility in Plovdiv. The remaining MiG-29s will be similarly upgraded by 2004, to become NATO-interoperable, with extended service lives through 2012.

France

AWACS upgrade

The French government has awarded Boeing a $133 million contract to upgrade the radar system of four E-3F airborne warning and control system (AWACS) aircraft operated by the Armée de l'Air (AdA). The update, which will be the most significant since the aircraft entered service in the early 1990s, will provide the French with the same surveillance capability as their NATO allies. The radar system improvement programme (RSIP) improves the AN/APY-2 radar by increasing its sensitivity, updating the electronic counter-countermeasures capability and increasing its reliability. Boeing will ship the modification kits to Air France Industries in mid-2004 and the subcontractor will modify the aircraft at its facility in Le Bourget, France. Boeing will flight-test the updated aircraft at the French AWACS main operating base in Avord, France. The installation will be completed during 2006. RSIP modifications have already been carried out on the E-3s operated by the USAF, Royal Air Force and NATO AWACS fleets.

South Africa

Upgraded C-130s into service

Formal acceptance took place at Waterkloof Air Base, near Pretoria, on 18 February, of the first three SAAF Lockheed C-130BZ tactical transports to receive glass-cockpit avionics upgrades by UK's Marshall Aerospace. Redelivered to South Africa in the second half of 2001, the three Hercules were handed-over to the South African government's Armaments Corporation (Armscor) and the SAAF after Marshall had designed, installed and certificated a complete Sextant (now Thales) Topdeck digital avionics package.

Coupled with a structural refurbishment, the upgrades, undertaken jointly by Marshall and South Africa's Denel Aviation, extend the C-130 operating lives by up to 20 years. Denel is incorporating similar upgrades in a further six SAAF C-130s in South Africa, where the first is due for completion in mid-2002. Seven SAAF C-130Bs, bought new and operated since late 1962, have been supplemented since 1997 by two ex-USN C-130Fs and three ex-USAF C-130Bs, transferred by the US Department of Defense from AMARC storage.

United Kingdom

Re-engined Jaguar

Formal delivery was made earlier this year by BAE Systems to the UK Defence Logistics Organisation of the first of 62 RAF Jaguar GR.Mk 3A ground-attack fighters upgraded by installation of uprated Rolls-Royce Adour Mk 106 turbofans, from a £105 million August 1998 contract. After completion of extensive trials programmes, the first uprated GR.Mk 3A was flown from BAE Systems' factory airfield at Warton on 24 January to the RAF's Jaguar operating base at Coltishall, Norfolk, for the hand-over.

While offering a small increase in take-off power from the earlier 34.14 kN (7,900-lb) thrust Rolls-Royce/Turbomeca Adour Mk 104 afterburning turbofans to compensate for increased Jaguar operating weights, replacement by the new 36.67-kN (8,245-lb) Mk 106s was aimed mainly at improving engine reliability and maintainability.

Displayed under a Tornado GR.Mk 4A of No. II Squadron, RAF, is the Raptor reconnaissance pod which gives the RAF's Tornados a medium-altitude dual-band capability, with datalink. Eight pods have been bought, with two ground stations.

Smart weapons under test at Edwards have included the AGM-154 JSOW (above), two of which were launched simultaneously at two separate targets from a B-52H on 8 February 2002, and the Wind Corrected Munition Dispenser (right), which was dropped from an F-15E for the first time on 25 March 2002 over the China Lake range.

United States

ICAP III testing
The first of two EA-6Bs modified as part of the Improved Capability III (ICAP III) programme has begun system performance testing with the Naval Air Warfare Center Aircraft Division (NAWC-AD) at NAS Patuxent River, Maryland. The testing, which involves both ground and flight segments, began when BuNo 159909/SD-536 was placed in the centre's anechoic chamber for electronic compatibility testing. Initial flight tests will be conducted at Patuxent River between April and July, and in August the aircraft will undergo further tests at Point Mugu, California, where the performance and accuracy of the AN/ALQ-128 electronic warfare receiver will be evaluated. The updated aircraft will undergo an operational assessment in September/October 2002. A low-rate initial production (LRIP) decision is expected in January 2003, and the aircraft will receive a Technical Evaluation (TECHEVAL) and a full operational evaluation (OPEVAL) between late 2002 and mid-2003. The ICAP III system is expected to attain initial operational capability (IOC) during 2005. The Navy is expected to order 8-15 ICAP III systems under the initial LRIP contract, and plans call for the conversion of 80-100 ICAP II Prowlers.

C-5 update continues
Under the terms of a $126 million contract, General Electric will deliver 14 CF6-80C2 propulsion systems to Lockheed Martin for the company's C-5 Galaxy reliability and enhancement and re-engining program (RERP). The engines will be installed on three C-5B Galaxies that will support the system development and demonstration (SDD) phase of the RERP. Each propulsion system includes the CF6-80C2 engine, thrust reverser and nacelle. Flight tests associated with this phase of the RERP, which is part of a comprehensive modernisation plan for the USAF's C-5 Galaxy fleet, are scheduled to begin in 2005. There is a further option for four additional propulsion systems to equip a fourth aircraft. Delivery of the propulsion systems is scheduled to begin in 2004

and a further four systems could be purchased under an option between the two contractors. At the conclusion of the six-year SDD phase the programme will enter a production phase that could require the manufacture of up to 490 additional CF6 propulsion systems. The RERP is designed to keep the C-5 fleet operationally viable until at least 2040.

New radar for Hawkeye
The US Navy has awarded Northrop Grumman a $49 million Pre-Systems Development and Demonstration (Pre-SD&D) contract for the E-2C Hawkeye Radar Modernization Program (RMP). The 12-month RMP Pre-SD&D effort will define the physical architecture of the next-generation E-2C mission system, produce the preliminary weapon system specification and provide associated programme plans. The development programme will integrate a number of new capabilities into the aircraft's new airborne early warning and control system. The result will be a solid-state, electronically steered UHF radar and the next-generation Hawkeye will also have theatre missile defence capabilities. Further improvements will include a new communications suite, new generators, improved identification friend or foe system, and an updated mission computer and software. In addition, a tactical cockpit will give the co-pilot the capability to function as a fourth mission system operator. The full SD&D programme contract could eventually lead to a multi billion-dollar programme for production of a next-generation Advanced Hawkeye.

Rafael has released this photo of a Chilean Navy MBB BO 105 helicopter equipped with a He-lite electro-optical payload. Derived from the Litening targeting and navigation pod, He-lite is a 117-lb (53-kg) payload comprising a TV camera (either b/w or colour) with a x20 zoom, a third-generation Forward Looking Infra Red (FLIR) sensor with three Field Of View (FOV) options, a Laser Designator (LD) with a range of up to 12 miles (20 km) and a Laser Range Finder (LRF) with an accuracy of 16 ft (5 m) at ranges of up to 12 miles (20 km). A laser spot tracker is optional, as is a datalink to a ground station.

Army special ops helicopters
Rolls-Royce Allison recently completed certification testing of the Model 250-C30R/3M turboshaft engine, which will be retrofitted to the US Army's Special Operations Command (USASOCOM) fleet of AH/MH-6 multi-role helicopters. The engine is developed from the 650-shp (485-kW) commercial Model 250-C30R/3 engine installed in the TH-67A Creek training helicopter, which is based on Bell's commercial model 206B.

The installation of the new engine is part of the Mission-Enhanced Little Bird [MELB] programme that will also provide the aircraft with a six-bladed main rotor, a canted four-bladed tail rotor, and uprated transmission system.

Digital cockpit for Hurons
The USAF recently announced plans to equip its C-12C/D/F transports with a new digital flight instrumentation system. The service's fleet includes 27 aircraft scattered around the globe primarily in support of the Defense Intelligence Agency and Defense Security Cooperation Agency, which use the aircraft for embassy support. The digital modifications to the cockpit will replace the old analog system with three to five multi-function displays, depending on the avionics suite selected. Although assigned to the Air Force Materiel Command (AFMC), the C-12 fleet is maintained using contract logistic support (CLS). The CLS contractor is responsible for engineering and programme management, operational safety, suitability

and effectiveness of the aircraft, and will likely also be responsible for modifying the Hurons at a number of facilities around the world. As part of the modification programme the aircraft will also be equipped with an enhanced ground proximity warning system (EGPWS) and traffic collision avoidance system (TCAS). The US Army has already carried out similar modifications on most of its C-12 fleet.

UH-1Y rolled out
Bell Helicopter rolled out the first engineering, manufacturing and development (EMD) UH-1Y destined for the US Marine Corps on 13 December 2001 at the company's Flight Research Center at Arlington Municipal Airport, Texas. The aircraft, known as 'Yankee 1', subsequently made its first flight on 20 December 2001 and the joint Bell/USMC crew conducted in-ground-effect hover manoeuvres. Modified to the new configuration as part of the H-1 Upgrade Program, 'Yankee 1' is scheduled to arrive at NAS Patuxent River, Maryland, for flight testing in May 2002. The Marines plan to spend approximately $4.5 billion over the next 10-12 years to upgrade 100 UH/HH-1N to UH-1Ys, and 180 AH-1Ws to AH-1Z configuration. The programme is about a year behind schedule and the estimated cost of rebuilding 280 helicopters has increased from $4 billion in the past year. Over the next few months Bell will deliver two more AH-1Zs and a second UH-1Y, and flight testing will continue into 2004, when operational evaluation testing will begin.

Europe's last 'Floggers'

In mid-December 2001 the Bulgarian air force (BVVS) decided to retain up to 12 MiG-23s – a mix of MLD and MF single-seaters and UB two-seaters. Under the Armed Forces Plan 2004, announced in 1999, these aircraft were due to be retired in 2001, but won a three-year reprieve and will serve until at least 2004. According to Minister of Defence Nikolay Svinarov, it was judged "unwise" to withdraw the still-useful fighter and close down the 1st Fighter Airbase at Dobroslavtzy, strategically located in the outskirts of the Bulgarian capital, Sofia.

Bord 204, a MiG-23MLD (export), taxis for a training sortie from its base at Dobroslavtzy. Air defence of the capital, Sofia, remains the 1st FAB's prime mission.

Until late 2000 the BVVS also operated MiG-23BN attack aircraft, which served with the 25th Fighter-bomber Airbase at Theshnnigirovo. A lack of airworthy aircraft put an end to the operational capability of this unit, although flying at the base continued using a pair of MiG-23UB two-seaters until September 2001, at which point the MiG-23BN was officially withdrawn from use.

Alexander Mladenov

PROCUREMENT AND DELIVERIES

Australia

Tiger contract signed
A \$A1.3 billion (\$672 million) defini-tive contract was signed on 21 December by the Australian Defence Department in Canberra with Eurocopter for procurement of 22 Tiger armed reconnaissance heli-copters. December 2001 also marked the 10th and last guided launch of MBDA's TriGAT-LP long-range fire-and-forget anti-tank missile from a French Tiger, to complete its heli-copter qualification trials.

Undertaken at the Captieux firing ranges in south-west France, nine of the 10 launches were completely successful, allowing qualification procedures to be completed later this summer, and delivery of the first production and combat-ready TriGAT-LPs for German army UHT (Unterstützungshubschrauber) Tigers. The first of these are due for delivery to the Heeresflieger in December. ALAT in France is due to receive its first Tiger HAP in July 2003. Each country will allocate 14 Tigers and

about 160 personnel to the joint Tiger training academy (EFA) now being established at Le Luc-en-Provence, in southern France.

RAAF orders new transports
The Australia Department of Defence has announced plans to lease two new Boeing 737s and three Bombardier Challenger 604s for use as government transports. The aircraft will be leased for a period of 12 years at a cost of \$255 million. Delivery of the first 737 will take place in May 2002 and the second aircraft will likely arrive in June or July, with the Challengers following later in the year.

Gripen proposed for Air 6000
Gripen International was among the first contenders to respond, on 1 February, to RAAF Requests for Information (RFI) for new aerospace combat capability within its Air 6000 project, issued last year. Air 6000 is considering various solutions, includ-ing manned and unmanned systems, in current concept development stud-ies for its next-generation combat aircraft. These will replace its current

Boeing F/A-18A/B Hornet fighters and GD F/RF-111A/C/G long-range strike/reconnaissance aircraft from around 2012, for estimated costs of up to \$20 billion.

Austria

US bids cheaper fighter package
Having lost out to the Gripen for new fighter requirements in the Czech Republic and Hungary, the US govern-ment and Lockheed Martin redoubled their promotion of surplus and new F-16s in other key Central European countries, including Austria, Poland and Slovakia. From the original \$1.74 billion package of up to 30 Block 50/52 F-16C/Ds offered to Austria last November, the US Commerce Department and Defence Security Co-operation Agency (DSCA), together with Lockheed Martin, came up with a much cheaper bid on 18 March.

The revised \$1 billion US proposal is based on 30 surplus ex-USAF F100-PW-220-powered F-16A/Bs, incorpo-rating both mid-life avionics upgrades and 'Falcon Up' airframe life-extension structural reinforcements. While considerably cheaper than the leading Gripen bid, the new US proposal might offer less scope to meet Austrian industrial offset requirements for a massive 200 per cent of the over-all contract value. An Austrian fighter decision is expected this summer.

In March 2002 F-14B squadrons VF-11 (illustrated) and VF-143 began combat operations over Afghanistan with the GBU-31 JDAM GPS-guided bomb, flying with CVW-7 aboard John F. Kennedy in the Arabian Sea. The final JDAM clearance and training work for the F-14 had been performed in early February at the China Lake ranges. A VF-11 crew conducted the first combat drop of a GBU-31 from a Tomcat on 12 March.

Bahrain

Primary trainers acquired
While continuing to rely on out-sourced training for its pilots, the Bahrain Emiri air force (BEAF) has ordered three T67M260 Firefly primary trainers, powered by 195-kW (260-hp) Textron Lycoming AEIO-540 piston engines, from the UK's Slingsby Aviation. Delivery is scheduled by the year-end.

Chile

F-16 contract finalised
Selection of the F-16 for its next-generation combat aircraft require-ments was confirmed on 1 February by the Chilean air force, with govern-ment signature of a letter of offer and acceptance (LOA) for six F110-GE-129-powered Block 50 F-16Cs and four two-seat F-16Ds. Chile's \$500 million Foreign Military Sales (FMS) agreement with the US government will include a USAF order for some \$400 million with prime aircraft contractor Lockheed Martin, for mid-2005 to 2006 delivery. Their GE engines are being procured under discrete commercial arrangements. Chile thus becomes the 22nd F-16 customer nation.

Dominican Republic

A-37B replacement order
The diminutive Dominican air force (FAD) became launch export customer for the single-turboprop EMBRAER EMB-314 Super Tucano or ALX light attack aircraft, from an August 2001 order for 10. Equipped with advanced Elbit digital avionics, these cost about \$5.5 million each, fly-away, and will replace four to six ageing Cessna A-37Bs light ground-attack aircraft currently operated by the FAD. They follow a \$380 million Brazilian air force (FAB) contract placed in mid-2001 for 25 single-seat EMBRAER A-29 Super Tucano ALXs and 51 two-seat AT-29 versions, plus 23 more options. Developed from EMBRAER's EMB-312 Tucano advanced turboprop trainer, with some 700 world-wide sales to date, the ALXs for Brazil and Dominica will be delivered between December 2003 and August 2006. Some two-thirds of the FAB's ALXs will be used for Amazonian (SIVAM) operations, and the remainder for advanced and tactical training.

Estonia

Ravens ordered
The Estonian Air Force has ordered four Robinson R44 Raven light heli-copters and will use the aircraft in training, liaison and observation roles. Delivery of the first of the four-seat aircraft is expected in April 2002.

France

New transports acquired
Delivery took place on 8 February to the French air force (AdA) of the first of two Airbus A319CJ corporate jets on order for government transport. With 50-seat interiors, the A319CJs

An Aéronavale E-2C Hawkeye from Flottille 4F prepares to launch from USS John C. Stennis *in March 2002. Both* Stennis *and the Hawkeye's 'homeplate' – Charles de Gaulle – were sailing in the Arabian Sea during combat operations over Afghanistan.*

have a maximum unrefuelled range of up to 10000 km (5,400 nm).

Two additional EADS-CASA CN-235 twin-turboprop tactical transports are scheduled for delivery later this year, as a follow-on to initial AdA procurement of 15 Series 100 versions, from 1991. The new French contract also includes options for a further three CN-235s.

Germany

More frontier patrol helicopters
Further re-equipment of the helicopter fleet of Germany's BGS (Bundesgrenzschutz) frontier patrol service is planned from a 68 million euro ($59.3 million) February order with Eurocopter for 11 liaison EC135s and two larger transport EC 155s, for delivery from 2002. These will follow initial orders in 1997 for 11 EC135s and 16 EC155s, to continue replacing SA 318 Alouette IIs and various Bell helicopters.

Greece

C-27J selected for MRTA
After competitive evaluation against the CN-235 and EADS CASA C-295, selection was announced in Athens on 1 March of the twin-turboprop Alenia/Lockheed Martin C-27J Spartan for its new medium-range transport aircraft (MRTA) requirement. HAF procurement is planned of 12 C-27Js, plus three options, from a $200 million Lockheed Martin contract.

Rolls-Royce will supply 24 to 30 3457-kW (4,637-shp) AE 2100 D2 powerplants, including Dowty R-931 six-bladed composite propellers for the HAF's C-27Js, from a $60 million contract. As an uprated Alenia G222 with enhanced performance, the C-27J also incorporates the C-130J's digital 1553B-based advanced avionics in an integrated glass cockpit suite, and completely new systems.

HAF air transport resources are also being expanded by acquisition from Lockheed Martin of two ex-Italian air force C-130Hs.

India

Su-30MKI deliveries delayed
Although the first examples of India's initial 32 Sukhoi Su-30MKI multi-role vectored-thrust combat aircraft were displayed last August by the Irkutsk Aircraft Production Organisation (IAPO) in Moscow, their delivery to the IAF was still awaited in late spring this year. All are due to arrive by

2003, to be followed in 2004 by the first of 140 more license-built by HAL at its Nasik plant.

The Rs200 billion ($4.16 billion) Su-30MKI programme will include progressive HAL production of every component, including Lyulka AL-31FP turbofans and their thrust-vectoring nozzles. HAL will also be involved in producing their avionics, initially supplied from French, Israeli and South African sources. Output is expected to build up to 10-12 aircraft per year, for completion by 2017.

The IAF's Vision 2020 long-term defence plan envisages the operation of at least 10 Su-30 squadrons, from a total of 55 first-line units. Force-multiplying strategic support for IAF operational units will also be supplied by six Ilyushin Il-78M 'Midas' tanker aircraft, reportedly being acquired as surplus from Uzbekistan. It is claimed that the Il-78Ms will confer a capability for the IAF's already long-range fully-armed Su-30MKIs to fly continuously for over 10 hours, with two aerial refuellings, on sorties covering a combat radius of more than 4500 km (2,428 nm).

Ireland

S-92 launch order challenged
Legal challenges were issued by Eurocopter in March, following Irish Air Corps selection in February of the Sikorsky S-92 helicopter in preference to the EC 725 Cougar Mk 2+, for SAR and military transport missions. A $62 million contract, including support services, was then being finalised for three S-92s, plus options on two more. Powered by twin GE CT7-8 turboshaft engines, the S-92 won out over the EH 101, having lost to the Merlin in the Nordic Standard Helicopter and Portuguese military SAR/transport programmes.

Eurocopter sought re-opening of the bidding because of alleged flaws and irregularities claimed in the selection process. S-92 development flying is now well advanced, with one static and four flight-test prototypes. Initial deliveries are planned to Canadian Cougar Helicopters, from the S-92's current civil orders for 18, in 2006.

Israel

IDF/AF F-16I options taken up
Following last September's signature by Israel of a US letter of offer and acceptance (LOA), Foreign Military Sales (FMS) contract finalisation was announced on 19 December by

Farewell, 'Mighty Wessex'

With a series of dinners, flypasts and other commemorations, the RAF's Support Helicopter Force bade farewell to one of its long-standing stalwarts at the end of March 2002. The Westland Wessex HC.Mk 2 entered service in 1963 and became the backbone of the RAF's transport helicopter force until more modern types such as the Puma and Chinook entered service. In 1969 No. 72 Squadron sent a detachment of Wessexes to Northern Ireland to assist the army and police force in anti-terrorist operations, and the whole squadron moved to RAF Aldergrove in 1981. The Wessex notched up 33 years of continuous operational commitments in the province, No. 72 Squadron claiming this as a record for the RAF. The squadron was tasking right to the last day of operations, much of the Wessex's work having been undertaken in the notorious South Armagh region. With the end of Wessex operations, No. 72 Squadron was disbanded on 31 March, its Puma flight being reassigned to No. 230 Squadron, also at Aldergrove. Ten Wessexes were airworthy for a tour of former haunts in the Province and on the mainland, which ended on 27 March when the formation left RAF Benson for the flight to Shawbury, where they entered storage pending disposal. Among the aircraft was XR497, the first Wessex to be delivered to the RAF in 1963. The retirement leaves No. 84 Squadron at Akrotiri as the last RAF Wessex operator, although its aircraft are used for SAR, rather than SH, tasks. They are due for retirement in 2003.

Above: Wessex at work – in the last year of its service, a No. 72 Sqn aircraft manoeuvres a slung load at Bessbrook Mill, the army's headquarters in the South Armagh region. The Wessex operated in this area right until its retirement.

Below: Eight of the 10 Wessexes depart RAF Benson in fine style on 27 March as the type embarked on its last flight as an SH helicopter in RAF service. The destination was Shawbury, where the helicopters entered storage.

Lockheed Martin Aeronautics, for 52 more two-seat Block 52+ F-16I Fighting Falcons. These had been listed as options in Israel's January 2000 Peace Marble V programme, involving an initial batch of 50 F-16Is. Their increase to 102 brings total IDF/AF Fighting Falcon procurement to 362 aircraft. Lockheed Martin's share of the new contract is $1.3 billion, from a total programme value of some $2 billion.

Lockheed Martin Aeronautics will assemble all 52 new F-16Is at its Fort Worth, Texas, facility, but there will be significant co-production of airframe and avionics components in Israel. Apart from the upgraded Northrop Grumman APG-68(V)9 multi-mode radar, providing a 33 per cent increase in air-to-air detection range and synthetic aperture (SAR) ground-

mapping, the Block 50+ F-16I two-crew mission system avionics will integrate many Israeli units. These include Satcom and L-band datalinks, an internal Elisra/Elta EW suite, incorporating missile warning and towed decoy provision. The Elbit DASH helmet sight will also be linked with an indigenous SAR reconnaissance pod.

Deliveries of the first IDF/AF F-16Is, which also have conformal long-range tanks containing 1360 kg (3,000 lb) of fuel, will start next year. Firm orders for the F-16 now total 4,347 aircraft, from which about 301 remained to be delivered at the end of 2001.

Grob 120A selected
Elbit Systems was nominated in February as the preferred contractor for the IDF/AF's private-finance initiative (PFI) programme, for a new

primary flight-training system. This will involve procurement, ownership and technical support by Elbit of Grob 120A two-seat composite trainers. These will replace the dozen or so venerable Piper PA-18 Super Cubs operated for many years by the IDF/AF for initial pilot screening. Elbit will provide their technical support from its Cyclone field base, near Carmiel.

Italy

Boeing 767 tanker selected

Development of a tanker/transport version of the Boeing 767-200ER was launched in late 2001 from its selection by Italy's Defence Administration for an Italian air force (AMI) requirement for four aircraft of this type, plus options for two more. Boeing and Alenia Difesa reached agreement on the $720 million joint development and production programme, with 100 per cent offsets, for planned deliveries

As part of Exercise Foal Eagle in March 2002, this KAI-built Sikorsky UH-60P from the Republic of Korea Army undertook deck-landing qualifications aboard the USS Essex (LHD 2).

from 2004, as replacements for the AMI's four Boeing 707-328B tankers. After prototype modifications by Boeing Wichita, further 767 tanker conversions would be undertaken in Italy by Alenia Aerospazio and its Aeronavali subsidiary.

As well as a three-point hose/drogue refuelling system, the AMI 767s would have provision for an alternative refuelling boom, to operate with its 30 leased GD F-16A/Bs. These will comprise 26 ex-USAF F100-PW-220E-powered Block 15ADF F-16As and four Block 10OCU F-16Bs, plus another four F-16Bs for spares, to replace Italy's leased Tornado ADVs from 2003.

Engine selection in the 267.75-kN (60,000-lb) thrust category was still

undecided for Italy's 767 requirement, from competition between the GE CF6-80C2B6, Pratt & Whitney PW4060 and Rolls-Royce RB211-524G turbofans. In both Italy and the UK, the Boeing 767 tanker/transport has been competing with EADS and its multi-role Airbus A330TT equivalent for lease/purchase contracts.

More EH101s

Italian navy contracts worth 145 million euros ($126.4 million) have been received by AgustaWestland for another four EH101s, increasing its overall total to 20, plus four more options. Earmarked for amphibious support, the new EH101s will supplement eight ASW, four utility, and four AEW versions now being delivered to Italian Naval Aviation (MMI).

Japan

Reduction in FY2002 approval

Previously-listed Japanese Defence Agency FY2002 tri-service procurement requests for 49 new aircraft were only slightly reduced in February, when approval was received for 46, costing $1.76 billion. Cuts were confined solely to second-line support aircraft, including a single Sikorsky/MHI UH-60J SAR helicopter from the JMSDF's requested 12 aircraft, now costing $395.36 million. The JGSDF's revised $315.37 million allocations for 12 aircraft resulted from deletions of one of four requested Bell/Fuji UH-1Js and one of two Beech LR-2 King Air 359 liaison aircraft.

JASDF requests for $1.05 billion to acquire 23 aircraft remained unchanged, and included funding for the first of four new tanker/transports, for which the Boeing 767, as expected, has now been selected for commonality with Japan's AWACS fleet. Operational requirements have been quoted for a range of more than 6500 km (3,507 nm) with a 30-tonne payload, and a capability to refuel up to eight aircraft on a single mission, using a flying boom system.

South Korea

F-15K selected for F-X bids

A decision on South Korea's long-standing F-X next-generation multi-role combat aircraft requirement was finally announced in April, in favour of the Boeing F-15K, after earlier delays, reports of illegal agency commission payments, and protests by Dassault. For flight safety reasons, the RoKAF limited its evaluations to twin-turbofan fighters, from which the Boeing F-15K and Dassault Rafale Mk 2 were eventually short-listed, after elimination in March of the Eurofighter Typhoon and Sukhoi Su-35. Initial RoKAF procurement is planned of 40 aircraft, with options for up to 40 more and deliveries from 2007.

RAAF Air Combat Group

A twilight ceremony at Williamtown on 7 February 2002 marked the formation of a new command structure within the RAAF, bringing change in the way the strike/fighter elements currently do business. Air Combat Group, commanded by Air Commodore John Quaife, replaces both Tactical Fighter and Strike & Reconnaissance Groups, and brings the 'fast-jet' squadrons under the same umbrella for the first time.

Designed to restructure the RAAF into the configuration required for the AIR 6000 introduction, reorganisation gained momentum under the previous Chief of the Air Force, Air Marshal Errol McCormack. The immediate benefit is the streamlining of command structure but, more importantly, the aim is to drive the most logical AIR 6000 solution, enabling tactical innovation, and configuring the Air Force in how it wishes to fight.

Core combat capability of the two offensive wings will be retained: No. 81 retaining Control of the Air and No. 82 continuing as the Precision Strike unit. The third wing, No.78, remains as the Operational Conversion and Training Wing. The three wings will therefore be role-specialised, with the flexibility of adding a 'shadow' capability by the transfer of squadrons into and out of the wings as deemed necessary. The new arrangement replaces the two 'contingency' shadow wings (No. 95 Contingency Strike, and No. 96 Contingency Air Defence), and will allow other assets such as tankers to be brought in as required. The Wedgetail AEW platform will make a broader contribution to the ADF than can be utilised by ACG, and therefore will remain outside the new structure.

The major benefit of this new arrangement will be the realisation of the Hornet's precision strike capability within 82 Wing, thereby providing the Tactical Fighter community with an appreciation of the capabilities of the Strike Squadrons, and vice-versa. The formation of a dedicated Forward Air Control Development Unit (FACDU) under the control of No. 82 Wing is designed to progress the FAC role beyond the Vietnam-era skills currently practised, and has seen the four FAC-configured PC-9/A(F)s transferred from within No. 77 Sqn.

FACDU will continue to train Hornet pilots in the 'Fast FAC' role, but most of the work will be for the Army. At the present time, the unit is training members of the SAS and Artillery Corps (Forward Observers) as FAC-G (Ground). Trials are also being conducted with the Army's Kiowas in order to develop the FAC-H (Helicopter) role, prior to the introduction of the Tiger ARH in 2004.

Stage One of the restructure (completed 02/02) raised HQ ACG and FACDU at Williamtown, and disbanded the SRG & TFG. Stage Two (complete by mid-2003) extends the training responsibility of No. 78 Wing into the F-111 community, most likely involving a division of responsibility – OC 78 Wing responsible for training and OC 82 Wing for the platform. Stage Three (complete by end of 2003) will make recommendations on how 82 Wing can harness the Hornet's precision strike capability. A series of attachments into 82 Wing for a longer period than previously exercised with the Contingency Wing concept will occur.

Nigel Pittaway

AIR COMBAT GROUP (HQ Williamtown, NSW)

78 Wing, Williamtown, NSW		Fighter Conversion/Training
76 Sqn	Williamtown, NSW	Hawk Mk.127
79 Sqn	Pearce, WA	Hawk Mk.127
2 OCU	Williamtown, NSW	F/A-18A/B

81 Wing, Williamtown, NSW		Air Control
3 Sqn	Williamtown, NSW	F/A-18A/B
75 Sqn	Tindal, NT	F/A-18A/B
77 Sqn	Williamtown, NSW	F/A-18A/B

82 Wing, Amberley, Qld		Precision Strike
1 Sqn	Amberley, Qld	F-111C/RF-111C
6 Sqn	Amberley, Qld	F-111C/F-111G
FACDU	Williamtown, NSW	PC-9/A(F)

In unit terms, the main change brought about by the creation of the Air Combat Group is the reassignment of the RAF's Pilatus PC-9/A(F) aircraft from No. 77 Squadron, whose markings this aircraft wears, to the new FACDU. The aircraft remain at Williamtown, but now come under 82 Wing.

Following previous delays in F-X evaluation processes, original best and final offers (BAFOs) from all four RFP respondents were rejected in Seoul in early January as being too costly for its planned $3.2 billion budget. Revised BAFOs called for by 4 February left little scope for price reductions, and were still considered excessive, although allowing possible improvements in industrial offset returns. Seoul had already sought offset increases from around 30 per cent to 70 per cent of the total contract value, although Eurofighter reportedly eventually offered up to 100 per cent in contractual returns.

With substantial US government support, Boeing's customised F-15K – powered for the first time by GE F110-129 turbofans, armed with Raytheon AIM-9X and AIM-120 AMRAAM air-to-air missiles and Joint Direct Attack Munition (JDAM) ASMs, in conjunction with a Boeing/Vision Systems Joint Helmet-Mounted Cueing System (JHMCS) – was always front-runner for F-X procurement. Boeing's $4.4 billion F-15K submission further incorporated Raytheon's advanced APG-63(V)1 fire-control radar, as a successor to the F-15E's Raytheon APG-70. The US also approved possible RoKAF receipt of the APG-63(V)2, with electronic phased-array scanning.

RoKAF procurement was additionally being sought from the US Defence Security Co-operation Agency (DSCA) of 45 278-km+ (150-nm+) range Boeing/MDC AGM-84H Stand-off Land Attack Missile-Expanded Response (SLAM-ER) ASMs, costing around $115 million, for its F-15Ks. DSCA approval was further received for an unsolicited proposal for RoKAF F-15K integration of MBDA's ASRAAM close-combat air-to-air missile.

Malaysia

Mi-171 procurement extended
Long-planned replacement of the RMAF's 30 or so Sikorsky S-61A-4 Nuri transport helicopters appears to have started from a March order placed by the Malaysian government with Russia's Ulan-Ude Aircraft Plant (UUAZ) for 10 'Hips'. These are likely to be Mi-171-1V export versions of the Mi-8AMT, with uprated (to 1556 kW/ 2,100 shp) Klimov TV3-117 turboshafts, for improved tropical performance. Two Mi-171-1Vs were delivered to Malaysia in 1999 from a previous $20 million UUAZ contract, as part of the company's sales in the past decade of over 150 'Hips' to such countries as Algeria, China, Ecuador, Mexico and Sri Lanka.

Mexico

Maritime patrol aircraft defined
Raytheon Systems has been selected to modify two EMBRAER ERJ-145 aircraft for maritime patrol missions for the Mexican navy. Equipment being installed in the aircraft includes the SeaVue surveillance radar, AN/APX-114 identification friend or foe (IFF) system and a communications intelligence (COMINT) system. Raytheon will also provide a communications suite for a single ERJ-145 that is equipped as an airborne early warning and control platform.

Nepal

Mil helicopters delivered
Nepal received two Mil Mi-17 helicopters purchased from India during December 2001. The aircraft will support Royal Nepal Army operations against communist rebels who are trying to overthrow the Himalayan kingdom's constitutional monarchy.

Oman

F-16 and Super Lynx orders
Gulf Co-operation Council plans for enhanced mutual defence, agreed late last year, were reflected in a 38 per cent military budget increase to OR926 million ($2.4 billion) by Oman in January. This was followed in March by Omani plans to acquire Lockheed Martin F-16s and AgustaWestland Super Lynx 300 scout helicopters, for further defence expansion.

In early October, the US Defence Security Co-operation Agency revealed confirmation of Omani plans to acquire 12 GE F110-engined Block 50/52 F-16C/Ds, with uprated APG-68(V)XM radar, costing $1.12 billion. Also included were requests for 14 each Lockheed Martin LANTIRN targeting/TFR navigation pods, in addition to a previously-listed FMS package of AAMs, ASMs and PGMs.

Negotiations were then advanced with Westland Helicopters Ltd for RAFO acquisition of the Super Lynx with requirements for 15-20 equipped for battlefield support, plus Omani police needs for six AW 109 and AB 139 light helicopters. On 19 January, Oman's Defence Affairs Minister, accompanied by Royal Air Force of Oman (RAFO) Commander Air Marshal Mohammed al Ardhy, signed an agreement with Westland to provide 16 Super Lynx 300s. With CTS800 engines and wheeled landing-gear, their delivery is planned from about 2004, to replace the air force's current fleet of 20 or so Agusta-Bell AB 205As.

Pakistan

First Chengdu F-7PGs arrive
Delivery started last December of the first of an initial 50 Chengdu F-7PG fighters ordered in late 2000, as replacements for the PAF's ageing Shenyang F-6 versions of the twin-turbojet MiG-19, which have been in Pakistani service since 1966 and equipped two squadrons (Nos 17 and 23) of 31 Wing at Samungli, Quetta. F-7PG is the PAF designation for the J-7E/F-7MG, first revealed at the November 1996 Zhuhai air show.

As a major MiG-21F-13 upgrade, the J-7E/F-7MG retains the central portion of its 57° clipped delta wing, but increased in area by 8.17 per cent by new tapered outer panels with only 42° of sweep, and a span increase to 8.3 m (27.23 ft). With manoeuvring slats on the new tapered outer wings, and combat flap settings, the increased wing area is claimed to improve combat agility by 45 per cent, plus gains in take-off, climb, ceiling and landing performance.

The J-7E/F-7MG retains the twin under-fuselage 30-mm cannon, each with only 60 rounds, of earlier F-7s, but features five external weapon stations. GEC's original Skyranger range-only fire-control radar was initially replaced in the F-7MG by an upgraded GEC Marconi Super Skyranger full-function lightweight unit, using coherent technology to achieve scan, look-down and shoot-down capabilities. Pakistan's F-7PGs, however, are reportedly equipped with a development of the Galileo (Alenia/FIAR) Grifo 7 radar, produced under licence at Pakistan's Kamra aircraft factory, among other Western avionics.

As a long-term customer for Chinese-built Soviet aircraft, the PAF is believed to have overall requirements for 80 or more F-7PGs, with reportedly early delivery availability. They will supplement the PAF's remaining Chengdu F-7P and two-seat Guizhou FT-7P versions of the MiG-21 from 105 and 15 delivered, respectively, from 1991. The Shenyang F-6 was retired on 27 March 2002.

Poland

More naval aircraft orders
Delivery began from PZL Mielec earlier this year of the first four of 10 M-28 Bryza light twin-turboprop transports, from a Zl200 million ($48 million) follow-on order from the Polish Defence Ministry, placed in mid-2001. Two of the original Antonov An-28TD transport/paradrop versions of the M-28 have been operated by the WLOP since 1994, and in January were supplemented by two An-28TDs and two An-28E ecological monitoring aircraft for the Polish navy.

The navy also received two Mil Mi-17s from Russian production, plus a PZL-Swidnik W-3RM Anakonda SAR helicopter, and is awaiting four refurbished ex-USN Kaman SH-2G ASW helicopters, for operation from two US surplus frigates. These are among major new military equipment procurement planned by the Polish government in a Zl105 billion ($26.5 billion) mid-2001 six-year defence modernisation programme. This represents 1.9 per cent of Poland's GNP between 2001 and 2006, but further funding will be required for the WLOP's planned $3.5 billion new combat aircraft programme.

Switzerland

Transport requirement
Earlier Swiss government plans for the acquisition of two twin-turboprop EADS CASA-295s for SAFAAC tactical transport requirements have been put in abeyance. Evaluations have been re-opened to include the Alenia/LM C-27J, pending redefinitions of new Swiss military mobility requirements.

Turkey

Seahawks ordered for Navy
Despite a $5.5 billion military debt to the United States, Turkey has confirmed plans to purchase 14 SH-60B helicopters. The $324 million Seahawk purchase, which will be covered by bank loans, still requires the approval of the US Congress. The Turkish Navy will operate the heli-

Mexican Mils

Mexico has increasingly turned to Eastern equipment in the expansion of its air arms. Escuadrón Aereo 303, part of 3º Grupo Aereo at BAM 1 Santa Lucia, operates a mixed bag of Mil helicopters, including the huge Mi-26T 'Halo' (above) and the PZL-built Mi-2 (right). Mi-8s and Mi-17s are also on charge.

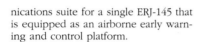

copters from its fleet of seven ex-US Navy 'Perry'-class guided missile frigates.

United Kingdom

Tutor deliveries complete

VT Aerospace, part of the Vosper Thornycroft Group, has placed the last of 99 Grob G-115 Tutor trainers, ordered as part of the Light Aircraft Flying Task (LAFT) contract, in service. In addition to the aircraft, the company provides engineering and logistic support to the RAF's University Air Squadrons (UAS) and Air Experience Flights (AEF) at 12 airfields around the United Kingdom, as well as at the Central Flying School (CFS) at RAF Cranwell. Under the terms of the LAFT contract, which was awarded to VT Aerospace in 1999, the RAF buys Tutor flight hours from VT Aerospace.

Airbus for VIP mission?

The United Kingdom's Ministry of Defence recently proposed purchasing a modified Airbus airliner for use by the Prime Minister. The A330 aircraft will receive a host of modifications designed to provide a secure mode of transportation at a cost of $114.5 million. Members of the UK parliament and the Royal Family were normally provided with transportation aboard Royal Air Force VC10s. However, the use of these aircraft was recently discontinued. The MoD, concerned about the safety of senior government officials, believe they would be safer aboard a dedicated aircraft operated by the military.

Super Lynx assessed for BLUH

A £20 million ($28 million) 18-month contract with the Westland Helicopters division of the AgustaWestland group, announced on 29 January by the MoD, confirmed selection of the company's Future Lynx project as having the best potential to meet the UK Army Air Corps (AAC) Battlefield Light Utility Helicopter (BLUH) requirement. In parallel with initial Future Lynx development by Westland, the MoD will commission

A recent delivery for the Peruvian army is this Cessna 208 Caravan, in amphibian configuration for use in the country's interior. The army's fixed-wing fleet comprises various light aircraft for communications duties, and An-32s for transport.

appraisal work on other aircraft options. Together, these will form the BLUH programme assessment phase, to select suitable successors to about 100 Lynx AH.Mk 7s and 25 AH.Mk 9s operated by the AAC. The Royal Navy also has a parallel Surface Combatant Maritime Rotorcraft (SCMR) requirement for 55-60 new ship-board helicopters, to replace some six Lynx AH.Mk 7s, 30 HAS.Mk 3s, and 40 HMA.Mk 8s.

A final BLUH decision is expected in late 2003, and an order for 70-80 Lynx replacements would be worth about £1 billion for Westland. Initial AAC service entry would follow in late 2006 for scout/utility, reconnaissance and utility roles, and continued operation possibly through 2030.

The multi-role Future Lynx will be developed from the company-funded Super Lynx 300 export project. In addition to an improved airframe with an increased 6260-kg (13,800-lb) take-off weight, and 25-year service life, the Super Lynx 300 has 40 per cent more powerful LHTEC CTS 800-4N powerplants and digital 'glass' cockpit avionics. Effectively a completely new design, Future Lynx would add a 10,000-hour airframe, increased-authority tail-rotor and more advanced mission system avionics, with state-of-the-art sensors, improved navigation systems and highly capable defensive aids.

United States

Airlifters ordered

Lockheed Martin has received a $356 million contract covering the delivery of five C-130J-30 airlifters to the USAF. The contractor also announced the USAF has assigned the military designation system (MDS) CC-130J to the stretched version of the airlifter. The 'Hercs' will be delivered to the USAF

in 2004, with two going to the California Air National Guard's 146th Airlift Wing, one to the Rhode Island ANG's 143rd Airlift Wing, and one to the Air Force Reserve Command's 403rd Wing at Keesler AFB, Mississippi. The final aircraft will be assigned to the new C-130J Formal Training Unit at Little Rock AFB, Arkansas. The service, which has a requirement for 168 C-130Js, has also included a multi-year acquisition plan for the C-130J in its proposed 2003 budget. A total of 40 aircraft has already been ordered.

The contractor has also received a $133.8 million contract covering the procurement of two KC-130J tanker aircraft for the US Marine Corps. The aircraft will be delivered by December 2003.

Gulfstream ordered

Gulfstream Aerospace has been awarded a $1.6 billion covering the delivery of up to 20 Gulfstrem V and V-SPs and 10 years of full or partial contractor logistics support (CLS) for each aircraft along with CLS for aircraft already in service. The USAF has already committed $43.4 million covering the purchase of one C-37A.

More Phantom drones

The USAF Air Armament Center at Eglin AFB, Florida, has issued BAE Systems in Mojave, California an $8.8 million modification to an existing contract to modify 12 F-4 Phantoms into QF-4 full scale aerial targets (FSAT). The target drones, which will be removed from storage at the Aerospace Maintenance and Regeneration Facility (AMARC) at

Davis Monthan AFB, Arizona, will be delivered by July 2004.

Targeting pod delivered

The Marine Corps accepted the first of 47 AN/AAQ-28 Litening II targeting pods for its AV-8B Harrier fleet from Northrop Grumman during December 2001. The Marine Corps is the first operator of the latest version of the Litening II, which incorporates a 640-by-512-pixel Forward Looking Infrared (FLIR) camera that will provide the Harrier with the highest performance targeting pod camera in the US inventory. The pod provides the Harrier with the ability to autonomously deliver laser-guided bombs, enhanced day and night target acquisition and improved low-level night flying capabilities. The service has a requirement for 56 of the pods.

Venezuela

Helicopters delivered

Enstrom Helicopters recently delivered four Enstrom 280FX helicopters to Venezuela's Guardia Nacional de Venezuela. The National Guard will primarily use the helicopters for flight training duties and will station them on Margarita Island.

'Civilian' helicopters operating in support of US forces involved in Afghanistan operations included two SA 330J Pumas (above), operating under contract from Fort Collins, Colorado-based Geo-Seis Helicopters to Military Sealift Command for vertrep duties, and an unknown number of Mi-8/17s which are operated by the CIA and other US agencies, ostensibly on charter from a US-based company. Here (left) US special forces help to evacuate wounded Afghan fighters from Mi-8MTV-1 N353MA during Operation Anaconda, a major campaign in March 2002 to flush out al-Qaeda fighters from the Arma mountains. The 'Hip' is one of at least two registered to the Maverick Aviation Corporation, of Enterprise, Alabama.

AIR ARM REVIEW

Canada

Last Silver Star overhauled

19 Wing's 414 Combat Support Squadron at CFB Comox, British Columbia, recently completed its last 300-hour periodic inspection on a Canadair CT-133 Silver Star. Better known as the 'T-Bird', the aircraft currently serves in the combat support role but will be retired beginning in March 2002. Developed from the Lockheed P-80 fighter and T-33 trainer, the Canadian aircraft were built locally by Canadair as the CL-30 and 656 were delivered to the Royal Canadian Air Force between 1952 and 1956. Today less than 30 remain in service. The 'T-Birds' are being retired as an economic measure and will be replaced by civilian aircraft that will operate under contract to the Canadian Forces. Canada officially retired the CT-133 on 31 March 2002, when operation of the aircraft formally ceased. As a result of the CT-133's retirement, 414 Combat Support Squadron (CSS) at CFB Comox and 434 CSS at CFB Greenwood will officially be decommissioned on 15 July 2002.

Hungary

Gripen pilots to train at NFTC

As the sixth customer for the $C3 billion ($1.88 billion) NATO Flying Training in Canada programme, the Hungarian Air Defence & Aviation Command (LeRP) signed a 17-year contract in March. Initially, seven LeRP MiG-29 pilots will undertake Phase III and IV advanced and tactical training on NFTC's BAe Hawk Mk 115s (CT-155s), to ease their transition to Hungary's 14 leased JAS 39 Gripens, due for delivery from 2004. Similar courses will also be completed by 13 LeRP pilots nominated for Gripen training, after graduating from the Aero L-139 Albatros.

With additional commitments from Denmark, Italy, Singapore and the UK, as well as Canada and Hungary, NFTC orders have recently been placed for two more Hawk 115s, for a total of 22. Another two Raytheon T-6A Harvard IIs (CT-156s) are being bought for basic Phase II training at CFB Moose Jaw, Saskatchewan, increasing NFTC's overall total to 26.

Peru

US aid for air power expansion

In early 2002, the US more than trebled its counter-narcotic aid to Peru to $156 million. This includes some $20 million to upgrade 10 of the FAP's 30 ground-attack Cessna A-37Bs, over $20 million to incorporate Huey II upgrade kits in 14 National Police Bell UH-1Hs and $2.7 million for eight P&WC PT-6A-25C turboprop engines for 25 FAP EMB-312 Tucano trainers. The FAP is also requesting eight Bell Textron 412EPs, together with additional sensors for its four twin-turboprop Fairchild/Swearingen C-26 electronic and optical surveillance aircraft.

Poland

Fighter requirements

NATO membership commits Poland to modernise its air defence forces and field 60 alliance-interoperable fighters to replace obsolete Soviet equipment by 2006. Some $3.5 billion was approved in 2001 for this programme, for which RFPs were issued by 15 June 2001. Initially, these envisaged short-term lease of 14 surplus multi-role fighters and two two-seat combat trainer versions, for late 2003 delivery, followed by procurement of 44 new-build aircraft by end-2006. In February, however, economies were sought by deleting initial lease plans, made possible by German MiG-29 transfers, while increasing new-build fighter procurement to 48 aircraft.

Poland's new combat aircraft decision was originally planned by last September. It was delayed, however, by investigations into procurement corruption allegations, and a government change last October. Since then, Germany's 19 MiG-29 'Fulcrum-As' and four two-seat MiG-29UB 'Fulcrum-B' combat-trainers, scheduled for withdrawal from service in 2003, have been transferred to Poland, for a nominal 1DM payment each. The ex-Luftwaffe aircraft were already equipped with Western avionics for NATO interoperability, unlike Poland's 28 MiG-29s and four MiG-29UBs, only 10 of which are currently listed as NATO-compatible. A $100 million upgrade for operational commonality among Poland's augmented MiG-29 fleet was being discussed with MiG Aircraft Product Support GmbH (MAPS) earlier this year.

Meanwhile, new Memoranda of Understanding (MoUs), signed in mid-February by SAAB and BAE Systems with the Polish Aviation Company (PZL) in Mielec, were planned to boost selection prospects for the multi-role JAS 39 Gripen submission for Polish air force (WLOP) fighter requirements. Work packages and tenders worth some $60 million have already been placed by SAAB/BAE with Polish industry. Swedish defence industry links also exist with Poland from a 10 million euro ($8.9 million) August 2001 contract, from Thales Netherland BV with SAAB Bofors Dynamics, for Polish navy RBS 15 anti-ship missiles.

SAAB-BAE's Polish submissions, with their governments' support, were based on leased Swedish air force Batch 3 Gripens, and follow-on purchases of similar new JAS 39C/Ds. Short-listed bidders from RFIs issued in November 2000 also comprised Dassault, with leased French air force Mirage 2000B/Cs and new Mirage 2000-5 Mk 2s, plus the US government and Lockheed Martin, with F-16s. These comprised 12 ex-USAF Block 15ADF GD F-16A and four Block 10OCU F-16Bs, with two more Block 10 F-16As for spares, with follow-on options for another 44 similar aircraft, or new multi-role Block 50/52+ F-16C/Ds.

A comprehensive US weapons, equipment and munitions package also included Raytheon AIM-120C AMRAAMs, Raytheon AIM-9M-2 Sidewinder close-combat AAMs, Raytheon AGM-65G Maverick ASMs, and Boeing/MDC 907-kg (2,000-lb) Mk 84 Joint Direct Attack Munition (JDAM) tail-kits, as well as 44 AN/ALQ-131 or ALQ-184 ECM pods, and 36 LITENING or LANTIRN navigation and targeting pods.

Russia

Air Force selects trainers

The Russian Air Force has announced that the Yakovlev Yak-130 has been selected as its new jet trainer, beating out the competing Mikoyan MiG-AT. Design work began on the Yak-130 in 1992 and the prototype made its maiden flight during 1996. The jet will be produced at the Sokol production plant in Nizhny Novgorod and the

Right: The ramifications of the 11 September terrorist attack in the US continue to be felt. Here an F-15 from the 125th Fighter Wing, Florida ANG, flies a CAP during the launch from Cape Canaveral, Florida, of the Space Shuttle Endeavour. US interceptors also kept up a constant watch over the Winter Olympic games in Salt Lake City, Utah.

Below: While US forces continued to fight pockets of resistance in Afghanistan, the war against global terrorism broadened to include operations in the Philippines. Here US special forces prepare to board a 160th SOAR MH-47E.

Exercise Cope Tiger 02 was held in Thailand in January 2002, with US and Singaporean forces participating alongside the Thais. Here Thai special forces ride into action in a RTAF UH-1H. The effectiveness of the mesh filter on the engine intake is noteworthy.

initial batch of four aircraft will be delivered in 2003. The Yak-130 will replace the air force's fleet of Czech-built L-29 and L-39 trainers beginning in 2005 and has a requirement for at least 120 trainers over a period of five to seven years. Subsequently, the Sukhoi Su-49 was announced as the preferred choice for a primary trainer.

Singapore

First NFTC graduates

Training to wings standards was completed in February of the first two

The Israel Defence Force/Air Force (IDF/AF) is planning to repaint its fleet of Sikorsky UH-60 Yanshuf (Night Owl) assault helicopters in a new scheme. The Israeli Yanshuf fleet is currently painted in the standard US Army olive drab scheme optimised for nocturnal operations. Two years ago an experimental scheme suited for daytime operations, known as the Syrian scheme, was introduced but was found to be unsatisfactory and a second scheme, known as the spotted scheme, was evaluated and selected for the repaint of the fleet. It is also planned to repaint the IDF/AF Boeing AH-64 Peten (Python) attack helicopters.

RSAF air force student pilots of six per year scheduled for graduation through the NATO Flying Training in Canada programme, from a 20-year contract. This involves Phase III advanced and Phase IV tactical and weapons training, on the NFTC's BAE Hawk Mk 115 lead-in fighters, which have digital cockpit and mission systems. Phase IV tactical training is now being completed by two more Singapore air force students, while four more are on the Phase III course, before moving to operational conversion units.

United Kingdom

RAF Tornado F.Mk 3 cuts
Planned disbandment by January 2003 of one of the RAF's five operational Tornado F.Mk 3 air defence squadrons will virtually coincide with deliveries of its first Eurofighters, as replacements. Initial deliveries of 13 Eurofighters to Warton-based No. 17 Squadron, the RAF's operational test and evaluation unit, are scheduled in 2003/04. No. 29 Squadron will also be re-formed at Warton, as the Eurofighter operational conversion unit in 2004.

According to the MoD's late January announcement, however, planned retirement of No. 5 Squadron's 15 F.Mk 3s, operated from Coningsby on stand-by alert to protect London since 11 September 2001, is due mainly to shortages of fast-jet pilots. Deficits of around 130 such pilots have been reported from requirements for 1,484.

Other quick-response F.Mk 3s, says the MoD, will be positioned "at an air base near London". Coningsby's Tornado crews will boost the operating strengths of the RAF's remaining northern-based first-line F.Mk 3 units, comprising Nos 11 and 25 Squadrons at Leeming, and Nos 43 and 111 Squadrons at Leuchars. Four Tornado F.Mk 3s have also been operated on detachment for some years by No. 1312 Flight, for Falkland Islands air defence. Annual savings of about £27 million are expected from No. 5 Squadron's disbandment, reducing annual flight commitments by about 4,500 hours.

Sea Harriers to be scrapped
Continuing its contradictory policies of cutting back government defence funding and equipment resources to pay for ever-increasing global military commitments, the UK government plans to withdraw the RN's entire force of nearly 50 BAe Sea Harrier FA.Mk 2 air defence/ground-attack V/STOL fighters in 2006. These entered service only in 1994, and their departure will leave the RN without organic air cover until planned JSF procurement from about 2012.

The MoD said in February that the FA.Mk 2 pilots and support personnel would be transferred to the recently-formed Joint Force Harrier's (JFH) main base at Cottesmore, in Rutland, where the RAF currently operates about 36 upgraded Harrier GR.Mk 7As. UK Armed Forces Minister Adam Ingram said that Sea Harrier retirement would move forward the JFH into the era of the planned Future Joint Combat Aircraft (FJCA) and Britain's two new aircraft-carriers (CVFs). "Recent commitments to the next phase of the Joint Strike Fighter (JSF) programme, the order for a further three Type 45 air defence destroyers, and service entry of new smart weapons, have given renewed impetus to Joint Force Harrier offensive roles", he added.

"We have concluded that JFH, [which combines the operation of some 75 RAF and 50 RN V/STOL fighters], should become an all-Harrier ground-attack/reconnaissance force, maximising investment in one aircraft type. The GR.7s will be further upgraded to GR.9 standard, to ensure maintenance of a credible expeditionary offensive capability". JFH is ultimately intended to operate a common aircraft type to replace the Harriers through the FJCA programme, for which the JSF is currently favoured.

Plans for a Harrier GR.Mk 7 mid-life digital avionics upgrade were revealed in mid-2000, following a December 1999 decision to replace their original 95.6-kN (21,500-lb) thrust Pegasus Mk 105s with uprated Mk 107 turbofans, similar to the 102.3-kN (23,000-lb) Pegasus 11-61s in the US Marine Corps AV-8B Plus Harrier IIs. Rolls-Royce delivered the first Pegasus 107, from an initial order for 40, plus options for 86 more, costing £350 million, on 30 November 2000, for installation in redesignated Harrier GR.Mk 7As.

A Mil Std 1760 stores management databus is the basis of the planned GR.Mk 7/7A avionics upgrade which, when contracted with BAE Systems, will allow the redesignated Harrier GR.Mk 9/9As to operate with smart air-to-surface and also air-to-air weapons. These will include Alenia Marconi Brimstone anti-tank missiles from 2006, plus MBDA ASRAAM close-combat AAMs. In addition to a new GD UK main computer and ADA software, GR.Mk 9/9A avionics will integrate BAE Systems multi-function colour cockpit and digital map displays, ground-proximity warning system and Northrop Grumman INS/GPS. No fire-control radar or gun installation is planned, but the former may be substituted by MIDS (Multifunctional Information Distribution or Data-Link 16).

Integration of ASRAAM in the Sea Harrier has now been terminated, after spending £1.2 million. Indian government enquiries have already been received by the MoD regarding future availability of the RN's surplus Sea Harriers. As part of its marketing strategy for the sale of these aircraft, the MoD's Disposal Services Agency "is following-up all expressions of interests, subject to normal export controls".

Nimrod cuts
Reductions in the RAF's long-delayed BAE Nimrod MRA.Mk 4 rebuild programmes from the originally-planned 21 aircraft to 18, also announced in February, are claimed by the MoD to have been possible without decreasing overall operational capabilities, while saving £360 million in support costs. This follows decreased submarine threats to the UK since the 1996 contract, markedly greater capabilities from MRA.Mk 4 sensor and aircraft performance than the 21 current Nimrod MR.Mk 2s, and potential BAE Systems in-service support improvements to maximise aircraft availability.

An MRA.Mk 4 first flight is now planned for September 2002, followed by two more aircraft in December and March 2003. Initial operating capability is planned following first MRA.Mk 4 deliveries in August 2004. Full mission specification will be met by the contracted time of the seventh aircraft delivery, and service entry at the Nimrod's Kinloss base, in Scotland, in March 2005.

United States

Coast Guard MH-68A
The USCG's Helicopter Interdiction Squadron (HITRON)-10 has taken delivery of the last of eight MH-68A helicopters from Agusta and, by early January 2002, the squadron had made three deployments. Although based on Agusta's A 109 Power helicopter, the MH-68A is equipped with a rescue hoist and carries a 0.50-in sniper rifle, grenades and weighted nets – all designed to stop drug-carrying ocean-capable speed boats or 'go-fasts'. The service has 42 'aviation-capable' ships, however, the MH-68s are more likely to be deployed as single-aircraft detachments aboard 370-ft (113-m) high- and 278-ft (85-m) medium-endurance cutters. These ships, unlike the older 210-ft (64-m) medium-endurance cutters, are equipped with hangar facilities. HITRON 10 is based at the former Naval Air Station at Cecil Field Airport in Jacksonville, Florida.

MH-60S enters service
The Sikorsky MH-60S Seahawk officially entered fleet service at NAS North Island, California, on 8 February 2002. Helicopter Combat Support Squadron Three (HC-3) will initially train pilots, aircrew and maintenance personnel on the aircraft, which enter operational fleet service with at HC-5 at Andersen AFB, Guam, in the summer of 2002. As part of its

On 1 January 2002 the Belgian Air Force ceased to exist, becoming the Air Component of a new, unified armed force structure. 8 Sm/Esc (F-16s) was disbanded on 8 March, although its aircraft had already been amalgamated with 31 Sm/Esc in January. Meanwhile, to celebrate its 60th anniversary, 349 Sm/Esc painted one of its F-16A MLUs in special colours (left). On 21 January 2002 the Air Component received its second and final ERJ-145 (below) which, along with two ERJ-135s, has allowed the Merlin and HS.748 to be retired. Official retirement date for both types was 20 March.

Helicopter Master Plan, the Navy will purchase up to 237 MH-60S helicopters, which will operate alongside the MH-60R. Under the plan the Navy will eventually reduce the number of helicopter airframes in service from seven to just two. Designed to fulfill multiple tasks, the MH-60S will initially conduct vertical replenishment (VERTREP), vertical on-board delivery (VOD), search and rescue (SAR) and remote site logistics missions. Its duties will be further expanded to include airborne organic mine countermeasures operations, combat search and rescue (CSAR) and special warfare support. Development of the MH-60S began in 1997 and the first production aircraft undertook its maiden flight in January 2000. The aircraft is currently undergoing operational test and evaluation (OPEVAL) at the Naval Air Warfare Center at Patuxent River, Maryland. Following the successful conclusion of OPEVAL, the newest Seahawk variant will be cleared for full-rate production. All training associated with the MH-60S will be carried out by HC-3.

USAF activates new squadrons
Air Combat Command reactivated the 12th Reconnaissance Squadron (RS) as part of the 9th Reconnaissance Wing at Beale AFB, California, on 8 November 2001. The squadron,

Long-term Luftwaffe F-4F operator JG 72 'Westfalen' has disbanded at Hopsten, to be replaced by the Zentrale Ausbildungseinheit F-4F (F-4F central training unit). The new unit is expected to operate Phantoms until 2006, by which time the Eurofighter will be entering service. JG 72 is scheduled to be the last of the four Luftwaffe wings to begin Eurofighter operations, in 2010. These two Phantoms were painted up to mark the passing of 1. Staffel, JG 72 (721 Squadron), which disbanded on 31 January 2002.

which last flew the RF-4C as part of the 67th RW at Bergstrom AFB, Texas, will operate the RQ-4A Global Hawk unmanned air vehicle.

The command also activated a third RQ-1A Predator squadron at Indian Springs Auxiliary Airfield, Nevada on 8 March 2002. The 17th Reconnaissance Squadron (RS), which joins the 11th RS and the 15th RS, is assigned to the 57th Wing at Nellis AFB, Nevada. The unit was created because of increased mission requirements based on Operation Enduring Freedom. Its initial cadre of personnel will be transferred from the existing squadrons.

The 558th Flying Training Squadron (FTS) was reactivated as part of the 12th Flying Training Wing (FTW) at Randolph AFB, Texas on 16 January

2001. Known as the 'Phantom Knights', the unit serves as the instructor training unit for the T-6A and is currently assigned 25 Texan II trainers. The squadron will eventually operate 35 aircraft but will then transfer 10 aircraft to other Air Education and Training Command (AETC) bases. The squadron was last active under the 12th FTW as part of the Specialized Undergraduate Navigator-Training Program (SUNTP) and was inactivated on 1 October 1996.

A new unit was assigned to the United States Air Forces in Europe when the 309th Airlift Squadron (AS) activated on 12 March 2002 at Chièvres Air Base, Belgium. The unit, which operates single examples of the C-9A Nightingale and the C-37A Gulfstream V in support of Supreme

Allied Headquarters Europe (SACEUR), is assigned to the 86th Wing via the 86th Operations Group (OG) at Ramstein Air Base, Germany.

Fifth Apache Longbow battalion
During December 2001, the US Army certified the 1st Attack Helicopter Battalion of the 101st Aviation Regiment as combat-ready with the AH-64D Apache Longbow helicopter. The unit is stationed at Fort Campbell, Kentucky, as part of the 101st Airborne Division (Air Assault). After initial training at Fort Rucker, Alabama, and Fort Eustis, Virginia, and subsequent training at Fort Hood, Texas, conducted by the 21st Cavalry Brigade, the battalion underwent a rigorous field examination, which included two live-fire exercises.

OPERATIONS AND DEPLOYMENTS

Canada

Auroras and Hercs deployed
14 Wing at Canadian Forces Base Greenwood, Nova Scotia, recently deployed two CP-140 Auroras, three aircrews and support personnel to Southwest Asia in support of allied naval forces operating in the Arabian Sea. A third aircrew was deployed from CFB Comox by 19 Wing's 407 Maritime Patrol Squadron. Once in theatre, the aircraft and crews supported Operation Apollo and the war on terrorism as part of the Long Range Patrol Task Force by conducting long-range sea surveillance missions. 8 Wing at CFB Trenton, Ontario, also dispatched to the region a tactical airlift detachment comprising three C-130s and 150 personnel. The

unit is supporting a Canadian Light Infantry battalion, humanitarian missions and other operations.

NATO

Additional AWACS to US
In response to a request from the United States, NATO approved the deployment of two additional Boeing E-3A airborne early warning aircraft in support of North American Air Defense Command (NORAD) operations. The E-3As joined five other aircraft already operating in support of US homeland security as part of Operation Noble Eagle. NATO Sentries began operations over the US on 9 October 2001 and are based at Tinker AFB, Oklahoma, alongside the USAF's fleet.

United States

Carriers deploy
USS *John F. Kennedy* (CV 67) departed its homeport in Mayport, Florida, on 7 February 2002 in preparation for final workups and a combat deployment. The Kennedy Carrier Battle Group (CVBG) initially conducted Phase II of Joint Task Force Exercise (JTFEX) 02-1 from 7-16 February 2002, and departed immediately for the Arabian Sea when this was concluded. The Kennedy CVBG relieved USS *Theodore Roosevelt* (CVN 71), which had been supporting combat operations over Afghanistan, on 7 March 2002.

The USS *Wasp* (LHD 1) amphibious ready group (ARG) deployed from Norfolk, Virginia, on 22 February 2002, with the 22d Marine Expeditionary Unit (MEU) embarked. HMM-261(Reinforced) is assigned as

the MEU's aviation combat element (ACE). The Wasp ARG relieved the USS *Bataan* (LHD 5) ARG, which had deployed in September 2001.

Sea Stallions deploy
On 7 February 2002 the 'Red Lions' of HMH-363 arrived at MCAS Iwakuni in support of the Marine Corps Unit Deployment Program (UDP). The event marked the first UDP deployment for a CH-53D squadron since 1991 and the first time a Sea Stallion unit has been attached to Marine Air Group (MAG)-12 at Iwakuni. While deployed, the squadron will support USMC operations through the western Pacific. The squadron's eight Sea Stallions were airlifted from MCAF Kaneohe Bay, Hawaii, to MCAS Futenma, Okinawa, and later flown to Iwakuni. The 'Red Lions' will return to Hawaii in April 2002 when another CH-53D squadron takes over.

This scene at Kandahar on 1 February 2002 shows a C-17 from the 437th AW, a C-141 StarLifter and a KC-130 from VMGR-352 'Raiders'. The latter unit deployed home to Miramar on 14 February.

On 28 February 2002 the French detachment of six Mirage 2000Ds arrived at the base at Manas, Kyrgyzstan, to begin operations over Afghanistan.

HC-85 'Golden Gators'

Naval Reserve's last Sea Kings

Around 50 H-3 Sea Kings remain in service with the US military, and Helicopter Combat Support Squadron Eight Five – the 'Golden Gators' based at NAS North Island, California – operates eight of these airframes. The unit was originally established as Helicopter Anti-Submarine Squadron Eight Five (HS-85) on 1 July 1970 at Naval Air Station Alameda, California. Equipped with the SH-3A, the squadron's mission was to provide inner-zone anti-submarine warfare (ASW) protection for the Navy's West coast Naval Reserve Carrier Air Wing 30 (CVWR-30). During the 1980s the squadron upgraded to the SH-3D, and in 1990 it completed its transition to the SH-3H model.

In conjunction with the disestablishment of CVWR-30, HS-85 moved to its current location at North Island and assumed the Target Launch and Target/Torpedo Recovery mission. This mission, formerly performed by an active-duty squadron (HC-1), required the reconfiguration of the SH-3H aircraft from ASW platform to the UH-3H utility and external load-capable configuration. That transition was completed in October 1994 and the squadron was redesignated HC-85 to reflect its new fleet support mission.

Commanded by Cdr Robert Johnson, the unit falls under the control of Captain Robert D. Howell, Commodore, Helicopter Wing Reserve, which is the rotary-wing element of the US Navy's Naval Air Reserve Forces. The unit comprises 23 reserve officers/pilots and eight full-time officer/pilots, making up 16 complete aircrew (each of two pilots and two enlisted aircrew men). The maintenance team comprises 115 reservists and 100 full-time personnel. Typically for a Reserve unit, the cadre of full-time officers and enlisted personnel provides continuity in the squadron's operations: they ensure that the squadron functions properly on a day-to-day basis in terms of its flight operations and maintenance, and that the resources are available to support the reservists when they report for duty.

Reservist crews at HC-85 are mainly ex-US Navy helicopter flight crew, most coming from the SH-60 community. Because the Sea King is no longer in front-line Fleet service, HC-85 effectively acts as a Fleet Replenishment Squadron for the helicopter, performing 'in-house' type conversion training for all its new pilots. With its location in San Diego and the type of mission it flies, HC-85 is a very popular 'billet' for reservist aircrew and whenever a vacancy arises HC-85 has no shortage of applicants.

Operating from NALF San Clemente, an HC-85 crew practises underslung load-carrying, using a slab of concrete. In the foreground are the baskets used to retrieve training torpedoes.

Primary mission for the unit is to provide Fleet Contributory Support for target launch, target/torpedo recovery and utility operations in support of the Southern California Offshore Range. To achieve its mission the squadron maintains a semi-permanent detachment of two or three aircraft on the small Naval Auxiliary Landing Field (NALF) on San Clemente Island, some 70 miles (110 km) off the coast of southern California. From San Clemente the squadron conducts all-weather flight operations and much of its regular training. HC-85 also has a sea- and shore-based search and rescue role,

Compared to its intended replacement, the SH-60F, the H-3 offers longer endurance, a larger cabin and amphibious capability. It remains very popular with its crews, and a useful asset in a variety of utility roles.

In addition to its transport, recovery and SAR roles, HC-85 also has a fire-fighting capability using an external bucket which can be filled from any convenient water source. This UH-3H is seen during a training sortie over the kelp beds off San Clemente Island.

and aircraft airborne on training missions can be re-tasked if a SAR call is made.

Although the unit rarely deploys away from its southern California base, it occasionally provides 'plane-guard' support for amphibious ships and aircraft-carriers. During the RIMPAC 2000 exercise in Hawaii, HC-85 provided VIP transport support around the participating fleet vessels.

As far as the future is concerned, the official word is that HC-85 should expect to keep its venerable Sea Kings for another three/four years before receiving the SH-60F. However, it would not come as a surprise to any of the 'Golden Gators' if they were still clattering over the Pacific in the Sea King at the end of the decade.

Richard Collens

CACOM-3

Colombia's Dragonflies

For many travellers, Colombia is an unknown territory – a land of myths, of cocaine, emeralds and guerrillas. Due to an apparently never-ending civil war and a reputation as the capital of the narcotics world, Colombia has a low international standing. An unemployment rate of 40 to 50 percent fuels the ranks of drug producers and rebels alike. Nevertheless, the ordinary people of Colombia remain admirably relaxed and easy-going, while the government is committed to the long fight against both the drug barons and guerrillas. At the heart of this battle is the Fuerza Aérea Colombiana base at Barranquilla and Comando Aéreo de Combate No. 3.

CACOM-3 history

During 1940 Colombia severed its relations with Nazi Germany, after two Colombian schooners were sunk by a German submarine. Following the sinking of a third in 1942, a fighting and reconnaissance squadron was established at Barranquilla, a large port town on the Atlantic coast, equipped with T-6 aircraft from Palanquero Air Base. The unit used this type of aircraft until the end of World War II.

On 1 December 1976 General Alfonso Rodriguez Rubiano became the new commander of the Fuerza Aérea Colombiana (FAC). He thought it his duty and first activity to establish an air force base in the Colombian Atlantic region to control the flow of marijuana from Guajira to the USA, which was being transported through the air and over the sea. Thanks to an effective intelligence service, the air force was well informed about the transportation of drugs.

General Rodriguez made a proposal to the Minister of Defence to obtain funds for establishing an air base, including the purchase of land and the construction of a runway, taxiways, ramps, air traffic control tower and lighting. The government gave the go-ahead for these funds. Due to economic realities, the existing airfield at Barranquilla was the only workable solution, necessitating joint civil/military use. Protests were raised by the director of the civil aviation authorities, but the overall board of aviation, with the commander of the FAC as a member, gave its approval. Consequently, parts of 'Ernesto Cortizo' airport were bought for the FAC. On 8 February 1977 contracts were signed in Barranquilla for the use of the land for a period of 99 years. These contracts were signed between the civil aviation authorities and the air force. The military area was situated near the end of runway 36 and near the main taxiway.

As air traffic with cocaine increased out of Guajira and as the government increased its crop-control operations against the secret growing facilities in the area, it became obvious that the FAC facilities in the region had to be increased as soon as possible. Through the intervention of the Rotory, the air force command received further funds for the expansion of the base. It acquired the ramp and administrative facilities of the airline Aerocosta, which had ceased operations some time before. Work was started to create an air group.

On 8 November 1977 the northern air group GANOR was established through decree 7077 of the air force command, the first commander being Major Santos E. Cueto. Initially, the northern air group used aircraft of the following types: Cessna T-41, de Havilland Canada U-6A, Douglas C-47, Hughes OH-6A, Bell UH-1H and a detachment of Lockheed T-33As from Apiay air base. The budget of 1978 contained an amount of 30 million Colombian pesos for the construction of a concrete ramp on the new air base, with two taxiways and a connection with the main taxiway.

To control the region the T-33As had to fly over sea during their intercepts of smuggling aircraft which flew north from Guajira. This was considered risky for the single-engined aircraft. During the 18th conference of air force commanders of the American continent, which was held in San Antonio, Texas, on 27 April 1978, General Rodríguez received an invitation

Colombia received a total of 32 A-37Bs in three batches, of which 14 are still active with Escuadrón de Combate 411 at Barranquilla. They are mainly used in the battle against guerrilla forces and narcotics traffickers in the north of the country.

CACOM-3's two operational units are Esc. 411 (above) which operates the A-37s, and Esc. 412 (right) which flies the Cessna O-2 and Bell UH-1.

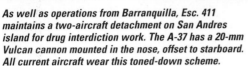

As well as operations from Barranquilla, Esc. 411 maintains a two-aircraft detachment on San Andres island for drug interdiction work. The A-37 has a 20-mm Vulcan cannon mounted in the nose, offset to starboard. All current aircraft wear this toned-down scheme.

to meet General Jones, US Air Force commander in Washington, in order to negotiate for some twin-engined jet aircraft. The meeting was held at the Pentagon, where the delicate matters of intercepting drug transports with single-engined, lightly armed T-33As was explained to General Jones.

Special attention was paid to the twin-engined A-37, which was being phased out by the USAF. General Jones promised to negotiate with his government to find a solution for this serious problem and, after a short time, the go-ahead was given by the US government for the delivery of 12 A-37s, together with two years' worth of spare parts. Training of Colombian personnel began immediately, in order to prepare supporting equipment and for the arrival of the aircraft. In December 1978 the first two aircraft (FAC2151 and 2152) were delivered to Palanquero Air Base. Shortly after, the A-37s were flown to their new base at Barranquilla, with the 10 others (FAC2153/2162) arriving by the end of 1980.

GANOR (Air Group North) was installed at Barranquilla on 13 November 1978 and, in September 1979, the headquarters of the FAC assigned the designation Comando Aéreo de Combate No. 3 (CACOM-3). Thanks to the good relationships with the USAF commander, the FAC Commander received support for an additional squadron of A-37s to complement the strength of CACOM-3. During the middle of 1984 12 additional A-37Bs (FAC2163/2174) arrived and, on 4 September 1989, the final eight A-37s (FAC2175/2182) were delivered.

On 11 November 1999, during the 80th anniversary celebrations of the FAC and 21 years of operations by CACOM-3 (including GANOR), the unit was honoured with the name of Major General Alberto Pauwells Rodriguez, commander of the FAC in 1953 and an officer who had done much to advance the stature and capability of the air force.

Operational unit

CACOM-3 is one of the units of the FAC which is on a constant alert and it flies operational missions against the National Liberation Army (ELN) settled in the Sierra Nevada, one of the most spectacular mountain ranges in South America. The highest peak, Ritacuba Blanco,

reaches 5330 m (17,487 ft) and there are more than 15 peaks over 5000 m (16,400 ft).

CACOM-3 consists of the Escuadrón de Combate 411, equipped with the Cessna A-37B Dragonfly (14), Escuadrón Aerotáctico 412, equipped with the Cessna O-2 Skymaster (1) and Bell UH-1H (6), and a transport flight with one EMBRAER EMB-110P Bandeirante, used primarily to transport VIPs and for flights to the detachment of Esc. 411 at San Andres island.

Fourteen of the original supplied A-37Bs are still in an airworthy condition. Around 12 can be found at Barranquilla, because the unit maintains a two-aircraft detachment on San Andres island, a Colombian territory just off the coast of Nicaragua and a 90-minute flight from Barranquilla.

There are still two Beech B80 Queen Airs at Barranquilla, but they were withdrawn from use some years ago and appear to be retained as decoys.

Although not confirmed, operations would appear to follow a pattern of the single Cessna O-2 flying surveillance missions. If a target is spotted, co-ordinates are radioed to the base, after which A-37s launch with rockets and/or Mk 82 bombs to attack guerrilla positions. The O-2 is also believed to provide forward air control for the Dragonflies.

Future aircraft

Due to insufficient funds, the FAC plans to use the A-37B for at least five more years. However, due to political decisions, for example those of the USA, it might be possible that replacements will be received sooner if funding arrives from outside. A likely candidate is the EMBRAER T-27 Tucano, already in service in some South American nations for armed anti-drug patrols. According to CACOM-3 pilots, the Fairchild A-10 is a possibility, with the 30-mm cannon replaced by a smaller-calibre weapon and sensor systems such as a FLIR.

Bob Fischer

The original FAC A-37 scheme was dark green and brown with full-colour markings. This aircraft carries four drop tanks and rocket pods.

KAI KT-1 Woongbee

Korea's turboprop trainer

Faced with the need to replace elderly Cessna T-37Cs and T-41s in the basic training role, and spurred by the rapid

economic growth and industrial development of the 1980s, the Republic of Korea embarked on a programme to develop and field its own training aircraft. The result was the turboprop KTX-1 (Korean Trainer Experimental) basic trainer, developed by Daewoo Heavy Industries, and the KTX-2 advanced jet trainer, led by Samsung and now known as the T-50.

Design of the KTX-1 was begun in February 1988 by Daewoo and the ADD (Agency for Defence Development), and resulted in a turboprop trainer in the mould of the Pilatus PC-9 and EMBRAER Tucano. Construction of nine prototype airframes began in June 1991, comprising five fliers (KTX-1-01 to KTX-1-05) and four static test airframes (KTX-1-001 to

Left: Two production KT-1s made the type's first international airshow appearance at the February 2002 Asian Aerospace exhibition at Singapore's Changi airport. As evidenced by the Indonesian purchase, this region is seen as an important potential market.

The cockpit of the KT-1 is designed to provide a lead-in for more advanced trainers (such as the Hawk T.Mk 67 and KAI T-50). Four small colour MFDs are provided, supported by traditional dial-type instruments. The KO FAC derivative will have a HUD.

KTX-1-004). The first flying aircraft was completed in commendably short time, being rolled out in November. It first flew on 12 December 1991, under the power of a 550-shp (410-kW) Pratt & Whitney Canada PT6A-25A engine. This engine also powered the second aircraft, which took to the air for the first time on 5 February 1993.

At the time, the ROKAF was also giving serious consideration to purchasing either the Shorts Tucano or Pilatus PC-9. The end of Tucano production effectively ruled it out of the running, while Swiss government insistence on PC-9s being delivered without any weapons capability effectively ended that aircraft's chances. These plans were dropped in 1994, the delay to the trainer replacement programme giving Daewoo time to sort out some problems encountered with the first two KTX-1s.

From the third aircraft, the powerplant changed to the considerably more powerful PT6A-62, which was adopted for all subsequent aircraft. This engine produces 1,150 shp (858 kW), but is flat-rated at 950-shp (709 kW). Prototype no. 3 flew on 10 August 1995. Still designated KTX-1, the aircraft was given the name Woongbee ('great flight') in November.

Aerodynamic problems with the first three aircraft were solved from prototype no. 4 by shortening the nose and relocating the tailplanes to a lower position further aft. No. 3 was subsequently modified to this configuration. The revised configuration was first flown by no. 4 on 10 May 1996, while the fifth prototype joined the test programme on 16 March 1998. This final test aircraft was a production-representative machine. In the course of 1,184 sorties and over 1,500 hours the KTX-1 design was fully tested, leading to the official end of development on 18 September 1998.

In production form the Woongbee is typical of the breed of turboprop trainers, with a wing of unswept but tapered planform. The aspect ratio of 7 and efficient airfoil sections give a good speed range and economic operations. The aircraft is fully aerobatic, conforming to FAR 23 stipulations, and the fuel system allows for at least 30 seconds of inverted flight. Spinning is predictable and easily recoverable, and the KT-1 is among the few aircraft able to enter and recover from inverted spinning safely. The dorsal fin, originally of straight design, gave way to a curved design, but returned to the straight-edged profile.

All primary flight controls are moved by mechanical linkage, although the tabs are actuated electrically. An airbrake under the central fuselage is actuated hydraulically, as are the trailing-edge split flaps. An automatic rudder trim system gives 'jet-like' handling qualities by

Specification – KAI KT-1 Woongbee

Wing span: 34 ft 9 in (10.60 m)
Length: 33 ft 8 in (10.26 m)
Height: 12 ft 2 in (3.70 m)
Wing area: 172.3 sq ft (16.01 m²)
Aileron span: 7 ft 7 in (2.31 m)
Flap span: 8 ft 2 in (2.49 m)
Flap area (total): 23.9 sq ft (2.22 m²)
Wheel track: 11 ft 7 in (3.54 m)
Wheelbase: 8 ft 5 in (2.56 m)

Powerplant: one Pratt & Whitney Canada PT6A-62 turboprop, flat-rated at 708 kW (950 shp), driving a four-bladed Hartzell constant-speed, fully-feathering propeller of 7-ft 11-in (2.41-m) diameter

Empty weight: 4,000 lb (1814 kg)
Maximum take-off weight: 5,500 lb (2450 kg) unarmed and 7,200 lb (3266 kg) with stores
Maximum fuel weight: 900 lb (408 kg)

Never-exceed speed: 350 kt (403 miles; 648 km/h)
Maximum cruising speed: 310 kt (357 mph; 574 km/h)
Stall speed: 70 kt (80.5 mph; 130 km/h)
G limits (never exceed): +7/-3.5
Take-off run: 850 ft (260 m)
Landing run: 1,300 ft (396 m)
Maximum rate of climb: 3,500 ft (1067 m) per minute
Absolute ceiling: 38,000 ft (11582 m)
Range: 900 nm (1,036 miles; 1666 km)
Endurance: 5 hours

automatically eliminating the varying effects of propeller torque

Internally the KT-1 is built on conventional lines, although the design is CATIA computer-driven for ease of assembly and perfect marry-up between components. The interspar voids in the wings are used to accommodate the integral fuel tanks (each of 72.8-US gal/275.5-litre capacity) and the mainwheels when retracted. The fuel tanks extend from the wingroots to a point roughly level with the flap/aileron interface, and can be augmented by external fuel tanks on the inboard hardpoints.

Both pilots sit on lightweight Martin-Baker Mk 16LF zero-zero ejection seats. The one-piece canopy hinges to starboard and incorporates Miniature Detonator Cord (MDC) which shatters the glazing prior to ejection.

Into production

Despite the dramatic economic down-turn which afflicted South Korea in the late 1990s, production of the KT-1 (the designation having dropped the 'Experimental' tag) got under way in 1999 at Daewoo's Sachon plant, following a 9 August contract award from the ROKAF for 85 aircraft. During the year – on 1 October – Korea Aerospace Industries (KAI) was formed to merge the aerospace assets of Daewoo, Samsung, Hyundai and Korean Air.

At Sachon AB on 7 November 2000 the Republic of Korea Air Force (ROKAF) took formal delivery of its first KT-1 to begin the replacement of the elderly Cessna trainers. Although only 85 are on firm order, with final delivery due by the end of 2003, the total ROKAF requirement is believed to stand at 105.

The KT-1 has provision for four underwing hardpoints, which can carry fuel tanks (inboard pylons), machine-gun pods (any pylon) or LAU-131 seven-round pods for 2.75-in (70-mm) rockets. This allows the type to undertake weapons training, light attack/COIN missions and forward air control. Under the designation XKO, and using three KT-1 development aircraft, KAI is working on an armed version with head-up display and weapon management system to answer a ROKAF requirement for a FAC aircraft to replace the Cessna O-2. Around 20 KOs (also known as KT-1C) are required, with an order expected some time in 2003.

KAI is marketing the KT-1 internationally as a potent but competitively priced trainer/light attack platform. It has also developed a complete support package, consisting of the Integrated Logistic Support (ILS) and Ground-Based Training System (GBTS) packages. The latter includes full-flight simulator, cockpit procedures trainer, a computer-based training system and electronic technical manuals. Potential customers can tailor the specification of the ILS/GBTS packages to their own requirements. In 2001 the Indonesian air force became the first foreign customer, signing for 10 KT-1Bs.

David Donald

Armed Forces of Malta

Malta Air Squadron

Today's Malta Air Squadron traces its origins back to 1971, when the newly-elected Labour Government under Dom Mintoff undertook to upgrade the island's armed forces, a move deemed necessary due to the imminent withdrawal of the British forces which had, by agreement, held responsibility for the island's defence since its independence in 1964. Although the original agreement covered a period of 10 years, the 1971 negotiations gave Britain continued use of the bases (at a cost) for a further seven years, although relieving it of the responsibility to defend the Island.

First results of the promised upgrade saw the introduction of equipment for coastal patrol duties, including patrol boats from the US and an offer of aid from West Germany in the form of four surplus Bell 47G helicopters. The first steps towards the introduction of these helicopters were taken in October 1971, when a selected group was sent to the WGAF base at Fassberg for training, pending delivery of the first aircraft. Two Bells arrived in May 1972 on board the same aircraft – a Transall C.160 – as that returning with the newly trained pilots. Two further aircraft were delivered five days later.

A decision was made to operate the Helicopter Flight from St Patrick's barracks as part of the 1st Regiment of the Malta Land Forces so, after arrival at Luqa, the aircraft were taken by road to their new home, where they were assembled by Maltese technicians with assistance from German personnel. Shortly after delivery, the first four took up their new identities of 9H-AAE to -AAH, these being allocated from the Maltese civilian register, a system that was to continue up until 1999.

In 1973 Libya donated a single Agusta-Bell 206A JetRanger. While greatly enhancing the flight's capability, this signalled a new influence over the island's affairs by the North African country. For the next six years Libya maintained a military presence on the island, a significant part of which was the stationing of a LARAF SA321 Super Frelon.

Malta's first military type was the Bell 47, of which two still remain in use after 30 years. Prior to their delivery they flew with the West German army.

Delivered on board a LARAF C-130 in June 1973, the JetRanger initially used its Libyan serial (81851), until allocated the 'civilian' registration 9H-AAJ. Like the Bell 47Gs, it operated from St Patrick's barracks. The arrival of the Libyan Super Frelon allowed the Bells to undergo a much needed overhaul. During 1975 all were air-freighted to Bergamo by the Italian Air Force, returning by mid-1976. By then, all wore the colours of the 1st Regiment: a horizontally divided red/blue roundel with a white '1' at its centre.

By 1978 the Libyan detachment had grown to include three Alouette IIIs, this necessitating a move to a larger and more practical operating base. With the RAF due to leave Luqa the following year, a temporary home was found at the former Royal Navy base at Hal Far. Early in 1979, the unit moved to its present base at Luqa, occupying the facilities vacated by No. 13 Sqn on the west side of the airfield. Organisational changes in 1980 led to the Helicopter Flight to be incorporated into a newly-formed Task Force, and to new markings consisting of a red and white roundel with a black 'TF' at its centre.

Disagreement between Malta and Libya in 1981 over oil exploration rights led to the sudden departure of the Libyan military. The three Alouettes were left behind, albeit in an unairworthy condition. These aircraft remained in limbo until 1992 when, having been officially 'donated' by Libya, they were sent by sea to Eurocopter in France for complete overhaul.

Libya's departure led to closer ties being forged with Italy, which already had a small military mission in place on the island. In 1982 a pair of Agusta-Bell AB 204Bs with SAR capability was deployed to Malta, later replaced by AB 212s. The Italian mission remains in place today, operating any longer-range search and rescue operations unsuited to the AFM's Alouette IIIs.

In 1992 the flight received two Nardi-Hughes NH500Ms from the Italian Guardia Di Finanza and, more significantly, introduced its first fixed-wing equipment with the purchase, via the USA, of five former Italian Army Cessna O-1E Bird Dogs. Shortly after, the Helicopter Flight was renamed as the Malta Air Squadron and became part of the 2nd Regiment of the Armed Forces of Malta. The insignia changed once again to its current format of a red and white roundel with a George Cross at the centre. This medal was famously bestowed on the island for outstanding valour during World War II.

The redoubtable Alouette III is the Air Squadron's principal SAR asset. Five are in use, split between three taken over from the Libyan Arab Republic Air Force detachment which operated from Hal Far and Luqa until 1981 (above), and two ex-Netherlands aircraft (below), which continue to wear their KLu camouflage. Long-range SAR cover is provided by a detachment of AB 212s from the Aeronautica Militare Italiana.

Fixed-wing assets consist of two Islanders (above) and five Bulldogs (right). The aircraft are used for training and coastal patrols. One of the Islanders (illustrated) has a nose-mounted search radar to aid its tasks. The Bulldogs came from the RAF, and replaced a similar number of Cessna O-1s used in utility roles.

In 1995, following a search for a suitable aircraft, a BN-2B Islander was added to the fleet as part of a deal involving the supply of four BAe 146s to Air Malta. The Islander required substantial repairs at its location in Malaysia before it could be delivered to Malta. It arrived in December 1995 to begin its task of coastal patrol. This first Islander now has a search radar fitted into a modified nose cone, although a second example purchased in 1998 has no such modification. Two further Alouette IIIs were purchased from the Netherlands Air Force in 1996. In the other direction a single Bell 47 and the Bell 206 were sold in November 1997, with a further Bell 47 following in 1999.

In February 2000 the purchase of four ex-RAF Bulldogs led to the retirement of the four remaining Bird Dogs, all of which were subsequently sold to civilian owners in the United States. The fifth Bird Dog, which was damaged in a landing accident in May 1993, had been used for spares recovery since that time and

has now been donated to the Malta Aviation Museum, situated on the former Royal Navy base at Ta'Qali.

With the arrival of the Bulldogs, the MAS introduced a new military serial system, having hitherto used the Maltese civil register. All of the aircraft ever used by the squadron have been allocated a military serial, even though many are no longer on strength. All serials are prefixed by AS (Air Squadron), followed by four digits The first two indicate the year of introduction, while the third and fourth indicate

the sequential number of aircraft operated by the squadron (for example, Bulldog 0020 is the 20th aircraft to be operated by the unit and entered service in 2000). The most recent arrival is a fifth Bulldog acquired in early 2001, which has been allocated '0124'.

Chris Knott and Tim Spearman

Malta's close ties with Italy led to the arrival of two ex-Guardia di Finanza Nardi NH500Ms (Model 369MC) for utility roles, including police support and medical evacuation.

Malta Air Squadron
aircraft operated since formation

Type	Military serial	Former civil registration
Agusta-Bell AB 47G-2	AS7201	9H-AAE
Agusta-Bell AB 47G-2	AS7202	9H-AAF
Agusta-Bell AB 47G-2	AS7203	9H-AAG
Bell 47G-2	AS7204	9H-AAH
Agusta-Bell AB 206A	AS7305	9H-AAJ
Cessna O-1E	AS9206	9H-ACA
Cessna O-1E	AS9207	9H-ACB
Cessna O-1E	AS9208	9H-ACC
Cessna O-1E	AS9209	9H-ACD
Cessna O-1E	AS9210	9H-ACE
Aérospatiale SA 316 Alouette III	AS9211	9H-AAW
Aérospatiale SA 316 Alouette III	AS9212	9H-AAX
Nardi-Hughes NH500M	AS9213	9H-ABY
Nardi-Hughes NH500M	AS9214	9H-ABZ
Aérospatiale SA 316 Alouette III	AS9617	9H-ADA
PBN BN-2B Islander	AS9516	9H-ACU
Aérospatiale SA 316 Alouette III	AS9617	9H-ADA
Aérospatiale SA 316 Alouette III	AS9618	9H-ADB
PBN BN-2B Islander	AS9819	9H-ADF
BAe (SAL) Bulldog T.Mk 1	AS0020	9H-ADQ
BAe (SAL) Bulldog T.Mk 1	AS0021	9H-ADR
BAe (SAL) Bulldog T.Mk 1	AS0022	9H-ADS
BAe (SAL) Bulldog T.Mk 1	AS0023	9H-ADT
BAe (SAL) Bulldog T.Mk 1	AS0124	–

Ka-50/52
Kamov's 'Hokum' family

In applying its coaxial rotor concept to the demands of the battlefield attack mission, Kamov produced a helicopter unlike any other. The Ka-50 also incorporated a modern integrated weapon system, allowing it to be operated by a single pilot. Blessed with outstanding performance and agility, the aircraft outflew and outgunned its Mi-28 rival, and has entered limited service with the Russian Army. It also forms the basis of a growing family of two-seat derivatives with even greater capability.

By the mid-1970s, the Soviet Defence Ministry leadership had come to believe that the Mil Mi-24 'Hind' attack helicopter (then the backbone of Soviet Army Aviation) did not meet Army requirements. The attempt to develop the type into a multi-role helicopter had led to increases in the aircraft's weight and size, and deficiencies in flight performance which, in turn, decreased its combat efficiency. In addition, late in 1972 the United States launched its Advanced Attack Helicopter (AAH) programme, resulting in such new combat aircraft as the Bell YAH-63 and the Hughes YAH-64. (The latter, designated Apache, was adopted for mass production and

The long-suffering 11th flying 'Hokum' is now in its third major incarnation, having originated as a standard Ka-50. It then became the Ka-52 prototype for the side-by-side two-seater, and is now in Ka-52K configuration, with nose-mounted sensor turret.

By the time the Ka-50 should have been entering large-scale production the Soviet Union was disintegrating and defence budgets were being slashed to lower than subsistence levels. Consequently, very few Ka-50s have been built and put into service, despite the obvious attractions of the type. Shown above is a production aircraft during flight test, while at right the sixth production aircraft (024) – in 'Black Shark' scheme – leads two later production machines in an air show flypast at MAKS'99.

in the mid-1980s entered the US Army inventory as its main combat helicopter.)

In light of this, on 16 December 1976 the Council of Ministers of the Soviet Union passed a resolution regarding the development of a new-generation combat helicopter that could be fielded with Soviet Army Aviation in the 1980s. The prospective helicopter's primary purpose was to destroy enemy materiel, particularly tanks on the battlefield and near the forward edge of battle. The resolution also provided for competing proposals by the Kamov and Mil design bureaux, ensuring one could later be selected for series production. At the time of these programmes, both developers already had 30 years experience in designing rotary-wing aircraft.

When developing its advanced army combat helicopter, known as the Mi-28, Mil drew on both Soviet and foreign experience in operating army aviation combat helicopters and opted for a single-rotor twin-seat machine. Flying the helicopter and employing the weapons were independent functions requiring a pilot and a weapons system operator, i.e., Mil adopted the same concept as the US AAH programme.

The Kamov design bureau – a dedicated developer of naval helicopters – had a wealth of experience from designing its complex Ka-25 and Ka-27 anti-submarine warfare helicopters,

Right: 010 was the V-80 prototype, first flying on 17 June 1982. No operational systems were installed and the aircraft was used for handling trials and envelope expansion. It was used to assess a variety of tail configurations, including one which featured small auxiliary side fins but without the extension above, effectively forming a T-tail. It also flew without wings.

Below: The second V-80 prototype, 011, featured a cylindrical fairing in the top of the forward fuselage – a mock-up of the intended Mercury LLTV installation – and the gun was fitted. The aircraft retained the vestigial tailplanes mounted high on the fin, which were removed from the third prototype, 012.

Left: The third flying prototype (012) was essentially similar to the second aircraft, and also had a Mercury fairing in the upper nose section. It introduced the definitive fin arrangement without auxiliary side fins. Note that the gun is not fitted here, revealing the cut-out into which the weapon's swivelling mount was attached.

lapping. The V-50's estimated speed was 400 km/h (249 mph). In 1975-1976, during research for its future combat helicopter concept, Kamov designed the V-100, which would have featured lateral rotor positioning and a pusher propeller for propulsion. Both the V-50 and V-100 projects were daring for their times, but neither was destined to be realised.

V-80 concept features

The Kamov Helicopter Plant initiated design of its advanced combat machine, designated V-80 (aka Product 800), on the heels of the government directive in January 1977. The programme was run by the head of the design bureau, chief designer Sergei Mikheyev, later Designer General.

Various aerodynamic configurations were considered but Kamov's traditional coaxial configuration was retained, since it offered undeniable advantages. By comparison with a single-rotor configuration, the substantial reduction in power loss resulting from the deletion of a tail rotor provided a hefty increase in main rotor thrust. Coaxial rotors resulted in a greater hovering ceiling than did single rotors developing the same power. The aerodynamic symmetry and lack of cross-links within the control system simplified the act of flying the helicopter. Such a helicopter also offered less restriction regarding sideslip angles, angular speeds and acceleration over the whole speed range. The helicopter's compact size also meant lower moments of inertia, providing rapid and effective control.

Another fundamental aspect of the design was that it was a single-seater, in which the lack of a weapons operator was to be offset by a highly-automated sight/navigation suite. The feasibility of a single-seat combat helicopter was confirmed by the experience of attack aircraft and fighter-bombers, whose single crew members acted as both pilots and navigators/systems operators.

Kamov designers believed that if flying, target detection and tracking were automated, combining the functions of the pilot and weapons operator in a single person would be possible and would not cause an excessive psychological and physical strain. By the late 1970s, the state of the national helicopter industry was such that building these automatic systems was possible; even the Ka-25 and Ka-27 featured an automatic submarine search capability, automatic navigation and flight modes, automatic data exchange among helicopters operating as a team, etc.

Using just a single crew member would reduce the helicopter's weight and increase its flight performances. It would also cut the expense of personnel training and reduce casu-

which featured an ingenious and reliable coaxial rotor configuration. The bureau also had some experience in developing Army helicopters: in 1966, in a competition for a transport/combat helicopter, Kamov developed the Ka-25F (F = *frontovoy*; front line). Derived from the shipborne Ka-25 version, it was armed with series-produced weapons including 23-mm rotating automatic cannon, six Falanga anti-

015, the fifth flying Ka-50, was the first built to representative initial production standards, complete with K-37-800 ejection seat. Both this and the preceding aircraft dispensed with the Mercury system fairing.

tank guided missiles, six rocket pods and free-fall bombs. The Ka-25F had a crew of two and could carry eight assault troops in its cargo cabin. However, preference was given to Mil's Mi-24 based on its proposed engines and sighting systems, and its new Shturm ATGMs.

In 1969, in the last stage of that competition, the Kamov team offered a radically new design in the shape of the V-50 combat helicopter. The aircraft would have had two longitudinally-positioned rotors that rotated counterclockwise in the same plane – blade synchronisation would have prevented the blades from over-

The fifth prototype (015) was used heavily on firing trials, the aircraft being seen here firing rockets. The coaxial rotor layout of the Ka-50 allows precise and rapid aiming of the fixed weaponry such as rockets, and is also used for initial aiming of the cannon. Fine-tuned aiming of the gun is handled by a limited gun traverse angle.

alties in time of war, making an overall substantial reduction in the costs of supporting army aviation.

Developed by the Tula instrument-making design bureau (designer general Arkady Shipunov), the Vikhr anti-tank guided missile system was chosen to be the V-80's main weapons system. The Vikhr ATGM's most distinctive feature was its laser guidance system which, coupled with the automatic target-tracking system, ensured high precision irrespective of range. The missile's range exceeded that of the foreign Chapparal, Roland and Rapier anti-aircraft systems, and its impact and proximity fuses – plus powerful shaped-charge/fragmentation warhead – enabled the missile to destroy both armoured vehicles and aerial targets.

In designing the helicopter, special attention was paid to the choice of cannon, which eventually settled on the 2A42 30-mm single-barrelled cannon developed by the team at Tula headed by Vassily Gryazev. The cannon was initially intended for infantry fighting vehicles, so Kamov's designers faced the problem of mounting it on the helicopter in such a manner that it would retain its main feature – high accuracy – and make up for its main deficiency – heavy weight, compared to dedicated aircraft guns. It was decided to mount it on the starboard side close to the centre of gravity – the strongest part of the airframe. This would reduce the impact of the recoil on the airframe and provide maximum precision. The restriction on the cannon's horizontal angle of rotation was overcome by the coaxial-rotor helicopter's ability to turn at any speed, with an angular speed matching that of present-day aircraft cannon. Thus, the rough horizontal aiming of the cannon was achieved by turning the whole helicopter rather than the gun.

In addition to the ATGM system and cannon, it was desirable to outfit the helicopter with a range of other weapons. As a result, the V-80's weapons suite was bolstered by rocket pods, UPK-23-250 cannon pods, bombs and KMGU small bomb pods, with provision to mount air-to-surface and air-to-air missiles in the future.

Daytime optics

The V-80's launch-and-leave Shkval automatic TV/laser sighting system was developed by the Krasnogorsk-based Zenith Optics Mechanical Plant. It was developed in two variants – for the V-80 combat helicopter and the Sukhoi Su-25T attack aircraft. The Leningrad-based Electroavtomatika scientific production association was tasked with the development of the V-80's Rubicon unified sight/navigation/flight system.

One priority was to enhance the helicopter's survivability; configuration and systems' arrangement were chosen, assemblies designed

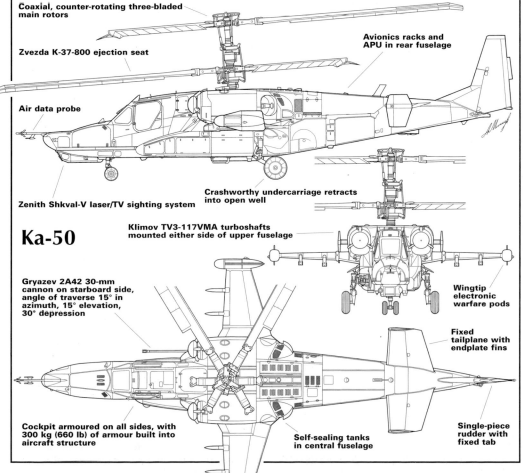

Coaxial, counter-rotating three-bladed main rotors

Zvezda K-37-800 ejection seat

Air data probe

Zenith Shkval-V laser/TV sighting system

Avionics racks and APU in rear fuselage

Crashworthy undercarriage retracts into open well

Ka-50

Klimov TV3-117VMA turboshafts mounted either side of upper fuselage

Gryazev 2A42 30-mm cannon on starboard side, angle of traverse 15° in azimuth, 15° elevation, 30° depression

Wingtip electronic warfare pods

Fixed tailplane with endplate fins

Cockpit armoured on all sides, with 300 kg (660 lb) of armour built into aircraft structure

Self-sealing tanks in central fuselage

Single-piece rudder with fixed tab

and structural materials tested all with this in mind. One advantage was that the helicopter lacked the very vulnerable tail rotor as well as the intermediate and tail reduction gearbox and control rods. The following measures were made to heighten survivability:

■ the engines were placed on both sides of the airframe, preventing them both being damaged with a single shot
■ the helicopter could fly on one engine in various modes
■ the cockpit was armoured and screened with combined steel/aluminium armour and armoured transparency
■ the hydraulic steering system compartment was armoured and screened
■ vital units were screened by less important ones
■ the self-sealing fuel tanks were filled with polyurethane
■ composites were used, thus preserving the helicopter's integrity if its load-carrying elements were damaged

■ the two-contour rotor-blade spar was developed
■ the diameter of the control rods was increased and most were positioned inside the armoured cabin
■ the powerplant and compartments adjacent to the fuel tanks were protected against fire
■ the transmission could operate faultlessly for 30 minutes with the oil system damaged
■ the power supply systems, control circuits, etc. were duplicated and placed on the sides of the airframe.

Armoured cockpit

The pilot, instruments, part of the control wiring and the sighting/navigation system were accommodated in the armoured cockpit. The armour consisted of spaced aluminium plates with a total weight of more than 300 kg (660 lb). The metal armour was fitted in the fuselage's load-bearing structure, which reduced the total weight of the helicopter. Tests at the GosNIPAS proving ground confirmed the

pilot's protection from shell fragments and shells of up to 23-mm calibre.

A unique feature of the V-80 was its rocket parachute ejection system. The helicopter emergency escape system, using the K-37-800 ejection seat, was developed by the Zvezda Scientific Production Association (chief designer Guy Severin). The pilot's safety was also increased by an undercarriage design capable of absorbing large loads in an emergency landing. The cockpit's internal volume was designed to shrink by no more than 10-15 per cent upon impact, while the fuel system incorporated fire prevention features.

A helicopter's combat effectiveness largely depends on its characteristics and ground maintenance facilities, issues that were considered at the early design stages of the V-80. Experts from the Defence Ministry NIIERAT Aircraft Operation and Maintenance Scientific and Research Institute actively participated in the work. While developing the helicopter's maintenance system, due consideration was given to

self-sustained deployment to unpaved airfields.

Thus, by the late 1970s, the Kamov design bureau had finalised the concept of its new combat helicopter – a coaxial single-seater with a wide variety of powerful weapons having a firing range that exceeded that of hostile air defence systems. It was expected to be fitted with functionally-integrated, highly-automated equipment that ensured high combat surviv-ability and the pilot's survival in an emergency, and offered long-term deployment from unpre-pared locations. The helicopter was intended to operate as part of a reconnaissance/attack system comprising aerial and ground recon-naissance, surveillance, and target designation components.

Testing

In August 1980, the USSR Council of Ministers Presidium Commission on Military Industrial Issues decided to build two V-80 and two Mi-28 experimental prototypes in order to hold comparative tests. That same year, the

Ministry of Defence issued a common perfor-mance specification for both experimental heli-copters.

The first V-80 prototype (Product 800-01, side number 010) left the Kamov Helicopter Plant in June 1982. On 17 June, for the first time, test pilot Nikolay Bezdetnov hovered the V-80, and on 23 July the helicopter made its maiden hori-zontal flight. V-80 no. 1 was designed to assess flight characteristics and try out the helicopter systems. In particular, it flew with tail assem-blies of various forms, and without the wings. It was fitted with two TV3-117V turboshaft engines.

In August 1983, the second prototype (800-02, side number 011) was completed to test the onboard equipment, avionics and arma-ment. It was powered by upgraded TV3-117VMA engines. For the first time, it was fitted with the Rubicon sighting/navigation system and NPPU-80 rotating cannon mount. The second aircraft flew for the first time on 16 August 1983.

As well as its strength against ground targets, demonstrated here in a rocket attack, the Ka-50 has a notable air-to-air capability, especially against other helicopters.

Just a little later than the V-80, the Mil design bureau began testing its two Mi-28 prototypes. The first helicopter (side number 012) made its first hovering on 10 November 1982, with the second prototype (side number 022) joining the trials in the autumn of 1983. Phase I of the V-80 and Mi-28 joint official comparative flight test programme began in December 1982 and ended in the autumn of 1984. Late in 1984, the Phase I assessment was made, which noted that many V-80 characteristics were superior to those of the Mi-28. In October 1984, the Minister of Aircraft Industry ordered V-80 series production preparations to be launched by the Arsenyev-based Progress Aircraft Plant in the Russian Far East, which at the time manufac-tured Mi-24 helicopters. Late in the year, flight trials under Phase II of the test programme assessed the V-80's and Mi-28's flight and

Ka-50 and Ka-52 specifications

	Ka-50	Ka-52
Engines	TV3-117VMA	TV3-117VMA
Take-off power, kW (hp)	2 x 1641 (2 x 2,200)	2 x 1641 (2 x 2,200)
Overall length, rotors turning, m (ft)	15.96 (52.36)	15.96 (52.36)
Rotors diameter, m (ft)	14.5 (47.57)	14.5 (47.57)
Height, m (ft)	4.93 (16.17)	4.93 (16.17)
Wing span, m (ft)	7.34 (24.08)	7.34 (24.08)
Empty weight, kg (lb)	7700 (16,975)	7600 (16,755)
Normal take-off weight, kg (lb)	9800 (21,605)	10400 (22,928)
Max take-off weight, kg (lb)	10800 (23,810)	11300 (24,912)
Max take-off weight for ferry, kg (lb)	–	11900 (26,235)
Combat load, kg (lb)	2300 (5,070)	2300 (5,070)
Internal fuel, kg (lb)	1460 (3,219)	1490 (3,285)
Fuel in drop tanks, kg (lb)	1720 (3,792)	1720 (3,792)
Max never exceed speed, km/h (mph)	350 (217)	350 (217)
Max speed in level flight, km/h (mph)	310 (193)	310 (193)
Cruising speed, km/h (mph)	270 (168)	270 (168)
Hovering ceiling, m (ft)	4000 (13,123)	3600 (11,811)
Service ceiling, m (ft)	5500 (18,045)	5000 (16,404)
Max rate of climb, m/s (ft/min)	10 (1,968)	8 (1,575)
Max *g* loading	3.5	3.0-3.5
Service range without drop tanks, km (miles)	455 (283)	455 (283)
Flight endurance without drop tanks, hours	2.4	2.4
Ferry range (with 4 drop tanks and 20-min reserve), km (miles)	1160 (721)	1120 (696)

Above: 22 was an early production Ka-50 assigned to the Torzhok centre. It was used for display work, for which a full 'Black Shark' scheme with jaws and shark fin tail markings was applied. It crashed in 1998, killing the centre's commander, Major-General Boris Vorobyov, who was also a renowned display pilot who had often demonstrated the Ka-50 at international shows.

Above: The cockpit of the standard 'daytime' Ka-50 is dominated by the large HUD, underneath which is a screen for displaying imagery from the Shkval-V system. Note the pull handles for the K-37-800 ejection seat.

combat performances in a more profound manner.

On 3 April 1985, test pilot Yevgeny Laryushin died in a crash of the V-80 no. 1. While descending quickly from low altitude to ground level in order to manoeuvre behind a terrain feature, his machine's main rotor blades crossed. The investigation revealed that the cause was the pilot exceeding the maximum allowed *g* load (it totalled -2*g*) rather than perceived deficiencies of the machine itself. Having poured over the investigation records, air force experts agreed to further tests.

The V-80's design was later modified to prevent the main rotor blades from crossing: the distance between the main rotors was increased, and the control system was fitted with a device increasing the load on the controls when the blades got too close. The tragic death of Yevgeny Laryushin dealt a heavy blow to the Kamov design bureau. His expertise, experience and instinct as a true aviator had greatly influenced the V-80's configuration.

In December 1985, in order to complete the flight performance evaluation programme, the third V-80 prototype (800-03, side number 012) was built with the Mercury low-level TV sighting system mock-up.

The Gorokhovets-based Main Rocket Artillery Directorate proving ground hosted the V-80 and Mi-28 flight trials in September 1985

to assess their combat capabilities within the framework of Phase II of the test programme. Flights were performed by test pilots of the Chkalov Air Force Scientific Research Institute, Colonel V. Kostin being the first military pilot to master the V-80.

These tests finished in August 1986. Results showed that the V-80 outperformed the Mi-28 in combat effectiveness owing to higher survivability, better flight characteristics (especially at high altitudes and temperatures), and wider weapons capabilities. During flights, the level of psycho-physical strain on the pilot was close to that on a fighter-bomber pilot but, in principle, it proved possible to combine the pilot's and navigator's functions.

The Defence Ministry Institutes concluded that the Kamov combat helicopter was more promising than the Mil. Nonetheless, a number of shortcomings were highlighted, the most serious being that the helicopter could not operate at night due to the limitations of the Mercury TV night-vision system. In accordance with the results of the comparative tests, it was recommended that Kamov improve the night-vision system, equip the V-80 with an airborne defence system, reduce the number of operations performed by the pilot while searching and attacking a target, and ensure the integration of the onboard equipment with ground and air reconnaissance systems.

Production launch

On 14 December 1987, the Council of Ministers of the USSR decided to halt development and to launch production of the V-80Sh-1 single-seat combat helicopter by the Progress plant in Arsenyev. The helicopter competition effectively ended. However, the same resolution provided for further Mi-28 development into the improved Mi-28A variant, production of which was launched at the Rostov Helicopter Plant (known today as Rostvertol JSC).

Under the resolution, the Kamov Helicopter Plant manufactured the fourth V-80 (800-04, side number 014) in March 1989, followed in April 1990 by the fifth (800-05, side number 015), which became the baseline version for future series production. Both aircraft were equipped for the first time with integral self-defence aids, i.e., the UV-26 passive countermeasures devices and a laser illumination warning system. The Rubikon system incorporated external target designation equipment. The analog weapons control system gave way to an advanced lightweight unit assembled

015 looses off a salvo of S-8 80-mm rockets (NURS) from the 20-round B-8V20A pods. Comprehensive firing campaigns have proven the Ka-50 to be a deadly accurate platform, at least in daylight.

Chechnya – baptism of fire

In December 2000, a strike/reconnaissance formation comprising two Ka-50 combat helicopters and one Ka-29 (for reconnaissance and target designation) arrived in Chechnya to join the ongoing Russian Army counter-terrorist operations in the northern Caucasus. The Ka-50s flew their first missions on 6 December.

Through January and February 2001 sorties against Chechen rebels included several weapons attacks – the first combat use of the Ka-50. The missions, which were largely accomplished in difficult mountainous terrain, reaffirmed the new machine's features, especially its high power-to-weight ratio and manoeuvrability, which exceed those of the Mi-24 by a considerable degree.

On 9 January, the pilot of a Ka-50, accompanied by a Mi-24, used S-8 rockets to destroy a rebel ammunition dump at the entrance to a gorge in the vicinity of the village of Komsomolskoye. On 6 February, operating in forested mountainous terrain south of the village of Tsentoroy, the strike/reconnaissance formation of two Ka-50s and the Ka-29 detected and eliminated with two Vikhr ATGMs a well-fortified rebel camp in a single pass from a distance of 3000 m (3,280 yards).

On 14 February the strike/reconnaissance formation conducted a search-and-destroy mission around the villages of Duba-Yurt and Khatunee. Operating over

Illustrating the effectiveness of the camouflage against a typical Russian background, a pair of Ka-50s follows a Ka-29 as they depart for the first combat deployment to Chechnya in November 2000. The Ka-29 operated as a reconnaissance/designation platform, in much the same role as that envisaged for the Ka-52.

difficult terrain, the team pinpointed and knocked out eight targets. During this operation the excellent magnification and resolution of the Shkval sighting system's TV channel was called upon.

While the two-month northern Caucasus stint of the two Ka-50s did little to tip the scale in favour of the federal forces in Chechnya, the deployment allowed the Ka-50 to demonstrate its capabilities in a real 'shooting' war. The combat experience gained has already proved useful for both refining the existing machine and in deriving new variants.

---

around a digital computer. With its K-37-800 ejection seat installed, V-80 no. 5 became the first rotary-wing aircraft in the world to boast a rocket/chute emergency escape system.

Four V-80 helicopters underwent development flight tests between July 1988 and June 1990. The no. 3 and no. 5 prototypes were used to hone the rotor system, control system, landing gear and external fuel tanks, assess flight performance and gauge the load factors affecting the machine in flight. Nos 2 and 4 prototypes evaluated weapons suites and fire control systems, avionics electromagnetic compatibility and the powerplant's gas-dynamic stability.

In 1990, the USSR Council of Ministers Commission on Military Industrial Issues authorised the manufacture of an initial batch of the helicopters, soon designated Ka-50, at the Arsenyev-based plant. The lead series-built helicopter (side number 018) took to the air on 22 May 1991 with test pilot N. Dovgan at the controls. The first stage of the Ka-50 state tests (the assessment of flight characteristics) was initiated in mid-1991 using prototypes nos 4 and 5. In January 1992, the lead production

Initial deliveries to the Russian Army were made to the Combat Training Centre at Torzhok, from where field trials began in November 1993. Further aircraft have been delivered to a regular unit in the Russian Far East.

Ka-50 was sent to the GLITs Russian Air Force State Flight Testing Centre for the second stage of state tests (the assessment of combat effectiveness), which started in February.

Soon, the Ka-50 entered the world arena. In March 1992, Designer General Sergei Mikheyev made a speech about the new helicopter at the international symposium in the United Kingdom, where the new designation of the helicopter – Ka-50 – was mentioned for the first time. In August 1992 the third Ka-50 prototype was revealed in flight during Mosaeroshow '92 at Zhukovskiy. The second production Ka-50 (side number 020) made its foreign debut at the Farnborough air show in September that year, where it topped the bill. Its rudder sported the image of a werewolf and an appropriate logo. By that time, the fifth prototype, painted black, had starred in a film called *Black Shark* – and that name has been associated with the Ka-50 ever since. The third production machine (side number 021) later sported a black shark on its rudder and wore the logo 'Black Shark'. Both it and the werewolf-adorned helicopter were showcased at the Le Bourget air show in France in June 1993. Following its debuts at Farnborough and Le Bourget, the Ka-50 was regularly seen at air and arms shows in Russia and abroad.

Ka-50s began field trials in November 1993 at the Torzhok-based Army Aviation Combat Training Centre, which had been provided with four production helicopters. The Centre's pilots and engineers made a substantial contribution to testing the Ka-50 and devising its tactics. Having mastered the new machine, the Centre's chief, Major-General Boris Vorobyov, flew the Ka-50 at international shows at Le Bourget, Dubai, Malaysia and Farnborough. He died in a crash in 1998, while testing a Ka-50 (side number 22) in the maximum allowable regime at Torzhok.

On 28 August 1995, the President of the Russian Federation authorised the fielding of the Ka-50 with the Russian Army. However, these helicopters have not yet replaced Mi-24 veterans. The collapse of the Soviet Union and a plunge in defence procurement resulted in a mere dozen machines built, compared to the several hundred planned to have been built by 2000. Three have been flown by the Torzhok-based Army Aviation Combat Training Centre. In 1999, a decision was taken to field the Ka-50 with a helicopter regiment in the Far Eastern Military District.

Night-capable Ka-50Sh

The Ka-50 single-seat combat helicopter was the forerunner of a family of army helicopters, of which the first production machines were able to operate only during the day. However, as far back as the late 1970s when the requirements of the Ka-50 were being devised, it was decided that the future combat helicopter should be fitted with aids enabling its employment around the clock and in adverse weather. Such aids could be night-vision goggles, stationary low-light television systems using special amplifiers and operating in very low light such as starlight, imaging infra-red (IIR) systems, or the only all-weather means – radar. Each system had its own demands.

Initially, the designers opted for the Mercury low-light TV system, mock-ups of which were installed in the second and third V-80 prototypes. There were plans to mount them on Mi-28 helicopters and Sukhoi Su-25T attack aircraft, as well.

At the same time, the debugging of the TpSPO-V helicopter IR imager was proceeding. It had been developed by the Geofizika Research and Production Association in Moscow and underwent trials on an Mi-24V

018 was the first production series Ka-50 (V-80Sh-1), but was subsequently modified to become the Ka-50Sh day/night demonstrator. In the first night-capable configuration to reach flying hardware stage, the Samshit-50 turret was mounted above the existing Shkval-V daytime optics suite.

The cockpit of 018 in its first Ka-50Sh incarnation was similar to that of a standard Ka-50, but reworked on the right-hand side to add a screen for Samshit-50 imagery.

prototype in 1986. A mock-up of its flight and sighting version, designated Stolb, was fitted to the fifth V-80 flight prototype under the protective glass cover of the Shkval-V system.

The first Soviet night-time surveillance and sighting systems were far from reliable, and their performance left much to be desired. The commission that assessed the Mi-28 and V-80 official comparative flight tests, which ended in August 1986, stressed that neither machine met the requirement for night-time operation set forth in the specification. Sixteen months later, when the government resolution dated 14 December 1987 announced the V-80 as the winner and authorised its series production, the Kamov design bureau was faced with upgrading the helicopter's avionics suite as soon as possible to make it capable of night-fighting.

However, it was roughly a decade before the emergence of the first night operations-capable version of the single-seat helicopter, designated Ka-50Sh. There were a number of reasons for the delay, the main one being that the Soviet Union lagged behind other industrial nations in night-time surveillance systems and sights for helicopters: for example, the US AH-64 Apache attack helicopter, fitted with the TADS/PNVS system featuring a FLIR flight and navigation sensor package, had been in production since 1984. This feet-dragging was exacerbated by the general economic crisis in the country and ensuing catastrophic slump in funding defence R&D efforts.

Refining the Mercury equipment, as well as developing other similar systems, was deemed pointless. Both domestic and foreign experience in this area proved that weight and size constraints prevented the night-vision TV systems from being installed in production tactical and army aviation aircraft – every extra kilometre of TV range would come at a prohibitive cost.

Infra-red imagers

This prompted the designers to shift their focus to developing and perfecting infra-red imaging systems. In 1993, the Kamov design bureau generated a draft design of the night-capable Ka-50. The Krasnogorsk plant launched development of the Shkval-N round-the-clock sighting system, which was to comprise an additional night-time capability wrapped around an infra-red imager designed by Geofizika (Moscow) and the State Institute for Applied Optics (Kazan). However, insufficient funding hampered progress, and in the mid-1990s Kamov decided on the temporary use of foreign-made equipment.

With Russian government approval, Kamov struck a deal with France's Thomson-CSF to buy several examples of its infra-red imagers for tests on the Ka-50. A pod-mounted imager was shown in 1995 under the wing of the tenth Ka-50 flying prototype (side number 020) during the MAKS '95 air show.

By the mid-1990s, the leading Russian manufacturer of optronics, the Urals Optico-Mechanical Plant (UOMZ) in Yekaterinburg, devised a concept of optronics for helicopters, aircraft and other vehicles, based on a line-of-sight stabilisation and control system using high-precision gyro sensors, ball bearings and a 3D torque engine. The UOMZ was able to quickly develop a range of efficient optronics systems, collectively known as gyro-stabilised optronics systems (the Russian acronym being GOES).

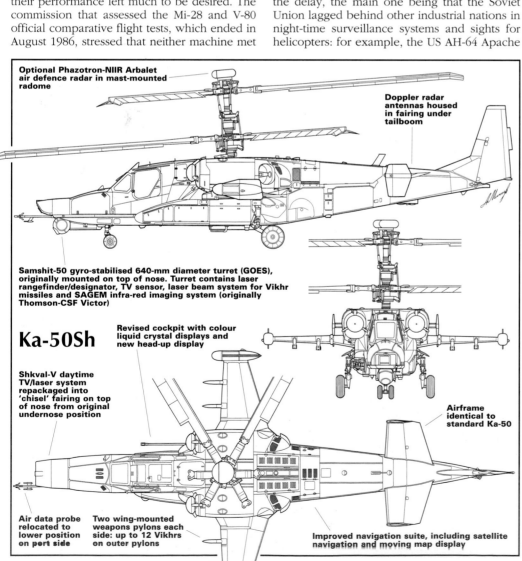

Optional Phazotron-NIIR Arbalet air defence radar in mast-mounted radome

Doppler radar antennas housed in fairing under tailboom

Samshit-50 gyro-stabilised 640-mm diameter turret (GOES), originally mounted on top of nose. Turret contains laser rangefinder/designator, TV sensor, laser beam system for Vikhr missiles and SAGEM infra-red imaging system (originally Thomson-CSF Victor)

Ka-50Sh

Revised cockpit with colour liquid crystal displays and new head-up display

Shkval-V daytime TV/laser system repackaged into 'chisel' fairing on top of nose from original undernose position

Airframe identical to standard Ka-50

Air data probe relocated to lower position on port side

Two wing-mounted weapons pylons each side: up to 12 Vikhrs on outer pylons

Improved navigation suite, including satellite navigation and moving map display

The GOES, which have several standard sizes (optical unit diameter of 640 mm/25 in, 460 mm/18 in or 360 mm/14 in), are designed as a ball moving in different axes. Inside the ball are one to five optronics modules operating as information channels: the daytime or low-light (dusk) TV channels, an infra-red imager, a laser target designator/rangefinder, a laser illumination warning device, a laser beam control system, ATGM infra-red detector, etc. Thus, the GOES offer surveillance and sighting systems with multi-channel operating capability while ensuring a single line-of-sight for all optronics modules, as well as a degree of spatial stability and the ability to traverse (±235° in azimuth and up to 160° in elevation) at high angular speed (up to 60° per second) and acceleration (up to 150° per second^2).

Samshit-50

One of the first UOMZ-built gyro-stabilised optronics systems was the Samshit-50, designed for the Ka-50. Early in 1997, it was fitted to the Ka-50's eighth flying prototype (the first production aircraft, side number 018) in its

The second Ka-50Sh configuration applied to demonstrator 018 reversed the positions of the primary sensors, so that the Shkval-V occupied a new chisel fairing in a reprofiled nose section. While this modification work was ongoing, the cockpit and navigation suites were also revamped with newer technology equipment. Note the four windows for the optical sensors in the Samshit-50 ball turret.

nose section above the optical port of the original Shkval-V daytime laser/TV system. The retrofitted machine was dubbed Ka-50Sh and its maiden flight was made by test pilot Oleg Krivoshein on 4 March 1997. Ten days later, the Ka-50Sh and a production Ka-50 (side number 22) from the Torzhok-based Army Aviation Combat Training Centre departed for the IDEX '97 arms show in Abu Dhabi (UAE), held between 16 and 20 March 1997. There, the daytime helicopter took part in demonstration flights and revealed for the first time its cannon and rockets, and made live anti-tank guided missile firings.

The Ka-50's Samshit-50 gyro-stabilised optronics system is a 640-mm (25-in) diameter ball with four optic windows and houses four

Right: At the 1999 MAKS show Kamov revealed yet another night-capable Ka-50 design with two nose-mounted GOES turrets. The smaller, upper turret houses optics to aid flying and navigation, while the larger, lower turret accommodates targeting sensors. The aircraft was the fourth prototype and was armed with four B-8 20-round rocket pods.

Below: Three colour LCDs dominate the cockpit of the later Ka-50Sh demonstrators (this is 014, the demonstrator for the two-turret configuration). The screens can be configured to display sensor imagery, navigation information or aircraft status data.

The Ka-52 Alligator was initially designed to perform a reconnaissance and attack-lead role, working with single-seat Ka-50s. This is the prototype (061), converted from an early production Ka-50. The aircraft was painted with 'Ka-50-2' titles during the initial phase of the attack helicopter proposal to Turkey.

major channels: a laser rangefinder/designator, a TV channel, a laser beam system to control Vikhr ATGMs and the Victor infra-red imager from Thomson-CSF. The equipment for night operations enables the machine to spot a tank-sized target at a range of 4 to 4.5 km (2.5 to 2.8 miles).

In addition to its optronics, the single-seat night-capable helicopter carried the Arbalet radar from Phazotron-NIIR Corporation, the antenna being housed in a mast-mounted fairing. Coupled with the upgraded self-defence suite, the radar alerts the pilot to possible enemy aircraft attack. The Ka-50Sh's improved flight and navigation suite was augmented with a satellite navigation receiver, while the original PA-4-3 automatic plotter with a paper map was ousted by a navigational LCD displaying a digital terrain map.

The aircraft was subsequently fitted with an advanced integrated open-architecture avionics suite based on multiplex data exchange channels, developed by the Ramenskoye Design Company. The cockpit management system was built around three colour LCDs and a modified head-up display (HUD). At the same time, the positions of the Shkval-V and Samshit-50 systems were altered, with the latter's ball moving downwards and the former's optic window moving upwards. The modified Samshit-50 system included an infra-red imager from SAGEM.

The updated Ka-50Sh was ready in June 1999 and displayed at the arms show in Nizhny Tagil, as well as at the MAKS '99 air show.

The Ka-50Sh's weapons suite matches that of the production daytime Ka-50, but the helicopter can fire its weapons (i.e., Vikhr ATGMs) round the clock owing to its night-vision surveillance and sighting capability. To counter aerial threats, the weapons suite was beefed up with R-73 or 9M39 Igla air-to-air missiles, while Kh-25ML semi-active laser-guided air-to-surface missile could also be carried.

To determine the best avionics suite configuration for the single-seat helicopter, in 1999 Kamov derived another prototype from the fourth Ka-50 flying prototype (side number 014), which was fitted with two gyro-stabilised optronics systems. Both are housed in the fuselage nose section, necessitating the removal of the Shkval-V system. The upper optronics system (GOES-520, with an optical unit diameter of 360 mm/14 in), has TV and IIR channels and is designed for flying and navigating, while the lower ball (GOES-330 with a 460-mm/18-in diameter) comprises the TV, IIR and laser channels and is intended to acquire, track and engage targets. To ensure night-time flight safety for all versions of the Ka-50, Kamov and the Orion Scientific Production Association suggested that crews be provided with OVN-1 night-vision goggles, which had been tested on Ka-50 helicopters in summer 1999 and been displayed at the MAKS '99 air show.

Ka-52 Alligator

In September 1994 at the Farnborough air show, it was announced that the Kamov company had produced a side-by-side twin-seat version from the baseline Ka-50 single-seat combat helicopter. As far back as 1984, when V-80 and Mi-28 comparative trials were in full swing, the Kamov design bureau proposed the development of a dedicated helicopter to conduct battlefield reconnaissance, provide target designation and support group attack helicopter operations. The system was supposed to be carried by the advanced Kamov V-60 helicopter that had also been designed in multi-role, utility and shipborne versions (it evolved into the current Ka-60 multi-role army aviation helicopter and the Ka-62 cargo/passenger variant). The system was intended to augment the combat efficiency of helicopter teams without introducing sophisticated and expensive surveillance and reconnaissance hardware.

The economic hardships that hit the nation in the late 1980s and dogged it throughout the 1990s significantly hampered the new helicopter's development programme. This prompted Kamov's Designer General to pursue mounting the reconnaissance and target designation system on a version of the production

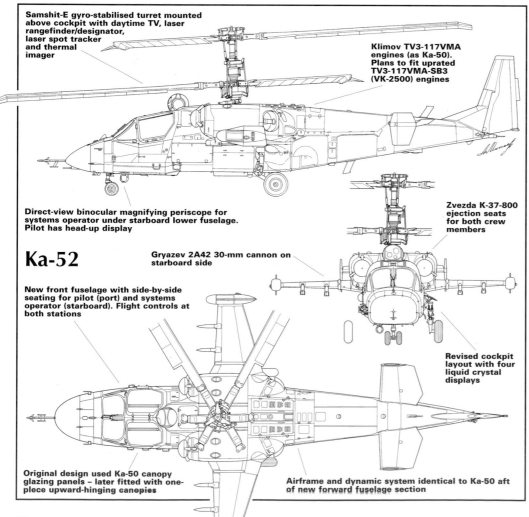

Samshit-E gyro-stabilised turret mounted above cockpit with daytime TV, laser rangefinder/designator, laser spot tracker and thermal imager

Klimov TV3-117VMA engines (as Ka-50). Plans to fit uprated TV3-117VMA-SB3 (VK-2500) engines

Direct-view binocular magnifying periscope for systems operator under starboard lower fuselage. Pilot has head-up display

Zvezda K-37-800 ejection seats for both crew members

Ka-52

Gryazev 2A42 30-mm cannon on starboard side

New front fuselage with side-by-side seating for pilot (port) and systems operator (starboard). Flight controls at both stations

Revised cockpit layout with four liquid crystal displays

Original design used Ka-50 canopy glazing panels – later fitted with one-piece upward-hinging canopies

Airframe and dynamic system identical to Ka-50 aft of new forward fuselage section

Ka-50 which, given that it retained its weapons suite, could not only conduct reconnaissance but also undertake independent combat missions, including those at night and in adverse weather.

A second crew member was needed to operate the optronics/radar reconnaissance suite. The systems operator was to be seated by the pilot's side, rather than in front of him as is the case with most army machines across the globe. The side-by-side seating arrangement was intended to facilitate co-operation between the crew in combat, especially in such demanding modes as terrain-following. It also meant that a number of instruments and controls did not have to duplicated.

Unveiled in September 1994 to the mock-up commission, the first version of the Ka-50 twin-seat reconnaissance/attack variant utilised as many of the production aircraft's components and systems as possible. The wide cockpit, with its angular outlines, reduced streamlining and did little to reduce radar signature. Therefore, the helicopter, a full-scale mock-up of which was displayed at the MAKS '95 show, was turned down by the mock-up commission; it suggested the twin-seater's fuselage nose section be reconfigured and the cockpit be provided with new glazing. This configuration was ready in mid-1996, when Kamov's experimental production facility began manufacturing the first Ka-52.

To this end, a decision was taken to employ the production Ka-50 (the 11th flying prototype) that used to have side number 021. Its nose section was removed up to frame 18 and work on assembling and attaching a new nose section began in the rig. Work was completed by November 1996, the machine having been fitted with several surveillance and sighting systems to select the best configuration. The Ka-52, which sported an all-black paint job and side number 061 (Product 806, no. 1), was

Above: The Ka-52 has slightly reduced performance compared with the Ka-50, but Kamov hopes to redress the balance by fitting uprated engines.

Right: The cockpit of the Ka-52 has four LCDs and a HUD for the pilot. The systems operator has a shrouded sight for the direct-view periscope system. Contrary to Western practice, Soviet tradition places the helicopter pilot in the left-hand seat. Kamov's Ka-60 is the first Russian helicopter to adopt a right-hand seat for the pilot.

unveiled to the media on 19 November 1996. On the eve of its presentation, the aircraft received a name of its own – Alligator – stencilled in large white letters on its left side.

Two-seat cockpit

According to the developer, the Ka-52 had 85 per cent commonality with the production Ka-50. The Alligator inherited the powerplant, rotor system, wing, empennage and landing gear intact from the baseline version. The central and tail sections remained unmodified. The main difference was a new nose section with a two-seat cockpit. The crew are seated side by side in K-37-800 ejection seats, the same seats as in the Ka-50, and enter the cockpit via the upwards-hinged canopy sections. Flight controls are provided at both crew stations. Instruments are updated, with most older needle-type instruments being replaced by four liquid-crystal displays (LCDs). Relevant flight and navigation data are fed to the pilot's HUD. The systems operator has a high-magnification optical binocular periscope

whose objective is housed by a spherical turret beneath the cockpit.

A 640-mm (25-in) diameter moving ball atop the upper fuselage forward of the rotor mast houses the Samshit-E gyro-stabilised optronics surveillance and sighting system. This comprises a daytime TV system, a French-made infra-red thermal imager, a laser rangefinder/designator and a laser spot detector. The first Ka-52 carried a swivelling mount below its chin, which accommodated the Rotor day/night surveillance and sighting system with two optical windows. One of the latter was used by the sensor for the Thomson-CSF Victor thermal imager. The Rotor was intended to be replaced in the improved radome by several modules of the Arbalet all-weather multi-mode day/night multi-function radar, including the antenna to detect ground targets. The other part of the system, designed to handle aerial threats, was to be mast-mounted in a pod above the rotors.

In addition to the entire weapons suite of the single-seat machine (i.e., the 30-mm 2A42

Designed for use on ground vehicles, the 2A42 cannon is heavy but extremely accurate. It is also fitted to the Ka-50/52's rival: the Mi-28 'Havoc'.

Two missile options for the Ka-50/52 are the R-73 air-to-air missile (left), and the Kh-25ML laser-guided air-to-surface missile (right).

This Ka-50 is armed with four Vikhr missiles on the lower mounts and two Igla-V AAMs. The Igla is a shoulder-launched SAM adapted for helicopter use.

The unusual mounting system for the Vikhr missiles is hinged at the rear, and can be lowered for missile launch. Note the caps which protect the missile launch tubes. The rocket pod is the standard B-8V20A.

Ka-52 ordnance options

The primary armament of 12 Vikhr ATGMs is carried on the outer of the two wing pylons. A range of other weaponry is available, such as rockets (including the massive 240-mm S-24), free-fall bombs (including FAB-100-120, FAB-500, RBK-250, RBK-500 and ZB-500), gun pods and Kh-25ML laser-guided missiles on APU-68-UM2 launchers. Defensive armament options are the R-73 air-to-air missile carried under the wing on an APU-62-1M rail or 9M39 Igla-V air-to-air missiles carried on dual launchers under the wingtip EW pods.

4 x 9M39 Igla-V AAM, 30-mm 2A42 cannon with 460 rounds

12 x Vikhr PTUR (ATGM), 4 x 9M39 Igla-V AAM, 30-mm 2A42 cannon with 460 rounds

12 x Vikhr PTUR (ATGM), 2 x B-8V20A pods with 40 S-8 80-mm rockets, 4 x 9M39 Igla-V AAM, 2A42 cannon

4 x B-8V20A pods with 80 S-8 80-mm rockets, 4 x 9M39 Igla-V AAM, 30-mm 2A42 cannon with 460 rounds

12 x Vikhr PTUR (ATGM), 2 x UPK-23-250 gun pods, 4 x 9M39 Igla-V AAM, 30-mm 2A42 cannon with 460 rounds

2 x UPK-23-250 gun pods, 2 x B-8V20A pods with 40 S-8 80-mm rockets, 4 x 9M39 Igla-V AAM, 2A42 cannon

2 x FAB-500 HE bomb (RBK-500 cluster bomb or ZB-500 FAE alternative), 4 x 9M39 Igla-V AAM, 2A42 cannon

4 x FAB-250 HE bomb (RBK-250 cluster bomb alternative), 4 x 9M39 Igla-V AAM, 30-mm 2A42 cannon

4 x KMGU weapons dispenser pod, 4 x 9M39 Igla-V AAM, 30-mm 2A42 cannon with 460 rounds

2 x B-13L5 rocket pods with 10 S-13 rockets, 4 x 9M39 Igla-V AAM, 30-mm 2A42 cannon with 460 rounds

4 x fuel drop tank, 4 x 9M39 Igla-V AAM, 30-mm 2A42 cannon with 460 rounds

control system, coupled with the side-by-side seating arrangement, also makes it well-suited to training pilots for Ka-50 single-seaters.

Compared to that of the Ka-50, the twin-seater's flight performance is slightly downgraded. The systems operator's seat and the introduction of new systems increased take-off weight from 9800 kg (21,605 lb) to 10400 kg (22,928 lb) which, using the same powerplant, reduced hover ceiling from 4000 to 3600 m (13,123 to 11,811 ft) and service ceiling from 5500 to 5000 m (18,045 to 16,405 ft). The machine's rate of climb and maximum allowable loading also dropped somewhat. In the future, Kamov plans to enhance the Ka-52's flight characteristics by installing more efficient VK-2500 (TV3-117VMA-SB3) engines, a joint development by the St Petersburg-based Klimov plant and Motor Sich JSC. The VK-2500 boasts 1865 kW (2,500 hp) for take-off and 2014 kW (2,700 hp) in emergency mode.

International debut

The first Ka-52 prototype was revealed at the Aero India '96 air show in Bangalore, even before it had begun flight trials. Upon its return home, preparations recommenced for its maiden flight, which was made by test pilot Aleksandr Smirnov on 25 June 1997. Following performance definition trials, the experimental Ka-52 was used to develop various surveillance and sighting suite configurations.

In 1999, it took part in demonstration and familiarisation flights in Turkey under the Ka-50-2 Erdogan programme to build an advanced helicopter for the Turkish Army (see

below), and in spring 2001 it was remanufactured into the Ka-52K version to compete in a bid for the future combat helicopter for the Army of the Republic of Korea. That country plans to buy 36 combat helicopters from 2004, under a contract valued at some $US1.8 billion. The Ka-52K faces competition from the US Boeing AH-64D Apache Longbow and Bell Helicopter Textron AH-1Z King Cobra.

The Ka-52K's surveillance and sighting equipment includes two gyro-stabilised optronics systems from the UOMZ, namely the nose-mounted GOES-342 fire-control optronics system with TV, laser and IIR channels, and the GOES-520 flight/navigation system housed in a spherical pod mounted under the cockpit, offset to port. Customers can order a weapons suite complemented by Western weapons, such as the French 20M621 750-round 20-mm cannon in an underbelly extendable turret instead of the indigenous 2A42 30-mm cannon, and standard NATO 2.75-in rockets in four 19-round pods (a total of 76 rockets) instead of the Russian-made S-8 rockets. In such a case, the Ka-52K's weapons suite retains the Ka-52's 16 Vikhr ATGMs and four Igla-V air-to-air missiles.

Korean authorities had planned to decide on a winner in November 2001, but the decision was delayed. The Ka-52 has one obvious edge over its US rivals, as South Korea has been operating 36 Kamov machines – multi-role Ka-32s – since the mid-1990s. However, given the strong US-Korea military and political links, the decision on the bid could prove to be politically motivated.

cannon with a 280-rpm rate of fire, Vikhr laser beam-riding ATGM system, 80-mm rocket pods, gravity bombs, gun pods, etc.), the Ka-52 can carry Kh-25ML semi-active laser-guided air-to-surface missiles as well as R-73 'dogfighting' or 9M39 Igla-V air-to-air missiles. The same weapons equip the night-capable Ka-50Sh single-seater. The Ka-52's fully-redundant

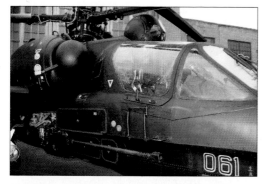

Mounting the Samshit-E turret on the Ka-52's roof reduces its 'look' angle, but raises its sightline (for better terrain masking capability) and, more importantly, leaves the nose free for other sensors. Note the 2A42 cannon installation.

Right and below: The broad nose contours of the Ka-52 Alligator can be used to accommodate a variety of sensors, including the ground target portion of the Arbalet radar system. Other options are the Rotor optical system.

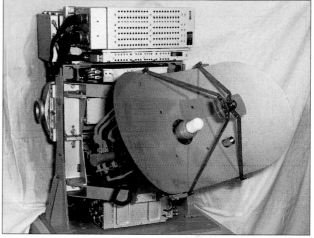

The Arbalet system consists of antennas located in a mast radome for air threats and nose radome (illustrated) for ground surveillance.

Meanwhile, the Progress plant in Arsenyev is prepared to launch the production of the Ka-52 for the Russian Army, and a few twin-seat airframes have been completed.

Ka-50-2 Erdogan

Another variant of the Ka-50 helicopter was developed for a tender issued by Turkey in late 1997. Under the ATAK programme, the Turkish Armed Forces expect to receive 145 modern gunships delivered up to 2010, the bulk of them licence-produced at a local TAI (Turkish Aerospace Industries) aircraft facility. Kamov is competing against the US Boeing AH-64D Apache Longbow and Bell AH-1Z King Cobra, joint French-German Eurocopter Tiger HCP and Italian Agusta A 129 Mangusta International helicopters.

Kamov responded to the tender with the Ka-50-2 two-seat multi-role helicopter, a derivative of the Ka-52 side-by-side all-weather combat helicopter. An important feature of the Ka-50-2 project was that it was developed in conjunction with the Lahav Division of Israel Aircraft Industries (IAI), which handles avionics development and integration.

IAI was chosen as a sub-contractor for several reasons. It has a wealth of experience in upgrading foreign-made – including Russian – aircraft by fitting them with up-to-date NATO-

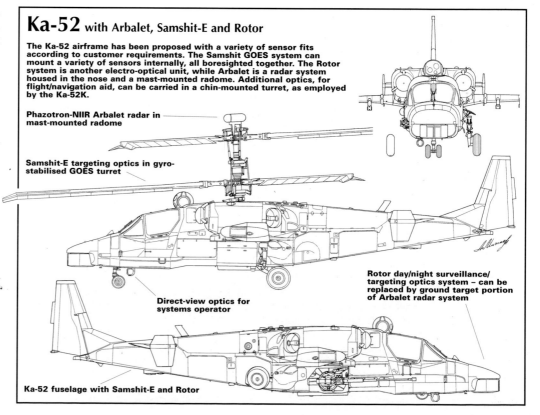

Ka-52 with Arbalet, Samshit-E and Rotor

The Ka-52 airframe has been proposed with a variety of sensor fits according to customer requirements. The Samshit GOES system can mount a variety of sensors internally, all boresighted together. The Rotor system is another electro-optical unit, while Arbalet is a radar system housed in the nose and a mast-mounted radome. Additional optics, for flight/navigation aid, can be carried in a chin-mounted turret, as employed by the Ka-52K.

Phazotron-NIIR Arbalet radar in mast-mounted radome

Samshit-E targeting optics in gyro-stabilised GOES turret

Direct-view optics for systems operator

Rotor day/night surveillance/targeting optics system – can be replaced by ground target portion of Arbalet radar system

Ka-52 fuselage with Samshit-E and Rotor

The large box containers carried by Ka-50/52s can be used for general cargo carriage for deployment, or for housing test and recording equipment during weapons trials.

compliant avionics. It is very familiar with the Turkish market, having already upgraded the Turkish Air Force F-4E and F-5A/B fighters. Finally, IAI agreed to share the financial burden of the project.

During the first stage of the tender, completed in March 1999, Kamov demonstrated the Ka-50/52's performance and revealed some elements of the future avionics suite developed by Israel. The helicopter proved its worth in the Turkish environment, with inherent high temperatures and mountainous terrain, as well as its day/night capability. All five Turkish crews that piloted the Russian helicopters appreciated the machine.

The second stage of the tender, which began in July 1999, compared the contenders' night performance and use of various weapons. The Ka-50-2 successfully completed demonstration flights in August. Its performance and avionics suite won high praise from local pilots who took an active part in its handling. Demonstration of the helicopter's firepower was another success – it fired the Vikhr anti-tank guided missiles, rockets and 30-mm cannon, hitting all of the targets.

Tandem cockpit requirement

At this stage, Turkey placed a number of additional requirements on the Ka-50-2, the most serious of which were reconfiguration of the side-by-side cockpit into a tandem one, replacement of the Russian 30-mm 2A42 cannon with a French 20-mm NATO-compliant flexible gun, and replacement of Russian 80-mm rockets with NATO-standard 70-mm

ones. Kamov prepared a number of proposals to meet these requirements, and by the September IDEF '99 exhibition a full-scale mock-up of a new helicopter had been built and delivered to Ankara. The new variant became known as the Erdogan (Turkish for warrior).

Kamov had been able to build this new machine so quickly due to the modular design of the Ka-50-2 – the nose part of the fuselage was simply cut off and replaced with a tandem one with a larger transparency area and less armour protection. Unlike in the Apache and Mangusta, the pilot was in the front seat and the operator in the rear, slightly above the pilot. The remainder of the helicopter fuselage, including the airframe, surfaces and assemblies, were left untouched. A 20-mm GIAT cannon was fitted to a mid-fuselage turret that was extendable after take-off. Many other requirements pertaining to the range of ammunition and equipment were also satisfied.

The Ka-50-2 avionics is of open architecture and is based on two R-3081 MDP processors and two Mil-Std-1553B-compliant databuses, one for the fire control system and one for the flight/navigation system. The main surveillance and sighting systems include the gyro-stabilised HMOPS consisting of FLIR and TV channels, a laser rangefinder and a laser ATGM guidance system, as well as NavFLIR and two IHS helmet-mounted target designators for each pilot. All information is displayed on four multi-function colour displays on the instrument panels.

The helicopter's flight/navigation system includes an INS/GPS navigation package and TACAN radio navigation system. Communications equipment comprises three VHF/UHF radios and a short-wave radio. Its electronic countermeasures suite consists of an Elint station, laser illumination sensors, an infra-red detector and a chaff/flare dispenser.

The Erdogan's primary weapons are its retractable belly-mounted 20-mm cannon in a turret, 12 Vikhr ATGMs (or 16 future foreign-made ATGMs), and 38 to 76 2.75-in rockets in two to four 19-round pods. Aerial threats could be handled by four Stinger AAMs.

The Ka-52K is a version of the Alligator on offer to the Republic of Korea Army to fulfil an outstanding attack helicopter requirement. The demonstrator is shown here armed with four Vikhr missiles on each wing. Alternative ATGMs could be integrated if a customer stated the requirement.

The main features of the Ka-52K demonstrator are the nose-mounted GOES-342 turret for the targeting optics, and the GOES-520 turret for flight aid/navigation under the port side of the forward fuselage. The demonstrator retains the 2A42 cannon, but Kamov can offer a French-built 20-mm cannon as an alternative.

The winner in the bid was determined in July 2000 when the Turkish government declared its intent to select the US-made King Cobra. Mention was made that the Russo-Israeli Kamov Ka-50-2 Erdogan remained on the short list and that the results of the tender were subject to reconsideration if the US failed to meet Turkish conditions.

The US and Turkey have yet to resolve their differences over technology transfer for licence-production of onboard computers and relevant software for the AH-1Z. In August 2001, the Pentagon announced that it would never allow the computer technology transfer and threatened to revoke the export licence. Ankara reciprocated with a sharp warning that the US stance could have an impact on both the AH-1Z acquisition and other joint military programmes. So, despite the official results of the tender, the Ka-50-2 still has some chances in Turkey. However, it is clear that Ankara's final decision will hinge on political expedience first and foremost.

Future Ka-50s

In 2001, the Kamov company announced its intention to develop another two-seat derivative of the Ka-50 helicopter. Like the Erdogan, it features the tandem configuration, but the cockpit design is believed to be different. Unlike the Ka-50-2, the new machine – designated Ka-54 – will feature a cockpit with much better armour protection, and the avionics and weapons suites will include advanced Russian-designed systems. The Ka-54 will be offered both to the Russian Army and foreign buyers.

The company is ready to develop other Ka-50 versions, as well, which could meet specific needs of demanding customers. The company guarantees that primary capabilities will be retained: unrivalled manoeuvrability, high reliability, flight safety and survivability, excellent combat efficiency. These qualities are grounded in the helicopter's unique coaxial rotor configuration, its ingenious and reliable design, and the top-drawer avionics and weapons suites, whose superiority has been proven by theoretical research, comparative trials and field operation.

Thomas Andrews

These two photos show Kamov's mock-up of the Ka-50-2 Erdogan proposal for Turkey. Unlike other gunships, the Ka-50-2's pilot occupies the front cockpit. The mock-up was displayed with standard Ka-50/52 weapons such as Vikhr and B-8V20A rocket pods, and also carried Vympel R-73 air-to-air missiles on wingtip launchers. If Turkey did adopt the type, Western equivalents would be substituted in most cases.

RSAF Republic of Singapore Air Force

Photographed by Peter Steinemann

Above: *Singapore's main combat capability is vested in the F-16C/D Block 52 fleet, which flies with No. 140 (Osprey) Squadron (illustrated) and No. 143 (Phoenix) Squadron, both located at Tengah. The F-16Cs are primarily roled for air defence, as displayed by these aircraft carrying AIM-7 Sparrows and AIM-9 Sidewinders, while the two-seat 'big-spine' F-16Ds have an air-to-ground tasking, for which AGM-65 Mavericks and GBU-10/12 Paveway II laser-guided bombs are important weapons. The Peace Carvin II/III/IV batches cover a total of 24 F-16Cs and 44 F-16Ds, the final 20 of which will be delivered in 2003.*

Left: *Mobility for the army has received a massive boost with the arrival of the CH-47D Chinook, which serves with No. 127 Squadron at Sembawang, home to all of the RSAF's home-based helicopter units. Type training is accomplished in the US at Grand Prairie, Texas, with G Company, 146th AVN.*

With a tiny island state to defend, the RSAF strives to maintain a qualitative edge over its neighbours through an aggressive procurement and training policy. The result is a highly professional air force with the best equipment it can acquire.

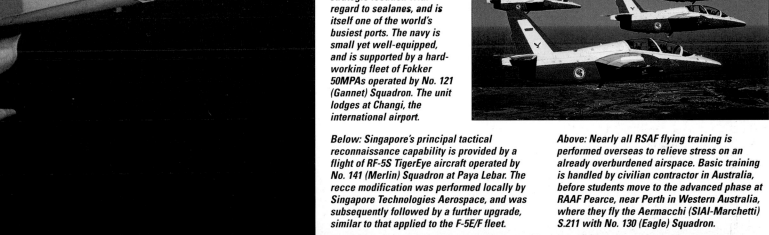

Above: Although territorial waters are small due to the small size of the island itself, Singapore occupies a strategic location with regard to sealanes, and is itself one of the world's busiest ports. The navy is small yet well-equipped, and is supported by a hard-working fleet of Fokker 50MPAs operated by No. 121 (Gannet) Squadron. The unit lodges at Changi, the international airport.

Below: Singapore's principal tactical reconnaissance capability is provided by a flight of RF-5S TigerEye aircraft operated by No. 141 (Merlin) Squadron at Paya Lebar. The recce modification was performed locally by Singapore Technologies Aerospace, and was subsequently followed by a further upgrade, similar to that applied to the F-5E/F fleet.

Above: Nearly all RSAF flying training is performed overseas to relieve stress on an already overburdened airspace. Basic training is handled by civilian contractor in Australia, before students move to the advanced phase at RAAF Pearce, near Perth in Western Australia, where they fly the Aermacchi (SIAI-Marchetti) S.211 with No. 130 (Eagle) Squadron.

Above: Northrop F-5s serve with three squadrons at Paya Lebar, in three variants (F-5S, two-seat F-5T and recce-configured RF-5S). This No. 149 (Shirka) Squadron F-5S prepares for a practice bombing sortie.

Right: Paya Lebar's F-5 units are represented here by F-5S Tiger IIs from Nos 141 (background), 144 and 149 (foreground) Sqns, together with an RF-5S from No. 141 Sqn. The upgraded F-5 is a cost-effective multi-role fighter, mainly used in the air defence role with a secondary attack tasking. Singapore is now actively studying a replacement, aircraft under consideration including Rafale, Typhoon, Su-30 and several US types.

Singapore's original Peace Carvin I F-16A/B Block 15OCU batch remains in service with No. 140 Sqn (below) at Tengah, albeit for second-line duties. The unit's 'A' Flight acts as an OCU for the F-16, taking students from the US-based F-16 conversion unit and providing local familiarisation training. In 2000 the squadron also provided two aircraft for the occasional 'Black Knights' aerobatic team (right), while another aircraft serves as an advanced cockpit technology demonstrator.

Below: Two of No. 140 Squadron 'C' Flight's F-16Ds show off their GBU-10 and AGM-65 armament. The RSAF's F-16C/Ds are Block 52s powered by P&W F100-PW-229 engines. In addition to the two Singapore-based units (Nos 140 and 143 Sqns), the RSAF has another F-16C/D squadron (428th FS/27th FW) flying at Cannon AFB, New Mexico, which it regards as an 'operational' unit rather than a training one.

Above: Singapore has a tiny land mass and airspace, so early warning of attack is imperative. As well as ground radars situated in the hilly interior of the island, radar warning is provided by No. 111 Squadron's four-strong fleet of E-2C Group 0 Hawkeyes, which were delivered in 1987. Despite having received various upgrades, replacement of these aircraft is seen as a priority for the RSAF in the future.

No. 121 (Gannet) Squadron operates all of the RSAF's Fokker 50 fleet, which is divided into four UTA utility transport aircraft (below), including one in a VIP transport fit, and five MPA Maritime Enforcer Mk 2 surveillance aircraft (bottom). The MPAs are readily identifiable by their large observation windows in the forward fuselage, underbelly search radars, wing- and tail-mounted ESM antennas, and undernose FLIR turret. The aircraft are also fitted with pylons on the fuselage sides and under the wings, from which a variety of stores can be carried, including AGM-84 Harpoon anti-ship missiles. Unconfirmed reports suggest that one of the Fokker 50s has an electronic surveillance function.

Below: F404-powered A-4SU Super Skyhawks remain in use with two units at Tengah. This aircraft serves with No. 142 (Gryphon) Squadron, the other unit being No. 145 (Hornet) Squadron. One A-4SU/TA-4SU squadron was transferred to Cazaux in France to provide advanced/weapons training as No. 150 Squadron.

Above: Singapore's F-5s are arguably the most advanced in the world, having undergone a major STAe/Elbit upgrade with new cockpit displays and HOTAS controls, multi-mode Grifo-F radar, defensive avionics and revised leading-edge root extensions, at the expense of one of the 20-mm cannon. Refuelling probes had been fitted in an earlier programme.

Above and right: Paya Lebar-based No. 122 Squadron is equipped with 10 Hercules, including four KC-130B tankers and a single KC-130H, used to provide refuelling support for the A-4/F-5 fleet. The remainder are in transport configuration, although reports suggest that one has a Sigint role.

Top right: The inability of the KC-130 to refuel the F-16 led to the purchase, in September 1999, of four ex-USAF KC-135R Stratotankers. After refurbishment and upgrade with Pacer Crag digital cockpits, the aircraft were delivered to McConnell AFB, Kansas, for crew training. In September 2000 the first aircraft arrived in Singapore for operation by a newly-formed No. 112 Squadron, which set up base at Changi International Airport. As well as the standard refuelling boom, for use with the F-16s, the aircraft have wingtip Mk 32B hose-drogue units for refuelling A-4s and F-5s.

Above: All RSAF Singapore-based helicopters are located at Sembawang in the north of the island. No. 124 Squadron operates six AS 350B Ecureuils for basic rotary-wing instruction, one of the few training processes which takes place in-country. These aircraft were originally flown by No. 123 Sqn, before being reallocated after the delivery of armed Fennecs.

Right: Principal army support helicopter is the AS 332M (AS 532UL) Cougar, which flies with No. 124 Squadron at Sembawang. A second Cougar squadron was established with AS 332M-1 aircraft, but this was subsequently relocated to the Australian Army Aviation base at Oakey, Queensland. Here No. 126 Sqn operates as the Cougar OCU, and also supports army exercises and training in Australia.

Above: In the anti-armour role the RSAF currently relies on No. 123 Squadron's fleet of AS 550C2 Fennecs, which are equipped with the HeliTOW system. From 2002 the RSAF will receive its first batch of eight AH-64D Longbow Apaches, although initial deliveries will be to a US-based training unit, probably at Marana, Arizona. At the time of the initial June 1999 order, an option was placed for 12 more AH-64Ds, and this was exercised in August 2001 to raise the RSAF Apache buy to 20.

Right: The elderly 'Huey' remains in service, flying with No. 120 Squadron at both Sembawang and a training detachment at Oakey, Queensland. The sizeable fleet includes both Bell UH-1Hs and Agusta-Bell AB 205s.

Below: The initial six-aircraft batch of Chinooks for Singapore were standard CH-47Ds, but the follow-on order was for the CH-47SD variant, with 'glass' cockpit, enlarged 'saddle' tanks and weather radar. Early deliveries were to the Dallas-based training establishment, the first Chinook arriving with No. 127 (Prancing Horse) Squadron at Sembawang in mid-1999.

Lockheed Martin
F-22
Raptor

Conceived in the free-spending Cold War era to be unassailable in air combat, the F-22 was born into a different world where the nature of air warfare is less certain, and where funding is far from assured. Adding an air-to-ground capability to its repertoire greatly aided the F-22 in its battle on Capitol Hill, and the programme has achieved Low-Rate Initial Production status with all nine flying and two static EMD aircraft delivered. Meanwhile, Air Force pilots are eager to get their hands on the F-22 and begin to meet the next big challenge: how to unlock the phenomenal capability of the aircraft in combat. The combination of high-speed supercruise, thrust-vectoring, integrated sensor-fused avionics, datalink and stealth is new to the fighter world and will require a complete revision of tactical doctrine. After all, as F-22 Chief Test Pilot Paul Metz puts it, "No-one's been invisible before."

A trio of Raptors flies over the Sierra Nevada mountains near the test base at Edwards AFB, California. To push through the aggressive development programme rapidly Lockheed Martin has built nine flying EMD airframes, of which six are fitted with integrated avionics, and one with structural measurement equipment. These aircraft now carry the brunt of the test work, the first two aircraft having tested aerodynamic and propulsion properties, initial stores separation, flutter loads and other non-systems related areas. Test work is performed by the F-22 Combined Test Force, jointly staffed by personnel from the Air Force and contractors. It is also known as the 411th Flight Test Squadron, part of the 412th Test Wing at the Air Force Flight Test Center.

Above: Current plans call for the F-22 to be declared operational in December 2005 with the first squadron of the 1st Fighter Wing at Langley AFB, Virginia. The wing will receive these aircraft from the third Low-Rate Initial Production (LRIP) batch, for which long-lead funding was provided in February 2002. Serials will be 03-4041 to 03-4067, provided the batch is authorised. Preceding them, and following on from the 17 EMD and PRTV test aircraft, the first two LRIP batches (Lot 1 01-4018/-4027 and Lot 2 02-4028/-4040) will equip the 325th Fighter Wing at Tyndall AFB, Florida, the designated training unit for the type. There will be no two-seater, despite earlier plans to build two F-22Bs as part of the EMD fleet.

Above right: As initially conceived, the F-22 was a pure 'air dominance' fighter to replace the F-15. Today the emphasis has shifted towards multi-role operations, in which the F-22 will carry bombs as well as missiles. Furthermore, a dedicated bomber version (FB-22) is under study.

Lockheed Martin's F-22A Raptor fighter has now been in the engineering and manufacturing development (EMD) stage for 11 years. Before EMD, there was a demonstration and validation phase that began in late 1986 and included the testing of two YF-22 prototypes and two examples of the competing Northrop YF-23. Serious studies of the Advanced Tactical Fighter (ATF) started four-to-five years before that. Some of the pilots who will start the first F-22 squadron at Langley will be training on a brand-new fighter that was being sketched on bar napkins when they were still toddlers.

Can an aircraft that has been around for so long still be revolutionary? Some critics say that it is not. Everest Riccioni, a retired USAF colonel and a founder member of the 'fighter mafia' whose activities paved the way for the F-16 and F/A-18, says that the F-22 "is really not a very spectacular increase in capability over current aircraft. It was maldesigned to an extent, conceived for a mission that no longer exists, and is totally irrelevant to modern warfare."

A different view comes from former USAF chief of staff General Ron Fogleman. In an interview in the USAF's *Airpower Journal*, published in early 2001, Fogleman said that the F-22's true capabilities have only been hinted at. "In the black world, the F-22 is a truly revolutionary airplane," Fogleman said. "On the surface, it looks conventional, like an F-15 with some stealth capabilities. But the combination of stealth, supercruise, and integrated avionics is a quantum jump. It will allow the United States to cease worrying about air superiority for the first 35 years of the next century."

Riccioni's charge of irrelevance nevertheless contains a core of truth. The F-22 was designed primarily as an air-to-air fighter, but the fighter threat to US and allied forces remains far smaller than anyone would have dreamed possible in 1991, when the F-22 EMD programme started. Since late 1999, the USAF has increasingly pushed the F-22 as an air-to-surface weapon – to the point where a Lockheed Martin proposal to modify the design into a medium bomber has gained serious consideration. F-22

critics argue that the USAF is simply attempting to make up a mission for the fighter. But it is also a matter of record that the technology of the F-22 was never solely aimed at defeating fighters.

Fighting in the SAM zone

When the USAF started the ATF programme in 1982, its mission was to perform air superiority missions above a dense, interlocking field of surface-to-air missiles (SAM) coverage zones in central Europe. Even public briefings in the early 1980s discussed the importance of speed, altitude and stealth in reducing the effective range of SAMs and preventing those missiles from inhibiting F-22 operations.

While the end of the Cold War caused the fighter threat to recede, it actually led the USAF to the realisation that the integrated air defence system (IADS) threat was in some ways worse than had been suspected. This happened because the US was able to conduct clandestine tests of Russian SAMs. "One of the side benefits of the end of the Cold War was our gaining access to foreign weapons," said former USAF chief General Fogleman. "We discovered that the SA-10s, -11s, and -12s are much better than we thought." In the 1996 Taiwan Straits crisis, says Fogleman, "we sent two carriers in and watched the Chinese move their SA-10s up." The SA-10 missiles would have severely inhibited strike operations by the carrier's air group. "Those two carriers did nothing more than make a political statement, which is fine as long as that is all that's necessary."

IADS can potentially prevent or limit successful attack operations. In early 2001, Air Combat Command chief General John Jumper (now the USAF's Chief of Staff) commented that "the greatest problem we have today with the F-117 and the B-2 ... is that they can't protect themselves from air-to-air and visually directed air-to-surface threats." Operating at medium-to-high altitude defeats the

latter class of threat, but the risk of a chance encounter with a fighter is what confines the B-2 to night operations.

This leads, in Jumper's view, to the concept of the Global Strike Task Force (GSTF), in which F-22s protect the bombers from the fighters and suppress the air defences. The F-22 performs two roles in the GSTF. If two F-22 squadrons are deployed, half the aircraft provide defensive cover for strike aircraft; non-stealthy intelligence, surveillance and reconnaissance aircraft such as Global Hawk unmanned air vehicles; and high-value targets such as AWACS. The other F-22s, armed with air-to-surface precision weapons, are tasked with destroying the most threatening SAMs and nodes in the IADS. States General Jumper: "We wouldn't be able to do it in all cases, but in certain cases we now enable daylight stealth with the F-22."

Programme slips

Despite the USAF's unwavering commitment, however, this is a perilous time for the F-22 programme. Last August, the Pentagon's Defence Acquisition Board (DAB) authorised the start of low-rate initial production (LRIP), a programme that will include more than 90 aircraft. But the start of operational testing slipped by eight months last year, from August 2002 to April 2003, due largely to late delivery of development aircraft. If it slips further, the Pentagon will not be able to authorise the start of full-rate production, a decision currently due for 2004. Congress's

Above: Raptor 02 flies past the Edwards tower. In addition to the flying EMD aircraft (91-4001 to 91-4009), there are two static test airframes: 3999 for structural load tests and 4000 for fatigue tests. By March 2002 Raptor 4000 had accumulated 5,232 simulated flight hours, over 65 per cent of the F-22's anticipated 8,000-flight hour first life cycle.

Far left: Lockheed Martin chose the plant at Marietta, Georgia, to be the final assembly location for the F-22. The site is traditionally the home of Lockheed's airlifters, having produced the Hercules, StarLifter and Galaxy. LM's other major fighter plant, Fort Worth, is still awash with F-16 orders, and is gearing up for F-35 production. It does, however, build the central fuselage for the F-22.

F-22 industrial team and defence industry consolidation

This table lists the main contractors on the F-22 programme, together with their primary areas of responsibility. Of the 14 listed, only three retain their original name, a graphic illustration of the effects of the consolidation – whether by merger or acquisition – that has affected the US defence industry during the lifetime of the programme.

Current team member	Original name	Role in programme
Lockheed Martin	Lockheed	Prime contract team leader, forward fuselage, final assembly
Lockheed Martin – Fort Worth	General Dynamics	Mid-body and systems
Boeing	Boeing	Aft fuselage, wings, avionics integration and training
Northrop Grumman	Westinghouse	Radar team leader
Raytheon	Texas Instruments	Radar team partner
BAE Systems	Sanders	ALR-94 ESM
BAE Systems	Sanders	AAR-56 Missile Launch Detector
BAE Systems	Tracor	ALE-52 countermeasures dispensers
BAE Systems	Lear Astronics	Flight control and vehicle management computer
TRW	TRW	ASQ-220 communications, navigation & identification system
General Dynamics	General Electric	M61A2 gun
Pratt & Whitney	Pratt & Whitney	F119 engine and nozzles
Raytheon	Hughes	Common Integrated Processor
Collins – Kaiser Electronics	Kaiser	Head-down displays

Above: Displaying a '100 Flights' logo below the cockpit, Raptor 4001 refuels from a 412th TW KC-135R. The first F-22 was grounded in late 2000 and dispatched to Wright-Patterson AFB, Ohio, to determine the effects of hits from rounds up to 37-mm calibre. The aircraft was placed under load for the tests, which also involved six jet engines blasting it from the front to simulate a 400-kt airflow.

Above right: Lt Col Bill Craig settles 4003 down at Edwards on 15 March 2000 after its cross-country delivery from Marietta.

General Accounting Office has expressed its belief that test targets will not be met. USAF Secretary James Roche issued a blunt public warning to an audience of defence industry leaders in February 2002. "If we have steady budgets, we can perform. At this point, we have nobody to blame but us," he said. "The prime is trying its best. Subcontractors, please know how critically important you are. You must deliver. You must deliver on time."

While there do not appear to have been any "show-stopper" problems in the programme, it has not been trouble-free. Quite early in development, Pratt & Whitney discovered a performance shortfall in the F119 engine, necessitating a redesign of the high-pressure turbine. There have also been structural problems. Before the first aircraft flew, the team realised that over-zealous weight-reduction efforts had left the airframes of the first two aircraft, 4001 and 4002, below the required strength. However, later tests and studies showed that the problem was worse than expected, with the result that much of the envelope expan-

sion programme and structural testing could not be completed until aircraft 4003 was available. This aircraft joined the programme nine months late, in March 2000.

One continuing problem has been the late delivery of development aircraft. Out of the nine EMD F-22s, only one – 4002 – made its first flight on time. The others flew an average of 10 months late; 4004 and 4005, the first avionics test aircraft, were 15 and 12 months late, respectively. There is no single problem responsible for the delays. Structural problems in the tail caused some aft fuselage sections to be delivered late from Boeing, and some avionics components were late. Generally, too, the contractors found themselves doing more scrap and rework than they had expected, with a larger number of parts that did not conform to specifications.

Attack from Congress

Further delays will be a serious political problem. In mid-1999, the F-22 came under fire from Congress, which had already capped the programme's cost. In July 1999, the House of Representatives voted to eliminate all $1.8 billion in F-22 production funds from the FY2000 budget. The cut would have prevented the USAF from ordering the first low-rate initial production (LRIP) batch of six F-22s and would have destroyed the complex system of incentives

F-22 Raptor history

The F-22 traces its origins to operational needs which were identified in the mid-to-late 1970s. At that time, the Soviet Union was close to completing the replacement of first-generation supersonic fighters with aircraft such as the MiG-23 and Su-24. Prototypes of the MiG-29 and Su-27 were observed in 1977-78. New SAMs were under development – and earlier SAMs had inflicted heavy losses on Israeli aircraft in 1973. If Soviet forces could prevent NATO aircraft from operating over their territory and attacking rear-area targets, NATO would have extreme difficulty resisting their advance on the ground.

In early 1980, influenced by the emergence of the new Soviet fighters, the USAF started to focus its attention on a new air superiority fighter, starting an Advanced Tactical Fighter (ATF) programme in April 1980. In May 1981, the USAF issued a request for information (RFI) for the ATF. The new fighter would replace the F-15, starting in the mid-1990s. A high-performance, air-to-air ATF, the USAF reasoned, could defeat advanced Soviet fighters and threaten the new A-50 'Mainstay' Soviet airborne warning and control system. In so doing, it would restore the low-altitude sanctuary, allowing the strike aircraft to enter hostile airspace below and between the interlocking SAM zones. (The secret F-117 would help by attacking the larger SAM sites and control centres.)

Responses to the RFI ranged from a Mach 2.8 'battlecruiser' by Lockheed to a Gripen-sized fighter from Northrop. By late 1982, however, after the RFI responses had been evaluated, the USAF had focused between these extremes, aiming at a supersonic-cruise fighter some 15-20 per cent larger than the F-15 Eagle.

In late 1982 the USAF issued a draft request for proposals (RFP) for the concept definition investigation (CDI) stage of the ATF programme. By this time, ATF was a real programme with money behind it. The USAF's fighter 'roadmap' called for the service to buy a strike version of the F-15 or F-16 first (McDonnell Douglas won this contest with the F-15E), and follow that with a fleet of 750 ATFs to replace the F-15, starting in the mid-1990s.

In May 1983, the USAF issued the final RFP for the

CDI phase – and amended it eight days later to place more emphasis on stealth technology. The secrecy surrounding the F-117 (which had flown in 1981) and other stealth programmes was so tight that even the ATF programme managers had not been made aware of technology that could allow them to combine supersonic speed, agility and stealth. In early 1983, however, the USAF's secret stealth programme office contacted the ATF team and brought them up to date on the new projects.

Concept definition contracts were awarded to Boeing, General Dynamics, Grumman, Lockheed, McDonnell Douglas, Northrop and Rockwell in September 1983. At the same time, the USAF issued contracts to Pratt & Whitney and General Electric, covering the design and construction of prototypes of the ATF's engine.

The next stage of the programme was to be demonstration and validation (Dem/Val), in which the USAF would pick two or more companies to test the riskiest ATF technologies at full scale. An RFP for this stage was released in October 1985 – and, once

With civilian registration N22YF, the General Electric YF120-powered YF-22 was the first of the two Dem/Val aircraft to fly. It did not fly again after the end of that phase, although the F119-powered aircraft flew in support of the EMD phase.

again, was amended almost immediately with tougher stealth requirements. The original RFP called for stealth in the frontal sector only, but studies at Lockheed and Northrop had shown that all-aspect stealth was attainable. Northrop was working on an extraordinary design which somewhat resembled a stretched, sleeker version of the Tacit Blue radar surveillance aircraft, while Lockheed's ATF studies had clearly evolved from the F-117.

In May 1986, just after the Dem/Val proposals were submitted, the USAF decided that Dem/Val should include a flight-test programme. The combination of stealth with supersonic cruise and agility had increased the risk of the project: the Packard Commission on defence procurement reform had just issued its report, heavily in favour of flight-

and milestones which the USAF, its contractors and their subcontractors had designed to keep the programme's cost within the limits that Congress itself had imposed. The Pentagon vigorously opposed the Congressional move, fearing with some justification that the project might go the way of the B-2, with development stretched and production cut back until the unit cost reached astronomical levels. By the end of the year, funds had been restored, although the six F-22s were described as PRTV II aircraft to save Congressional face.

The second F-22 is seen during its delivery flight from Marietta to Edwards, a 4.5-hour trip undertaken on 26 August 1998 by Lt Col Steve Rainey.

testing; and the USAF believed that two competitors, Lockheed and Northrop, were far ahead of the others. (General Dynamics placed third and Boeing fourth. McDonnell Douglas placed fifth, hampered by conservative programme leadership, and Rockwell and Grumman trailed the pack.) The contractors responded by forming teams: Boeing and Lockheed with General Dynamics, and Northrop with McDonnell Douglas. Around the same time, the USAF and Navy announced an agreement: the Navy's Advanced Tactical Aircraft (ATA) would be designated as a replacement for the F-111, and an ATF variant would replace the F-14. In theory, this added up to a joint requirement for well over 1,000 ATFs, but neither service's commitment to the other's aircraft was guaranteed, and the requirements were far in the future.

Lockheed's YF-22 and Northrop's YF-23 were declared the winners on 31 October 1986. At the same time, P&W and GE were both awarded contracts to develop and build flight-test engines. At this stage, there was no engine competition; one of each pair of prototypes would be fitted with the P&W YF119, and the other with the GE YF120.

ATF Dem/Val was the largest fighter competition in history, lasting more than four years and costing almost $2 billion – the $691 million contracts covered only part of the cost. In addition to the two prototype aircraft, the winning teams built complete avionics systems, which were tested in ground-based and flying laboratories. Full-scale RCS models were built and tested, and each design underwent thousands of hours of wind-tunnel testing.

For Lockheed the contest started badly. The initial design, with sharply swept leading and trailing edges and a straight-edged glove that ran all the way to the nose, would not fly. The rotary weapon bay was too fat and heavy for a fighter. By June 1987, the Lockheed team was forced to scrap its original design completely. The YF-22 design was not frozen until May 1988.

YF-22 Dem/Val - Demonstration/Validation

Having had to scrap its original ATF configuration, the Lockheed team redesigned the YF-22 in some haste to meet the proposal deadline. The aircraft emerged with a simple wing planform which was mirrored in the horizontal tails. The nose was long and pointed, and the vertical tails were huge and set well forward, where they would not be blanked even at very high Alpha. The forward fuselage was diamond-shaped, with engine intakes matching the alignment of the lower fuselage sides. The number of angular alignments was kept to a minimum in the interest of low RCS.

F-22A EMD - Engineering and Manufacturing Development

The final F-22 design is based on that of the Dem/Val aircraft, but introduces some notable changes. The wing planform is more complicated, and is no longer matched by the horizontal tails. Leading-edge sweep is reduced, to 42°. The vertical tails are smaller, and the nose shorter and less pointed. The intake/chines end further back, improving the view from the cockpit.

Painted with a spurious serial and 57th Wing codes, the first YF-22 (N22YF/87-700) appeared at the April 1997 celebrations at Nellis AFB, Nevada, on the occasion of the USAF's 50th anniversary. While the second YF-22 (N22YX/87-701) was used for productive work as a ground antenna testbed, the first went on display in the USAF Museum at Wright-Patterson AFB, Dayton, Ohio.

2003 rather than August 2002. Developmental testing – which was due to be completed in August 2002 – will now continue beyond the start of DIOT&E. Only those developmental tests that are essential before DIOT&E will be completed by April 2003. Overall, the number of test points to be completed before operational testing starts has been reduced by 31 per cent. The 2001 DAB review also delayed full-rate production and cut 36 aircraft from the USAF's planned force. Full-rate production (36 aircraft per year) will start in FY2006 rather than FY2004. The first LRIP batch includes 10 aircraft, with 13 the following year, ramping up to 35 aircraft in FY2005.

The 1997 Quadrennial Defense Review had set F-22 production at 339 aircraft. This was nominally reduced to 331 aircraft because eight LRIP aircraft (two ordered in December 1998 and two in December 1999) were re-labelled as PRTVs. The 2001 plan calls for a minimum of 295 aircraft (again, in addition to eight PRTVs). However, the USAF and Lockheed Martin still believe that, once full-rate production starts, they will be able to implement cost savings that will make it possible to build 331 for the same money.

Production decision

The next key milestone is the full-rate production decision, due in 2004. A vital element in this decision will be the successful completion of DIOT&E. But several other milestones have to be passed before DIOT&E can start, and there is not a lot of margin in the schedule. Aircraft deliveries are crucial. DIOT&E requires four primary aircraft and a back-up, all of them close to a production configuration. The plan is that the four primary aircraft will be 4008 and 4009, the last two engineering and manufacturing development (EMD) aircraft, and 4010 and 4011, which are the first two 'production representative test vehicles' (PRTVs). The

A boomer's-eye view of SRA-equipped Raptor 4002 as it nudges in towards a KC-135R. The refuelling receptacle is covered by unusual 'butterfly' double doors. The precise closure of these, as with all external openings, is crucial to maintaining the overall low radar cross-section of the aircraft, although the top sides are less critical than other aspects.

At the same time, as part of an agreement with Congress, the Defense Acquisition Board set specific criteria that had to be met during 2000 and 2001 in order to enter LRIP. The 2000 targets included the first flight of Block 3.0 software on the F-22, missile separation tests for the AIM-9 and AIM-120, and the start of RCS testing. Goals for 2001 included the first guided AMRAAM launch and first tests of radar performance. By August 2001, the DAB was sufficiently satisfied with the programme's performance to continue with LRIP. Because of the impact of delivery delays, however, the USAF has decided to extend the test schedule.

Under the new test plan, dedicated initial operational test and evaluation (DIOT&E) starts eight months later, in April

---

Northrop was first in the air, on 27 August 1990. The first YF-22, powered by the GE F120 engine, flew on 29 September, piloted by Lockheed chief test pilot Dave Ferguson. The flight-test programme was short and intense. The goal was to test a strictly limited number of key data points, identified by the contractors in their original Dem/Val proposals, as quickly as possible. This was not a fly-off programme

With an F-16B flying chase, Paul Metz guides 4001 – the first flying F-22 EMD aircraft – through its maiden flight from Dobbins AFB, Georgia (site of the Marietta facility) on 7 September 1997. Originally scheduled for May, the first flight was delayed by various technical problems.

in which the best-performing prototype would win. The winning ATF contractor would be selected on the basis of its proposed development and production programme, and the flight-test results were intended to validate the projected numbers in those proposals.

The first YF-22 went supersonic on 25 October 1990. The second, F119-powered, aircraft flew on 30 October with Tom Morgenfeld at the controls. Between them, the two F-22s made 74 test flights in three months. In early November, the first YF-22 sustained Mach 1.58 without afterburner while with afterburner it exceeded Mach 2. During December, it demonstrated its low-speed manoeuvrability, performing 360° rolls at a 60° Alpha (angle of attack).

Unarmed AIM-120 and AIM-9 missiles were fired from the side and centre missile bays. The YF-22 flight-test programme was completed on 28 December.

Proposals for the engineering and manufacturing development (EMD) stage of the programme were submitted on the last day of 1990. The decision hinged not just on what the contractors promised, but on the customer's confidence in their ability to deliver. McDonnell Douglas, at the time, was in the Pentagon's 'doghouse' because of the collapse of the A-12 programme; Northrop was in trouble on the TSSAM missile. Lockheed's F-117, meanwhile, had become the hero of the Gulf War. In several key areas, Lockheed had gone further in its demonstration

Having earlier been Northrop's YF-23 test pilot, Paul Metz joined the Lockheed Martin team as F-22 chief test pilot. Here he poses on the steps of Raptor 01 at Marietta around the time of the first flight.

PRTVs were ordered in FY2000 in a compromise that kept production moving, despite a Congressional decision to delay LRIP. Only one of these aircraft is flying today; 4007 is the back-up aircraft. Further delays will be a major problem: "We think we're over that," says USAF programme director Brig. Gen. Jay Jabour, "but it's difficult to recover the runway behind you."

The F-22 programme is streamlining its manufacturing and delivery process. Now, the aircraft are flown first in primer, and then return to the sophisticated robotic paint shop at Marietta for finishing. In the past, the aircraft were finished before they flew – and the inevitable post-first flight maintenance meant that radar-absorbent material (RAM) coatings and seals had to be removed and restored.

Some technical issues are still being cleared up – "the reason that we do EMD", says General Jabour. There have been some structural problems with the vertical tails, for the same reason that the F-15 and F/A-18 have had problems: the very hard-to-model interaction between forebody

vortices and the canted vertical tails. "The problem is a lot lower in magnitude than the F/A-18, but the onset of loads started at a lower angle of attack than we predicted." Structural fixes will be incorporated from aircraft 4018, the first LRIP aircraft, and some of the loads are being reduced by using flight-control software to toe-in and toe-out the rudders as necessary. Another problem concerns the canopy (which already caused one break in flight-testing when cracks were found in one test aircraft). A 'howl' has been traced to a canopy-to-fuselage seal, and is being fixed.

Above: This view highlights the bulges required to house the actuators for the ailerons, flaperons and horizontal tails.

Above left: Thrust-vectoring and large control surfaces make the F-22 highly manoeuvrable in any regime, including post-stall flight.

---

programme. The YF-22 had gone to high Alpha and had fired missiles; and it had flown with a prototype advanced cockpit. The Lockheed team's avionics demonstration was more comprehensive. Of the two proposed EMD designs, Lockheed's was more like what had flown in Dem/Val.

Lockheed, Boeing and GD's team was declared the winner in April 1991. The formal EMD contract was awarded in August, after a final review of the entire programme. The contract called for the construction of 11 EMD aircraft, including two F-22B two-seaters. The first aircraft was due to fly in August 1995. As it stood, the EMD schedule was later than that planned in 1986. The decision to build prototypes meant that the Dem/Val programme was longer, moving IOC forward to 1999. Plans were revised yet again under the Pentagon's Major Aircraft Review (MAR), published in April 1990, which delayed ATF production by two years and cut the annual production rate of the USAF version from 72 to 48 aircraft. When the EMD contact was awarded, the F-22's initial operating capability (IOC) was set for 2001. A year later, USAF Secretary Rice announced that total production would be cut from 750 to 648 aircraft.

Lockheed had announced before the EMD contract was awarded that, if it won, it would locate the new programme's headquarters in Marietta, north of Atlanta, Georgia, where the forward fuselage would also be built. Marietta, where Lockheed operated a massive production facility, offered lower costs and a larger labour force than California.

The engine competition was equally intense. There was a fundamental difference between the two engines: General Electric's F120 was the first 'variable-cycle' engine. For maximum power, supersonic acceleration and supercruise, the engine was a pure turbojet; but at subsonic cruise the bypass ratio could be increased for greater efficiency. The GE-powered Dem/Val prototypes were faster, but the GE engine was not as mature as the P&W engine by the time Dem/Val ended. Pratt & Whitney argued that it could scale up the F119 to match GE's performance in a design which might be less complex and expensive, and which presented fewer technical risks. The F119 was accordingly selected for the F-22.

The F-22 EMD schedule was to slip by another four years between contract award, in August 1991, and

the aircraft's first flight. Budget cuts moved the first flight from August 1995 to 7 September 1997, and delayed IOC from 2001 to 2005. The Pentagon reduced the planned F-22 fleet twice – from 648 to 442 aircraft, in the 1993 'bottom-up review' of US defence plans, and then down to 339 in the mid-1997 Quadrennial Defence Review. This review also cut the peak production rate from 48 to 36 aircraft a year.

In July 1996, the USAF deferred development of the F-22 two-seater to save money, and eliminated the two F-22Bs from the test programme. This was not a painless decision, but the fighter's carefree handling and straightforward flying qualities should make it easy and safe to fly, while recording devices and the debriefing functions built into the Boeing-developed training system allow a pilot's performance to be reviewed on the ground.

The second YF-22 resumed flying in October 1991

After two flights from Dobbins AFB, aircraft 4001 was shipped on 6 February 1998 by C-5 to Edwards AFB, California, from where it flew again on 17 May.

to support the EMD programme. In April 1992, the aircraft entered a series of oscillations during a pass along the runway at Edwards, crashed, slid 8,000 ft (2440 m) along the runway and burned. The aircraft could not be economically restored to flying condition, but was repaired externally and moved to Rome Air Development Center at Griffiss AFB in New York, where it was mounted on a pedestal and used to test antenna designs for the F-22A. The first YF-22, used as an engineering fixture during development, was restored in 1997 and is now on display at the USAF Museum in Dayton, Ohio. It has now been fitted with prototype Pratt & Whitney F119 engines, rather than the GE engines with which it flew.

A focus area is structural envelope testing, one of the tasks which must be completed before DIOT&E can start. In 1994, the designers discovered that the F-22 structure would have negative margins in some areas; but by that time it was too late to fix 4002, which was supposed to be the first structure-test aircraft. Consequently, the F-22 programme has only one fully instrumented structures aircraft, 4003. "If we have a mishap, we can't make it up with another aircraft," says Jabour. Flight testing is proceed-

ing with utmost caution. The structural tests are complex, mainly because of the weapon-bay doors. "The doors contribute to the structure," says Jabour. "There is a different loadpath when the doors are open or closed." The door actuators are actually pre-loaded to keep the doors closed (ensuring that they do not 'gap' and cause RCS problems), and some critical conditions are encountered with the doors closed rather than open, as might be expected.

Raptor test fleet

Of the other F-22s, the first EMD aircraft – 4001 – was retired from flying in late 2000. It was delivered in November to Wright-Patterson AFB and is being used for live-fire testing. 4002, the intended structures aircraft, is being used for weapon-separation tests and also carries the spin-recovery parachute. 4003 is continuing envelope expansion, and 4005, 4006 and 4007 are the avionics workhorses in 2002. 4004 was being modified in the early part of the year in preparation for climate testing in the McKinley Laboratory test chamber, located at Eglin AFB in Florida. At any time, too, one aircraft is dedicated to logistics testing – principally, ensuring that the computer-based system of technical orders is correct, and that maintainers will be able to support the aircraft by following them. The final EMD aircraft, 4009, was delivered to the test team on 15 April 2002, but was due to remain at Marietta for around six weeks while its operational maintainability was thoroughly tested, before moving to Edwards.

The F-22 should be delivered to the DIOT&E programme with a working set of avionics and software that reflects the initial operating capability (IOC) standard. General Jabour is "very optimistic" about avionics testing, which many observers expected to pose the biggest problems. "We invested a lot of resources in the flying test bed and ground testing," he says. Hardware and software were first installed in the avionics integration laboratory (AIL) at Boeing Field near Seattle, an operator-in-the-loop system which included a complete tower-mounted set of antennas. Representative emitters were set up on the hills opposite the airfield, and the radar was used to track air traffic using Seattle-Tacoma airport.

After the AIL, the system was installed on the FTB – the first Boeing 757. The FTB uses F-22-type processors, incorporates a cockpit simulator and has a forward wing to

757 FTB – the 'Catfish'

Boeing leads the avionics integration team for the F-22. To assist its work and to 'down-load' the F-22 test fleet, it converted the first Boeing 757 to become the Flying Test Bed (FTB) for the F-22's system. Initially it flew with a representative radome for the APG-77 radar grafted on to the nose, but subsequently gained vestigial chines either side of the nose, dummy 'sensor wing' sections above the flight deck (below) and an undernose radome. The unusual configuration gave rise to the nickname 'Catfish'.

Above: Having been used during the Dem/Val process in unmodified form as the Avionics Flying Laboratory, N757A flew again as the FTB in June 1997 with just the radar nose fitted. The full 'Catfish' configuration was flown on 11 March 1999. By February 2002 the FTB had flown over 925 test hours.

Test and recording equipment, together with operator work-stations, occupy much of the FTB's cabin, although the aircraft retains some vestiges of its airliner ancestry, notably in the provision of airliner-style seats in rest areas and the retention of some overhead luggage bins (below left, view looking forward). The most important station is the fully functional representative F-22 cockpit situated behind the flight deck on the starboard side (below right).

Below: Despite the delivery of (by April 2002) fall six avionics-equipped EMD F-22s, the FTB remains busy on systems-related work, especially trials of new avionics software releases prior to their installation in the F-22 fleet. The excrescence on the rear fuselage is part of the cooling system for the onboard equipment.

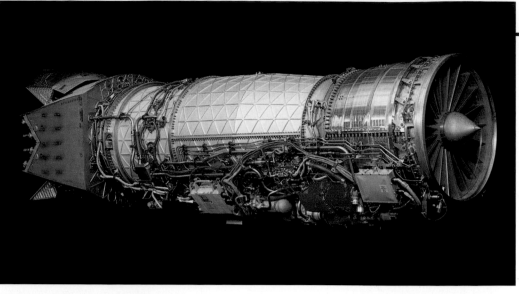

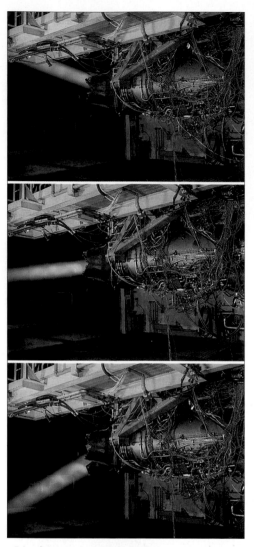

Above: The F119 builds upon P&W's considerable experience with the F100 engine, but is much larger and incorporates many areas of new technology. Advanced metallurgy – such as single-crystal technology, Alloy C burn-resistant titanium, 'blisks' (in which blade and disk are made as a single piece) and linear friction welding – allows the engine to run hotter, contain significantly fewer yet stronger parts, and be more durable and powerful than previous fighter engines.

Left: As well as vectoring thrust, the SCFN (Spherical Convergent Flap Nozzle) also flattens the exhaust plume, reducing its heat signature.

Below: Engines are slid into the F-22 airframe from the back. Once in situ, most of the external plumbing, wiring and accessories are easily accessible, and are colour-coded for ease of maintenance. Engine control, by FADEC, is integrated with the flight control system.

represent the location of the F-22's wing-mounted antennas. The FTB has been flown extensively, both against dedicated target aircraft and other military aircraft throughout the western US. "When we fly the avionics in the F-22, they work the way that they did in the FTB," says Jabour. Where there were problems, they concerned the nuts and bolts of aircraft integration, "the peculiarities of power sequencing and start-up," he comments. The FTB clearly does not replicate the agility of the F-22, which imposes a different set of dynamics on the radar and electronic warfare (EW) system. So far, though, the system seems to be performing well at moderate *g* levels.

According to Lockheed Martin programme manager Bob Rearden, there are two software releases due before the start of DIOT&E. Version 3.1.0 has been issued. In the summer, 3.1.1 will be installed on the F-22, and will be used to train pilots for DIOT&E. Finally, 3.1.2 will be the standard for DIOT&E and initial service, and will include all major radar, electronic warfare and communications, navigation and identification (CNI) functions.

Aircraft performance has not been a major issue. "The supercruise mission is exceeding the design goal," says

Jabour. A highlight has been the performance of the Pratt & Whitney F119 engines. "They have worked flawlessly," says Jabour. "We have not had a single unscheduled shutdown in flight."

Weapon testing has gone smoothly. A great deal of testing to support the AIM-120 AMRAAM uses an Instrumented Test Vehicle (ITV), a non-releasable missile simulator carried in the weapon bay. "In many ways, it gives us better data than the live missile." The AIM-120's autonomy makes missile integration rather different in nature from the same effort on a semi-active missile. The ITV contains electronics that read the datalink transmissions from the fighter to the missile; what is important is that these signals are correct, because the rest is up to the missile. So far, there have been two guided AMRAAM launches, both successful. In one case, the missile guided close enough to the target to have killed it if it had a warhead, and in the other the AMRAAM skewered the BQM-74 drone and killed it by impact. (Lockheed Martin programme manager Bob Rearden does not claim extra credit for the latter shot – again, it was the missile and its active radar that performed the final intercept.)

Guided 'Winder

The first guided AIM-9 shot will take place later this year, but there is clearly a consensus around the F-22 programme that the AIM-120 will be the primary weapon. Significantly, the programme office and the user have accepted a delay in incorporating the Joint Helmet Mounted Cueing System (JHMCS) on the F-22, because of integration problems between the JHMCS tracker and the active noise reduction system in the F-22 helmet. The JHMCS has been deferred until the system is mature. "This aircraft is designed to shoot a guy down before he gets close," says General Jabour. "It should mean that guys never get close to the merge." JHMCS is therefore less critical for the F-22.

The first production F-22, Raptor 4018, is due to reach the USAF in 2003. The first unit to receive the aircraft will be the 325th Fighter Wing at Tyndall AFB, Florida, which will be the 'schoolhouse' or training wing for the new fighter. L3 Communications' Link Simulation and Training division is due to have delivered two full mission trainers and four weapon tactics trainers to Tyndall by February. The first operational F-22 base will be Langley AFB in Virginia, adjacent to Air Combat Command headquarters. The first F-22s are due to arrive at Langley in September 2004. By 2009, three squadrons with 78 aircraft – 24 aircraft per squadron and six back-ups – should have replaced F-15s at Langley.

Pilots working on new-generation fighters believe that the process of tactical change – which is traditionally slow except in combat – should be accelerated by the use of full-

Above: Raptor 4002 taxis at Edwards. This Block 1 structures aircraft was scheduled to be retired along with 4001, but was kept flying to cover shortfalls in the test programme caused by delays in subsequent aircraft delivery and the cockpit crack grounding. It was retired in 2002.

Below: Optimised for high-speed supercruise at high altitude, the F-22 also has blistering pace at lower levels. Afterburner need only be used sparingly for rapid accelerations.

Left: A rare rainstorm at Edwards AFB in March 2000 allowed Raptor 01 to perform wet runway tests at 30-, 60- and 90-kt speeds. The F-22 employs a redundant, computer-controlled Brake Control System (BCS), which compares inertial velocity information with wheel rotation speed to command brake pressure. In initial tests there were some deficiencies, such as abnormally hot brakes or a perceived lack of response. Hardware and software changes were incorporated from Raptor 03 to provide a system which retains the advantages of 'brake-by-wire', but also provides traditional 'feel' for the pilot.

Flying the F-22

As pilots build up time on the F-22, its handling characteristics are beginning to become more familiar. Lockheed Martin chief test pilot Paul Metz comments on the fighter's high-speed behaviour: "The Raptor is always in a combat configuration. The aircraft has no external stores, so drag remains low and Ps [specific excess power, a measure of the fighter's ability to accelerate or climb at its current flight condition] stays high.

"Overall drag is at a minimum near the design speed of Mach 1.5 at 40,000 feet [12192 m]. This airframe is at its best at supersonic speeds. Conventional fighters start with a subsonic climb to the tropopause [about 36,000 ft/10973 m]) and then perform a pushover to supersonic speed and climb supersonically from there. The Raptor can dispense with this complex profile and blast off supersonic from the deck.

"Level acceleration in military power or less is sprightly at all altitudes, but downright astounding in full afterburner. Approaching Mach in military power, the acceleration reduces slightly as drag rises, but the aircraft punches on through easily. Accelerating through Mach in military power in the Raptor feels similar to accelerating in full afterburner in an F-15. When testing… we use afterburner to boost the aircraft to the point and then pull it out of afterburner to cruise at our test condition.

"The real secret to supercruise is thrust minus drag. Although the F-22 uses a fixed inlet design, the overall engine and airframe are optimised for high supersonic speeds. Acceleration and Ps are phenomenal at the right-hand side of the flight envelope. The Raptor can easily exceed its design speed limits, particularly at low altitude. We have incorporated max speed cues and alerts to remind pilots when approaching the limits."

Test pilot Jon Beesley discusses the fighter's low-speed qualities. "Many aircraft define the beginning of high AOA [angle of attack] flight at about 30°. On the F-22, flight test points are routinely held stable at a positive 60° AOA and a negative 40° AOA.

"Our initial tests showed sideslip excursions at 30° AOA, which were greater than we desired. We addressed it in the first software update. Through the magic of software, the airplane now passes through this AOA range with no apparent change in handling qualities. We also found that the rudder pedals offer the most intuitive lateral directional control at high AOA.

"Lateral-directional control remains good above 50° AOA, but the horizontal tails are working hard to provide that control, moving differentially like the webbed feet of a duck paddling through the water," says Beesley. "Differential tail is the primary yaw control device. Although these excessive differential tail movements make for great video, they indicate different aerodynamics than predicted."

Above: In early flight tests the F-22 showed a noticeable 'nose wander', or 'roll jerk', when rolling. This was corrected through changes to the flight control software.

Right: The F-22 cockpit is dominated by the central tactical display, which presents a 'god's-eye' view of the air battle. The imagery is second-generation: data from the aircraft's sensors – or other offboard sources received via the IFDL (Intra Flight Datalink) – having been fused and prioritised by the aircraft's 'brain', the Common Integrated Processor (CIP), to produce 'track files' of other aircraft. Either side of the main screen are slightly smaller displays for more detailed information, while below is another screen usually used for aircraft status. The F-22 uses a 'dark cockpit' principle, in which displays only appear if there is a problem with the aircraft. If the cockpit is 'dark', the aircraft and its systems are fine.

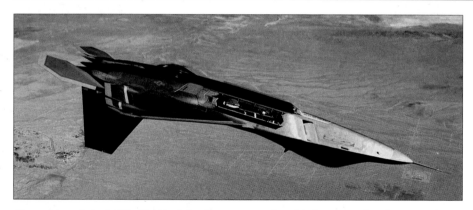

Above: A key element of the flight tests has been manoeuvring with the weapon bay doors open, as displayed here by Jon Beesley in Raptor 02, to assess the effects on flight performance, weapons, bays and doors. Aggressive rolls and sideslips were performed at varying speeds.

Below: During the course of envelope expansion flying the F-22 has recorded sustained, controlled flight at +/-60° Alpha, resulting in some unusual manoeuvres and spectacular imagery. In 2002 the aircraft began flying 9-g manoeuvres.

mission, multiple-player simulators. "Traditionally, you fly the new aircraft like the old one until you figure it out," remarks a former F-15 pilot who now works on an advanced fighter team. "It took us 10 years to employ the F-15 like an F-15, and not like an F-4." With simulation, "we'll do the same in four to five years." The development of tactics will be a key to exploiting the 'black world' capabilities alluded to by former USAF Chief of Staff Fogleman. Recently released details of the F-22 and its subsystems help to put together a picture of those capabilities, based on a combination of speed, stealth and avionics.

Supercruise

High performance has been crucial to the F-22 since the earliest days of the programme. The F-22's sustained supersonic speed (without afterburning) is now confirmed as Mach 1.7 – 80 per cent greater than any previous fighter. This confirms that the F-22 has a power reserve for manoeuvring and acceleration at high supersonic speeds – unlike today's fighters or JSF. The F-22's operational altitude, formerly quoted as 'above 50,000 ft (15240 m)' is now confirmed as 60,000 ft (18288 m). This is higher than other fighters, which are limited to 50,000 feet because of physiological limits: if cockpit pressure is lost, the pilot will lose consciousness before descending to an altitude where he or she can function on oxygen. However, the F-22's anti-*g* ensemble, including positive-pressure breathing, functions as a pressure suit. The USAF has stated that the F-22 has a maximum speed of Mach 2.0, using afterburner. This limit is probably set by structural temperature considerations in the airframe and engines.

F-22 chief test pilot Paul Metz notes that launching an AIM-120 Advanced Medium Range Air-to-Air Missile (AMRAAM) at supersonic speed, with an altitude advantage over the target, increases its range by 50 per cent. If a hostile fighter launches an MRAAM against the F-22, it can perform a supersonic turn or 'cranking' manoeuvre, presenting the missile with a rapid and unpredictable change in line-of-sight and thereby decreasing its effective range.

According to USAF test pilot Lt Col David Nelson, super-cruise is an advantage in attack and defence. "A subsonic airplane is easily targeted by fighters from the stern, aft quadrant, beam and front. Attackers approaching from the stern have a relatively good stand-off capability, since their missiles don't need to go so fast to run down a subsonic fighter.

"As the targeted airplane pushes past the Mach, however, it greatly reduces any stern firing opportunities," Nelson continues. "Similarly, aft-quadrant shots are denied because they lack the energy to complete the fly-out. Beam shots fail, because missile seekers exceed gimbal limits when they attempt to pull the proper amount of lead for an inter-cept. Front quarter shots are limited, because of the higher line-of-sight rates created by a high-speed target. All of these shots are complicated by thin air at high altitude, where the F-22 flies comfortably."

The USAF has quoted the ferry range as 1,800 nm (3333 km; 2,071 miles), but has not stated a combat radius. However, the USAF has disclosed that the aircraft has an internal fuel capacity of 18,348 lb (8323 kg), equivalent to a fuel fraction of 0.28 – that is, 28 per cent of the mission take-off weight is fuel. This is not unusually high, and the F-22 – unlike other fighters – does not carry external fuel tanks on combat missions. Based on these numbers, critics such as Riccioni allege that the F-22's range will be disap-pointing. According to Riccioni, the ATF requirement called for a 900-km (560-mile) mission radius, of which 740 km (460 miles) would be flown at supersonic speed. "I've heard numbers as low as 50 miles in and 50 miles out,"

Left: Raptor 4004 is seen in primer after roll-out at Marietta, and before its 15 November 2000 first flight. Following the first three EMD aircraft, which were allocated to aerodynamic, structural and weapons separation duties, were three 'wired' F-22s with integrated avionics and sensors installed. Of these, 4005 is focusing on weapon system work, while 4004 and 4006 are involved in LO trials. 4004 is also to be used for climatic tests.

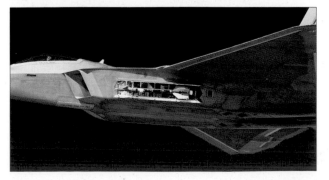

The F-22's side bays are large enough to accommodate one AIM-9 each. The missile is mounted on a launch rail which deploys outwards into the airstream, canted at an angle. Sidewinders launch 'from the rail', and the canting provides sufficient clearance for the missile, as well as allowing the missile's seeker head a 'look' ahead of the aircraft.

Riccioni remarked in June 2000. "But you can do that with the current F-15."

In a 1994 TV interview, Air Combat Command commander General Richard Hawley stated that the F-22 "will be able to cruise for … 30 minutes out of an hour-long mission at supersonic speeds." Assuming a supercruise speed of Mach 1.7 and a subsonic cruise speed of Mach 0.9, this would equate to a 700-km (435-mile) unrefuelled mission radius, with a 450-km (280-mile) supersonic segment. The USAF has also repeatedly told the General Accounting Office that the F-22 is exceeding its specified combat radius. The actual numbers are probably between Riccioni's gloomy projections and the 900-km target (which represents an ATF goal rather than EMD specifications). The F-22 does not fly or fight in the same way as a conventional fighter. It does not have the additional fuel carried in external tanks – but neither does it have the drag of tanks or pylons. Neither do weapons add to the fighter's drag (except indirectly, through greater weight). In a typical mission, too, the F-22 will make less use of the afterburner than a conventional fighter, making a massive difference to specific fuel consumption.

The F-22 is the first aircraft to combine high performance with stealth. The F-22 is designed for all-aspect stealth over a large bandwidth. Critics say that the F-22's beam-aspect radar cross-section (RCS) is much higher than its head-on RCS. They are right, but this is an intentional and logical design choice. The F-22 has what is known as a 'bow-tie' RCS pattern, with its smallest RCS at the nose and tail and a higher – but still reduced – RCS to each side. In most engagements, the side-on RCS is exposed only transiently to radar, and missiles have a hard time intercepting targets with high crossing rates. Stealth and speed are complementary: in a head-on engagement, where a missile has the greatest chance of reaching an F-22, the fighter is least visible.

The infra-red (IR) signature has not been ignored. The F-22 is covered with a Boeing-developed IR-suppressing topcoat. Moreover, according to F-22 critic Riccioni (who

does not identify his source), the F-22 also uses fuel to cool its leading edges; certainly, the USAF's work on heat-tolerant fuels comprising JP-8 with a chemical additive was inspired in part by the cooling needs of the F-22. IR tests started last year over Point Mugu, California, and showed that the F-22 will have a "low all-aspect IR signature under sustained supersonic conditions," according to the USAF.

Neither are IR search and track (IRST) systems as all-powerful as stealth critics like to suggest. Very long detection ranges claimed for IRST systems generally assume that the target is in afterburner, but the F-22 will use its thrust augmentors only transiently. The advanced IRST under development by EuroFIRST for the Typhoon, for instance, has a range of more than 90 km (56 miles) against non-stealthy targets such as Tornados and MiG-29s.

'First look, first shot'

Stealth protects the F-22 against detection and gives it the 'first look, first shot' advantage in an air-to-air engagement. By reducing radar detection range, it frees the F-22 to manoeuvre over a defended area and exploit its speed and mobility.

The third key element of the F-22 is its sensor system, spearheaded by the APG-77 radar with its active electronically scanned array (AESA). A report in late 2001 by the Pentagon's Defense Science Board credits AESA radars with "10-30 times more net radar capability" versus mechanically scanned systems. A unique advantage of an AESA is that its design reduces radio-frequency (RF) losses. An AESA can offer several times greater sensitivity than a conventional radar, partly because the receiver is coupled almost directly to an amplifier in each module. As a result, there is very little opportunity for interference or noise to enter the signal before it is amplified, resulting in a very clean signal to the processor.

One less desirable characteristic – at least in a conventional nose-mounted radar installation with a fixed array – is that the radar field of regard is restricted to 60° either side of the boresight. This is a matter of geometry: as the

Main picture and above: Flying at a speed of Mach 0.7 at 20,000 ft (6096 m), Raptor 02 fired the first AIM-9 Sidewinder from an F-22 on 25 July 2000. The aircraft carried a small test fixture below the fuselage. The initial tests were unguided and used an AIM-9M airframe without a warhead. In service, the F-22 will employ the next-generation AIM-9X with high off-boresight capability.

Raptor 02 undertook this Sidewinder firing, designed to demonstrate firing while manoeuvring. Weapons separation tests will continue to expand the manoeuvre/attitude envelope for missile launches. In 2002 the first guided Sidewinder launches will be made – another of the critical milestones for the F-22 CTF to reach.

Below: Armourers load an AIM-120C into the weapons bay of Raptor 02. The AMRAAM is carried on an LAU-142 launcher which extends down from the roof of the bay for loading. Many of the F-22's AMRAAM-related tests do not employ a missile at all, using a bay-mounted instrumentation pod instead.

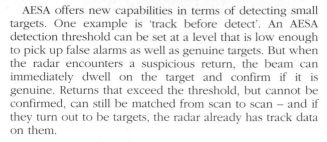

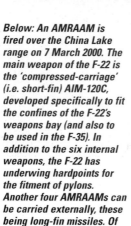

Above, above right and opposite: This sequence shows a calibrated AIM-120 separating and launching from 4002. The AMRAAM is thrust away from the F-22's bay by the Vertical Ejector Launcher (VEL) and falls well clear before the rocket motor is ignited.

Below: An AMRAAM is fired over the China Lake range on 7 March 2000. The main weapon of the F-22 is the 'compressed-carriage' (i.e. short-fin) AIM-120C, developed specifically to fit the confines of the F-22's weapons bay (and also to be used in the F-35). In addition to the six internal weapons, the F-22 has underwing hardpoints for the fitment of pylons. Another four AMRAAMs can be carried externally, these being long-fin missiles. Of course, external carriage of any store greatly compromises the aircraft's stealth properties.

off-boresight angle increases, the projected area of the antenna in the direction of the beam decreases and so does its gain. The F-22 has space, weight and cooling provision for side radar arrays in the forebody, but the arrays themselves were deleted in the demonstration-validation phase and have not been reinstated.

APG-77 specifics are still classified, but it has been stated that the Raytheon APG-79, under development for the Super Hornet, can track targets at ranges in excess of 185 km (115 miles), almost twice the range of some of today's radars. The APG-77 array is larger than that of the APG-79 and has more transmit/receive (T/R) modules (numbers in the DSB report suggest that the APG-77 has 1,500 modules, versus 1,000 in smaller radars) and more power. The AESA does not have to waste energy on sky that has already been found clear of targets and can concentrate on tracking targets that have already been detected. "Track quality throughout the entire field of regard of this radar is better than the weapon-quality track-while-scan volumes of current weapons," comments a USAF pilot.

AESA offers new capabilities in terms of detecting small targets. One example is 'track before detect'. An AESA detection threshold can be set at a level that is low enough to pick up false alarms as well as genuine targets. But when the radar encounters a suspicious return, the beam can immediately dwell on the target and confirm if it is genuine. Returns that exceed the threshold, but cannot be confirmed, can still be matched from scan to scan – and if they turn out to be targets, the radar already has track data on them.

Stealth advantages

The AESA is important to stealth in two ways: indeed, no known stealth aircraft has been built or even proposed with a mechanically scanned radar. The first advantage is reduced RCS. A mechanically driven antenna has a complex shape and is difficult to shroud. The APG-77 array is installed with a slight upwards tilt which deflects the head-on main-lobe reflection upwards and away from any likely receiver. There is still backscatter from the edges of the radar array, which is masked by the use of radar absorbent material (RAM) on the perimeter of the array.

If AESA technology has one 'killer application' on stealth aircraft, it is low probability of intercept (LPI). AESA greatly expands the number of LPI techniques that can be used to ensure that a stealth platform does not betray its position and identity through radar emissions. The agility of the

radar makes it possible to reduce peak power adaptively – as the target gets closer, the radar power can be decreased rapidly to a point where an intercept receiver cannot detect it. An AESA radar can also search simultaneously with multiple beams. Because each beam searches only a small sector, it dwells longer on a given spot and can therefore achieve the same probability of detection with less power.

Finally, given sufficient processing horsepower, an AESA radar can confuse an interception system by varying almost every characteristic of the signal, apart from its angle of arrival, from one pulse integration to the next. For example, it can change its pulse width, beam width, scan rate and pulse-repetition frequency (PRF), all of which are essential identifying characteristics for interception systems.

Passive cueing

Another important LPI technique is to leave the radar turned off. One system that helps do this is the BAE Systems (formerly Sanders) ALR-94, which has been described as "the most technically complex piece of equipment on the aircraft." The ALR-94 is a passive receiver system, with far greater capability than the simple radar warning systems fitted to most fighters, and includes multiple antennas covering several bands. It can detect sidelobes as well as main lobes, and can accurately locate and track any emitting target. It is likely that it uses the radar's AESA as a sensitive and accurate adjunct receiver.

Information from the ALR-94 and other non-emitting sources – such as the fighter's datalink – is used automatically to cue the radar, cutting down on electronic emissions. High-priority emitters – such as fighter aircraft at close range – can be tracked in real time by the ALR-94. In this mode, called narrow-band interleaved search and track (NBILST), the radar is used only to provide precise range and velocity data to set up a missile attack. If a hostile aircraft is injudicious in its use of radar, the ALR-94 may provide nearly all the information necessary to launch an AIM-120 AMRAAM air-to-air missile and guide it to impact, making it virtually an anti-radiation AAM. "The enemy," observes a Lockheed Martin manager, "dies relaxed."

Another important aspect of the F-22's avionics suite is its ability to identify targets in a beyond visual range (BVR) engagement. "If you impose visual identification [VID] on

an F-22, it deprives you of 90 per cent of what the aircraft can do," says Ben Lambeth, a fighter analyst with the RAND Corporation. Also, analysts agree, within-visual range (WVR) combat is becoming increasingly dangerous. "An F-5 or a MiG-21 with a high-off-boresight missile and helmet-mounted display is as capable in a 1-v-1 as an F-22," comments a former Navy fighter pilot, now a civilian programme manager. If the WVR fight is dangerous, the key is to avoid it by destroying as many hostiles as possible in BVR, preferably at extreme range.

Rules of engagement (ROEs) set the conditions under which an unseen target can be attacked. They vary from situation to situation, under political constraints, but a basic principle is that the target should be 'declared' by multiple independent channels. The F-22 has at least four channels available: the identification friend-or-foe (IFF) functions in the CNI suite, offboard ID provided by the datalink, the ALR-94 and the radar.

The ALR-94 should be able to identify a target rapidly if it uses radar. The APG-77 radar, with its ability to generate

Below: In September 2001 the F-22 Combined Test Force achieved a major milestone with the first guided AIM-120 launch. The launch platform was Raptor 05, the second of the 'wired' F-22s, and the pilot was Major Brian Ernisse from the USAF. The missile did not have a warhead fitted, but was adjudged to have passed well within lethal distance of the target drone flying over the Pacific Missile Test Range. A second guided AIM-120 test hit the target and destroyed it by kinetic energy. Raptor 07, which joined the CTF on 5 January 2002, will also be engaged on weapon trials.

Extensive testing of both real aircraft and mock-ups in special facilities allows accurate assessment of LO properties, emission control and interference between the aircraft's various systems.

Carrying weapons internally is essential to the F-22's stealth properties, but also has beneficial effects on performance. Opening the bay to fire the weapons means breaking the 'LO bubble' but, it is argued, in any foreseen engagement the F-22 will have reached its 'first launch' position before it can be tracked.

a very narrow, focused beam, should be able to deploy several non-cooperative target recognition (NCTR) modes. One NCTR technique – jet engine modulation (JEM), which relies on detecting characteristic patterns in the radar pulse caused by rotating compressor blades – has been in service since the 1980s on the F-15 and F/A-18 radars, and was employed operationally in Desert Storm. More advanced NCTR is based on High Range Resolution (HRR). The radar measures the signature of range cells along the length of the target, and compares the signature to a template. The exact range at which a target can be recognised has not

been revealed. However, it is a general principle that identification at greater distances will require more energy.

The use of a datalink, though, may be the biggest single change to offensive fighter tactics. The F-22's Intra-Flight Datalink (IFDL), which uses narrow pencil beams to defeat interception, ties the sensors of several F-22s together. Fighters in a datalinked flight no longer need to stay in visual contact and can spread out across many miles of sky. Another basic use of the datalink is 'silent attack'. An adversary may be aware that he is being tracked by a fighter radar that is outside missile range. He may not be aware that another, closer fighter is receiving that tracking data and is preparing for a missile launch without using its own radar.

Again, the way in which the F-22 is designed to exploit this datalink is classified in detail. But the potential uses of

Affordable stealth

Lockheed Martin claims that the F-22 will represent a considerable advance in 'affordable stealth', due to changes in design philosophy, development processes and materials.

While the F-117 was almost entirely covered with radar absorbent material (RAM), most of the F-22's surface is covered with a paint-like conductive metallic coating which prevents radar energy from penetrating the composite skin. The F-22 uses RAM around the edges of doors and control surfaces, and radar absorbent structure (RAS) on the body, wing and tail edges.

Because of this multi-layer, multi-spectral camouflage system, different F-22s at different points in the flight-test programme appear in different colour schemes. Avionics test-bed Raptor 04 made its first flight with a patchwork of grey-painted RAS edges, yellow-primered skin, metallic coating and sprayed-on RAM.

A number of principles underlie the low observables (LO) design of the F-22. The designers wanted to reduce the need to 'break the LO bubble' – that is, to perform maintenance actions which require the removal of seals and coatings. One element of this approach has been to make components as reliable as possible.

Where components require regular access, or where the technology to ensure long inspection intervals does not exist, the F-22 has frequent-access panels. Like weapon-bay and landing-gear doors,

these have frames, latches and gaskets which ensure that they can be opened and closed without compromising the LO 'bubble'. Those systems that need frequent attention are arranged so that the smallest number of frequent-access doors cover all the systems that need routine inspection.

Another fundamental principle of the F-22's approach to LO materials was to minimise their number. In the F-117 or B-2 programmes, where LO performance goals were paramount, designers tended to select the material which was best suited to the different electromagnetic, thermal, vibration and structural requirements in any particular location – with the result that different types of material proliferated. On the F-22, the individual merits of each material type were balanced against a requirement to reduce the number of materials, with the result that the F-22 uses about one-third as many different LO materials as earlier LO aircraft.

The F-22 LO development programme exclusively used full-scale models, following problems in earlier programmes where data from sub-scale models proved unreliable as a predictor of full-scale results. The radar cross section (RCS) test programme started with partial models, including an inlet, a radome and a dual-engine afterbody model. This culminated in tests of a high-fidelity full-scale model, which started in 1999. The model included a radar, all doors and even the first two stages of the engines. All control surfaces could be actuated by remote control.

The radome is one of the most complex components of the F-22. It is a 'bandpass' or frequency-selective type, which reflects signals at all frequencies except the wavelengths used by the F-22 radar. It evolved from a design developed for the F-117, and tested experimentally on the fifth full-scale development aircraft with an experimental Westinghouse radar. Design considerations included stealth, radar performance and resistance to birdstrikes, rain erosion and radar. In 1999, people without the appropriate security clearance were not

This head-on view highlights the common alignments of fuselage, intakes and vertical fins. The intake ducts snake up and inside to ensure that the fan blades are invisible to radar directly, and incorporate RAM to dissipate radar energy which bounces down the duct.

linked radars have been outlined by the Swedish developers of the JAS 39 Gripen, the most datalink-intensive aircraft in service today. For example, if the radars or ESM systems on two fighters detect the same target, they can locate it instantly by triangulation. The datalink results in better active radar tracking with fewer signals. Usually, three plots (echoes) are needed to track a target in track-while-scan. The datalink allows the radars to share plots, not just tracks: even if none of the aircraft in a formation gets enough plots on its own to track the target, they may do so collectively.

Each radar plot includes Doppler velocity, which provides the individual aircraft with range-rate data. However, this data on its own does not yield the velocity of the target. Using a datalink, two fighters can take simultaneous range-rate readings and thereby determine the target's track instantly, reducing the need for radar transmission.

Driving the aerodynamics of the F-22 was the requirement to be manoeuvrable across a wide range of speed/altitude conditions, and the ability to transition from one to another with great rapidity.

even permitted to touch the radome. Most probably, this is because the outer layer of the structure is a frequency-selective surface, incorporating a fine pattern of carefully shaped apertures.

Two new facilities have been built at Marietta to support F-22 production: a coatings hangar with a track-mounted robot applicator, and an indoor RCS measurement facility in which the aircraft can be suspended from the ceiling and rotated so that its RCS can be checked against specifications. While earlier stealth aircraft have undergone inflight RCS checks, the plan is that the indoor range will be used for acceptance of production F-22s.

To avoid the need to repair LO seals and coatings in the field, the F-22 has almost 300 specially designed access points. These include quick-access panels, featuring positive locks, seals and gaskets. Many service points are located inside the weapon bays, landing gear doors and other openings. The goal is to ensure that 95 per cent of all maintenance actions in a 30-day deployment can be performed without removing material.

In an 'integrated combat turn' – the equivalent of a pit stop in motor racing, in which the aircraft is refuelled, rearmed and sent back into combat – the goal is that no material should be removed. The F-22 allows for simultaneous gun ammunition and missile reloading, a process that normally goes in sequence. The Raptor has single-point refuelling and a single point where other consumables such as oil, chaff and flares can be checked.

The F-22 started its in-flight RCS tests on 31 January 2001, confirming that the RCS goals are being attained. LO testing continues, mainly aimed at maintenance issues. Progress in maintainability is being monitored by 50-hour block tests of two kinds. The first is a 'stability over time' test in which the aircraft is brought up to its LO configuration and operated with normal maintenance. At the end of the test, LO variations are measured. In the other test, the aircraft is kept up to LO specifications and the

Detail shaping around joints and necessary airframe protrusions is highly important to keeping RCS low. Note the shape of the notches at the ends of the control surfaces, and the subtle blending around the M61 cannon installation in the starboard wing root.

work involved is documented in detail. LO maintainability does not have to meet requirements before DIOT&E. The goal is to meet them by 'maturity', which is defined as 100,000 flight hours and will be reached around 2008.

A major challenge in the B-2 programme has been adapting an LO design originally intended for the nuclear attack mission to the conventional role. Some materials on the B-2 cannot be repaired or replaced except under controlled temperature or humidity, and

efforts to develop a deployable shelter have not been successful so far, with the result that all combat missions have to be launched from the air-conditioned barns at Whiteman AFB, Missouri. The F-22 "will not normally need an air-conditioned shelter, with a couple of caveats," says General Jabour. Some processes still have to be performed on a standard day, with a temperature between 60 and 100° F and humidity between 5 and 95 per cent. "You can't do it out in the rain," says Jabour.

and 8 MB is the standard for a $100 pocket organiser rather than a $100 million fighter. Moreover, the semiconductor industry is no longer interested in producing chips in the relatively tiny quantities required by defence programmes. The problem of diminishing manufacturing sources (DMS) affects most military electronics programmes.

The answer to DMS is to switch, as far as possible, to commercial off the shelf (COTS) components. COTS is a process rather than an end-state, because commercial memory and processor products are constantly changing. The goal is to accommodate COTS components in modules or cards which perform two functions. First, they protect the commercial chip from vibration, heat and cold, electromagnetic pulse (EMP) and other features of the fighter environment. Second, they act as an intermediate system layer that allows the ever-changing COTS chip to work with the fixed core of the fighter's avionics system, avoiding difficult and expensive changes to power, cooling and data functions.

Former F-22 programme manager Brigadier General Mike Mushala commented in 1999 that he did not expect to build two F-22s to a common configuration, because the design of the modules would be constantly changing. However, the differences would be entirely invisible to the pilot and maintainer. The first low-rate production batch of

Above: The heart of the F-22 CTF is this test operations room at Edwards, where telemetry data from test flights is monitored and recorded.

Above right: Boeing operates the Avionics Integration Laboratory at Seattle. At the centre is a dummy cockpit, surrounded by operator consoles and screens. Representative sensors are located outside the facility, operable from the lab's cockpit.

Improvements to this formidable capability are already being studied – and could lead to an F-22 that can hit multiple ground targets with precision weapons. The first wave of changes to the design, however, is intended to deal with a mundane but knotty problem: out-of-production components.

In the late 1980s, just before the YF-22 and YF-23 prototypes were flown, the USAF was sponsoring advanced-technology programmes to develop 384-kB memory chips, and it was a matter for wonder and amazement that the new fighter might need onboard data storage measured in megabytes! By the time the programme started, technology had reached a point where the Common Integrated Processor (CIP) could use commercially available Intel I-960 chips. Those processors are now out of production

For test work at the outer edges of the envelope Raptor 02 was fitted with Spin Recovery Apparatus (SRA), first tested on the runway at Edwards in July 1999 (illustrated). If the aircraft departs from controlled flight and ends up in a flat spin or tumble, the SRA is deployed to 'up-end' the aircraft, putting the nose down to an attitude from where recovery is routine. It is mounted high and aft to avoid fouling the twin fins.

F-22s will have a new CIP with fewer modules and a different backplane design, better adapted to the power and cooling needs of modern processors. A more tactically important change arrives with the first full-rate production F-22s, in 2005-06. These aircraft will be delivered with a new radar, using similar technology to the F-35 Joint Strike Fighter and the APG-80 for the Block 60 F-16.

From the outset, the JSF programme aimed to deliver a multi-role AESA for JSF at a fraction of the weight and cost of the F-22's radar. The Pentagon's Defense Science Board reported in 2001 that these goals have been achieved: "Factors of three to five or more in weight and cost reductions can be supported [relative to F-22 technology], along with innovations in mechanical design to simplify manufacturability and maintenance. Transmit/receive modules are approaching commodity status," the DSB said.

Gallium-arsenide microwave monolithic integrated circuit (MMIC) chips of the kind used in the T/R modules are now being produced for commercial communications applications, ranging from satellites to broadband wireless modems for computers. Another factor is the use of COTS processor technology. The current F-22 radar uses specially developed processors, but the APG-80 uses Mercury Computer Systems' RACE system, based on multiple commercial-type processors. The new F-22 radar will work better and cost less than today's system, and opens the way to an important expansion of the F-22's mission.

Attack weapons for the F-22

The USAF has three priority upgrade programmes for the F-22, says Lockheed's Rearden. The first is a synthetic aperture radar (SAR) mode, to be introduced with the new radar. The second is a transmit mode for the Joint Tactical Information Distribution System (JTIDS) datalink, and the third is the incorporation of new weapons – exploiting the other two changes.

The USAF decided to build a secondary air-to-surface capability into the F-22 in 1994, directing Lockheed Martin to incorporate a pair of 450-kg (1,000-lb) class Boeing GBU-32 Joint Direct Attack Munition (JDAM) guided bombs into the aircraft. The JDAMs replace four of the AMRAAM missiles. However, the USAF now has its sights on a new weapon for the F-22, the Small Diameter Bomb (SDB).

The SDB is a product of a mid-1990s project called the Miniature Munitions Technology Demonstration (MMTD). The goal of MMTD was to exploit the fact that a small, accurate bomb can be as effective as a larger weapon with a greater miss distance. Specifically, the researchers set out to show that a weapon weighing around 250 lb (110 kg), with GPS/inertial guidance, a modern warhead and fuse design, and precise control of impact angle, could defeat hardened targets that could previously be attacked only by bombs in the 2,000-lb (900-kg) class.

F-22 test pilots have all praised the Raptor for its air refuelling qualities, citing high stability yet precise control, and low pilot workload during the process. The use of tankers from both the Edwards trials unit and front-line units greatly extends the useful work that each F-22 test sortie can achieve. F-22s are qualified to refuel from both the KC-135R (above and top) and, from 19 November 1999, the KC-10A (above left).

Right: In March 2002 weapons and avionics test vehicle Raptor 4005 took time out from the busy flight test schedule to pose alongside a NASCAR racer which advertises the US Air Force as part of the service's recruitment campaign. In order to complete sufficient developmental testing for DIOT&E to begin, the F-22 CTF needs to fly around two flight test sorties a day. One F-22 is also required for ground logistics and supportability tests.

4006 joined the flight test fleet on 5 February 2001, and was soon at Edwards engaged on avionics integration work. It later became the first F-22 to fly with software release 3.1 – close to the initial operational standard. In April 2003 the DIOT&E phase begins, employing the last two EMD aircraft (4008 and 4009) and the first two PRTV (production-representative test vehicle) aircraft (4010 and 4011). Raptor 4007 will act as a back-up aircraft for DIOT&E, leaving 4003 to 4006 to handle any residual developmental testing.

On 25 January 2002 the CTF certified the F-22 for 'hot' refuelling with engines running, demonstrated here by Raptor 4005. Hot refuelling is an essential element of the integrated combat turn, in which a fighter is returned to the battle with the minimum of delay. Certification of the process was due for completion in April 2003 as part of the DIOT&E phase, but was brought forward to keep the test programme on track should inflight tankers not be available.

From the outset, it was seen that the MMTD could be very useful for stealth aircraft with internal weapons bays. The standard Mk 80 low-drag general purpose (LDGP) bomb was designed in the 1950s for external carriage, and its elliptical profile gives it a large maximum diameter. A cylindrical shape is more efficient for internal carriage. The SDB will resemble the MMTD prototypes in shape and size. It will also have wings – as demonstrated by Boeing in 2001, under the Small Smart Bomb Range Extension (SSBREX) programme – and it will be cleared for supersonic release from the F-22. Boeing's SSBREX demonstrator incorporated the Alenia Marconi Systems DiamondBack wing design and Russian-type lattice tail surfaces, which fold forward for internal stowage. The USAF's minimum requirement is that the smart rack should carry four SDBs, but both competitors are trying to meet the service's goal of six weapons per rack. This would give the F-22 the

ability to carry 12 SDBs, in addition to a pair of AMRAAMs and a pair of AIM-9s. Launched from the F-22's cruising speed and altitude, the SDB would have a range of 56 miles (90 km).

Boeing and Lockheed Martin were awarded contracts to develop SDB designs in September 2001. One of the two will be chosen to produce the weapon in September 2003, and it is due to become operational on the F-15E in 2006. The F-22 will be the second aircraft to carry the SDB. The first SDB version will have a jam-resistant GPS/INS guidance system. The second version, to be operational from 2010, will have a seeker (using radar or infra-red technology) and will be able to search a defined area for targets. The SDB will be a complete, integrated weapon rather than being built up in the field like JDAM. An integral part of the system will be a 'smart rack', compatible with the F-22, F-35 Joint Strike Fighter and external carriage, which will incorporate non-pyrotechnic ejectors and a databus that will connect the SDBs to the aircraft's avionics system.

In Lockheed Martin simulations, the F-22 uses its EW system, combined with off-board targeting information, to detect and locate SAM sites. Synthetic aperture radar scans the site and detects launchers, radars and control vehicles. The pilot selects multiple targets, releases SDBs at maximum range and breaks away in a 5-*g* supersonic turn.

Speed advantage

While the F-35 has a similar weapons and avionics suite, the F-22's unique advantage in this mission is its speed, high-speed manoeuvrability and altitude performance. Speed and altitude increase the range of the gliding weapon, while speed, altitude and manoeuvrability combine to reduce the effective range of the SAM. An F-22, therefore, can win a shoot-out against a SAM, where a subsonic, lower-flying F-35 would have to venture within the weapon's lethal radius in order to attack and could not escape as quickly.

Lo
91

F-2
Air

Air d
As be
three
carry
three

Acco
The p
of the
fighte
accon
USAF
take t
only a
The p
aircra
down
piece
Sierra
visibili
helme
fusion
joins t
piece.
side t
aircrat
elimin
collisic
briefly
discov
canop
the m

Speed and weapon range also allow an F-22 to respond more quickly to a new threat than a slower aircraft. For example, an air component commander may want to ensure that his forces can target any hostile emitter within a fixed time after its first signal. In that case, an F-22 can cover a much larger area than a subsonic aircraft. With a JTIDS transmit function, too, the F-22 will distribute information from its sensor suite to other aircraft in the theatre. For example, the F-22 will be able to confirm that it has located targets in the SAM site and launched weapons against them. Bob Rearden of Lockheed Martin describes a JTIDS-equipped F-22 as "a miniature Rivet Joint" able to feed an enormous amount of signals intelligence to other users.

Integration of the SDB may not be the end of the F-22's metamorphosis into a bomber. In early 2002, Lockheed Martin briefed the USAF on a radically modified version of the fighter, with a delta wing, longer body and greater range and payload. A company-funded study of the so-called FB-22 will continue until the end of 2002.

The FB-22 is clearly a bomber. Operations over Afghanistan drove home the fact that bases close to many potential targets may never be open to US combat aircraft. Apart from extreme-range operations by F-15Es, USAF fighters have been relatively little used in the theatre, and much of the burden has been carried by B-52 and B-1 bombers. However, the USAF has no bomber replacement plan and previously

As the development programme expands, so the avionics software packages of the aircraft have been improved. Each new release follows successful testing in the 757 FTB. Block 3.0 was the first software release to introduce full sensor fusion, and in 2002 Block 3.1 is the standard, with JTIDS-receive and JDAM capability. Block 4 is the proposed package for operational aircraft, with helmet-cueing, AIM-9X and JTIDS-transmit functions.

Above: Boeing builds the F-22's wings, which are 42 per cent titanium by weight. To improve battle-damage tolerance, every fourth spar is made of titanium, the others being composite sine-wave spars made by Dow in Wallingford, Connecticut.

Above: The aft fuselage is also built by Boeing. Weighing around 5,000 lb (2268 kg), the structure is 67 per cent titanium by weight, much of which is in electron-beam welded 'boom' sub-assemblies. Here a completed section is lifted from its assembly fixture for shipment to Marietta.

Above: An early EMD aircraft is sprayed in the paintshop at Marietta. Aircraft are now first flown in primer, before being painted by a robotic system.

EMD/PRTV F-22s proceed down the line at Marietta. Of note is the extensive use of sawtooth edges where components made of different materials are joined together, and similar edges to any openings and doorways. The aircraft in the foreground already has one engine installed.

expected to defer the service entry of a new bomber until the 2030s. This depends on some very optimistic projections concerning the survivability and operational lifetime of the B-52 and B-1. Some members of Congress have pressured the USAF to buy more B-2s, but the USAF does not want to do this because the bomber is expensive to buy and to maintain. One reason that the service may consider the FB-22 is that it demonstrates that the service is taking the need for a new bomber seriously, despite its lack of interest in more B-2s.

While some details of the FB-22 are considered proprietary and some have not been settled as yet, Lockheed's Rearden describes the aircraft as "very different, truly a derivative" of the F-22. To some extent, the FB-22 recalls two earlier Fort Worth projects: the F-16XL and the delta-winged F-16 proposed to the United Arab Emirates in 1995. Both these aircraft were based on the F-16, but with no separate tail, a much larger wing and a stretched fuselage. The result was a very different, heavier aircraft with a much bigger fuel fraction. Despite the airframe changes,

however, many elements of the aircraft – avionics, cockpit and systems – remained very similar to the original.

The FB-22's longer body accommodates larger weapon bays for nearly 30 SDBs, according to some reports. This would suggest that it could carry an alternative load of six to eight JDAMs, or a mix of SDBs and larger weapons. The FB-22 would have no horizontal tail and a more sharply swept, longer-chord wing – the span would not increase very much, says Rearden, because Lockheed Martin still wants the aircraft to fit in a 48-ft (14.6-m) wide hardened aircraft shelter. But with a longer chord and deeper section, the FB-22 wing would accommodate much more fuel, and there would also be more tankage in the stretched fuselage.

According to Rearden, there has been "quite a bit of aerodynamics work" done on the new wing, which would use complex camber variation to combine low drag with high volume and good low-speed characteristics. Another possibility is that the FB-22 could be completely tailless: Fort Worth has carried out a good deal of research into tailless designs with three-axis controls built into the wings. Whether the FB-22 would retain the same supercruise performance as the basic F-22 remains to be determined,

Above: The prominent SRA attachment identifies Raptor 02. The first two F-22s differed structurally from the production standard, weighed more and were slightly weaker.

according to Rearden. The engine might be "tuned to match the wing", suggesting that the bomber might use the more powerful, more efficient (at subsonic speed) and markedly cheaper F135 and F136 engines from the JSF. Some reports credit the FB-22 with an operating radius in the 2,175-mile (3500-km) class.

The last word on this remarkable aircraft may go to former USAF chief Fogleman. He resigned shortly after the release of the Clinton administration's second Quadrennial Defense Review (QDR) in 1997, and now says that the cutback in F-22 production in the QDR, from 448 to 339 aircraft, was a factor in his resignation. Fogleman added: "There are only two revolutionary weapon systems in the entire DoD budget: the F-22 and the airborne laser. There are no others."

Bill Sweetman

Above: F-22 4003 launches from Edwards for a test sortie. This aircraft is crucial to the test programme as it is the only instrumented structures airframe. Work with this aircraft will structurally clear the F-22 for the DIOT&E phase, which involves two EMD (91-4008 and -4009) F-22s and the two PRTV aircraft (99-4010 and -4011). The next six aircraft from the line are the PRTV-II machines (00-4012/-4017), slated for Force Development Evaluation (FDE) at Nellis AFB, Nevada. They will be used to devise and evaluate new tactics for the aircraft.

Left: Raptors 01, 02 and 03 gather for a family portrait under a Mojave Desert dusk sky. When it enters service in December 2005, the F-22 will restore the USAF's significant technological edge over rival fighters, while providing a basis for a new stealthy attacker.

Armada de Mexico

Photographed by Chris Knott and Tim Spearman

Mexico's naval air arm has recently undergone major organisational and structural changes, which have also seen the introduction of several new types to the inventory and the relocation of the Escuela de Aviación to La Paz. Also, in an unusual move for a military air arm, the Navy has also undertaken construction of its own aircraft at a small facility at Vera Cruz.

Above: Since the mid-1980s the CASA 212-300 has been the Navy's workhorse. Used for medium-range patrol work, primarily against drug traffickers, 11 remaining aircraft are distributed between four squadrons at Campeche, Chetumal, La Paz and Tapachula. This example wears the traditional blue/yellow scheme worn by patrol aircraft.

Two Beechcraft types in use are the Baron (left) and the F33C Bonanza (above). The Barons are used for multi-engine training by the Escuela de Aviación Naval at La Paz, while the Bonanzas are mainly used for inshore patrol duties, particularly by the Sixth Patrol Squadron at Tampico.

Above: Seen at its home base of La Paz, Baja California, Cessna 402B MP508 is one of two operated in the varied inventory of the Segundo Escuadrón de Patrulla. In the hangar in the background is a Piper PA-31 Navajo, also used for patrol by the squadron. Another lightplane type in use is the single-engined Cessna 182.

Below: The Navy's last airworthy Grumman HU-16 Albatross, and the only example to wear the latest blue and yellow livery, MP429 has not flown for some months and is likely to take up 'gate-guard' duties outside the impressive new naval aviation facility at La Paz in the near future.

A total of six Antonov An-32s is in use with the three patrol squadrons and, although used for low-level missions up to 300 miles (482 km) from the coastline, the aircraft are not ideally suited to the task and are more valued in the transport role. It is likely that a dedicated patrol aircraft, possibly in the form of surplus US P-3 Orions, may be acquired in the near future.

Below: The Navy's sole example of the DHC-5 Buffalo, MT220, is based at La Paz with the Segundo Escuadrón de Patrulla. It recently adopted this three-tone grey scheme. In February 2002 the navy took delivery of a Bombardier Dash 8-Q202, registered MTX-05, for service with the Escuadrón de Transporte.

Above and right: Maule MX-7s are used primarily for training by the Escuela de Aviación Naval at La Paz, Baja California (above), although several are operated by patrol squadrons on inshore coastal patrol duties. The latter are distinguishable by the predominantly blue colour scheme, as seen on ME108 (right), which flies with the Primero Escuadrón de Patrulla at Chetumal.

Left and below: The Navy operates two versions of the Lancair, constructed under licence in its own workshops at Vera Cruz. MT230 (left) is a Lancair IV-P flown as a 'hack' by the Centro de Mantenimiento at Vera Cruz, while ME121 (below) – a Lancair Super ES – is one of two flown by the Primero Escuadrón de Patrulla from Chetumal.

Above: Finnish-built Valmet L90TP Redigos are operated by two tactical reconnaissance squadrons at Tapachula and Guyamas. The latter unit has recently been detached to Campeche due to heightened security around oilfields in the Gulf of Mexico, following the 11 September terrorist attacks.

Above: A small number of Rockwell Turbo Commanders are used for light transport and communications duties. MT222 of PRIESCPAT is seen visiting Vera Cruz for maintenance from its base at Chetumal.

Right: A single Rockwell Sabreliner 60 is operated on VIP/staff transport duties by the Escuadrón de Transporte from the Navy facility on the south side of Mexico City International Airport. The aircraft replaced an earlier example of the same type in 1998.

Below: Single examples of three different versions of the Lear Jet are in operation, including this Lear Jet 25 MTX03. All are flown by the Escuadrón de Transporte at Mexico City.

Right: Acquired in 1999 and primarily used by the Secretary of the Navy, Lear Jet 60 MTX01 is the flagship of the VIP fleet. It replaced a Lear Jet 24 which carried the same serial.

Above: Acquired in the early 1990s, the MBB (Eurocopter) BO105C (above) and BO105CB are used for shipborne patrol operations, equipping two squadrons with land bases at La Paz and Vera Cruz. The BO105s have search radar mounted in the nose

Below: Two Robinson R22s were purchased in December 2000 for use by the Escuela de Aviación, and both operate in the bright yellow training colours worn by most of the unit's aircraft.

Left: Three elderly Hiller UH-12s are still used by the Escuela de Aviación Naval for basic helicopter training at La Paz, although the type is due to be replaced shortly by the Robinson R22 Mariner, which can be equipped with floats. The Hillers replaced Bell 47G/Js, the Navy's first rotary-wing type which was first delivered in November 1958.

Below: In use since March 2000 is a single four-seater Robinson R44 Clipper. Serialled ME-017, it is seen at its home base of La Paz, Baja California, where it flies with the Escuela de Aviación.

Below: Four MD500Es have been in service for more than 12 years. Used for advanced helicopter training by the Escuela de Aviación, all now wear the predominantly dark blue colour scheme normally associated with patrol aircraft.

Below: Built alongside the Lancairs at Vera Cruz were three Rotor Way Exec 162F helicopters. Acquired in kit form from the United States, the type was evaluated to replace the UH-12s of the Escuela de Aviación until one was lost in a crash in December 2000.

Above: Four Mi-2s are operated in the inshore search and rescue role by the Primer Escuadrón Busqueda y Salvamento. Nominally based at Salina Cruz, the aircraft are routinely deployed to other locations. Over 5,000 of these utility helicopters were built at PZL's Swidnik plant.

Below: Since 1994 around 20 Mil Mi-8MTVs have entered service. Used in the light transport and reconnaissance roles, the aircraft are regularly utilised in conjunction with fixed-wing units to provide rapid troop deployments during anti-narcotics operations. Most now wear the three-tone grey tactical camouflage, as seen on MR357 at Chetumal. Mexican Navy serials have a prefix denoting a basic role: 'MR' is for Marina Rescate (naval rescue), 'ME' for Marina Entrenador (naval trainer), 'MP' for Marina Patrullero (naval patrol) and 'MT' for Marina Transporte (naval transport).

Armada de Mexico – units, bases, aircraft

Transport/patrol

PRIESCPAT (Primer Escuadrón de Patrulla)	Chetumal	An-32, CASA 212, Lancair ES, Maule MX-7, Rockwell Commander 690
SEGESCPAT (Segundo Escuadrón de Patrulla)	La Paz	An-32, CASA 212, DHC-5, Cessna 402, Piper PA-31, Cessna 182
TERESCPAT (Tercer Escuadrón de Patrulla)	Vera Cruz	An-32
CUARESCPAT (Cuatro Escuadrón de Patrulla)	Tapachula	CASA 212, Maule MX-7
QUINESCPAT (Quinto Escuadrón de Patrulla)	Campeche	CASA 212
SEXESCPAT (Sexto Escuadrón de Patrulla)	Tampico	Beech F33, Cessna (various)

Tactical reconnaissance

PRIESCAREC (Primer Escuadrón de Reconocimiento Táctico)	Guyamas (detached to Campeche)	Redigo
SEGESCAREC (Segundo Escuadrón de Reconocimiento Táctico)	Tapachula	Redigo

Search and rescue

PRIESCBUSALV (Primer Escuadrón Busqueda y Salvamento)	Salina Cruz	Mi-2

Transport

PRIESCAMET (Primer Escuadrón de Mobil Exploración y Transporte)	Tampico	Mi-8MTV
SEGESCAMET (Segundo Escuadrón de Mobil Exploración y Transporte)	La Paz	Mi-8MTV
TERESCAMET (Tercer Escuadrón de Mobil Exploración y Transporte)	Chetumal	Mi-8MTV
CUARESCAMET (Cuatro Escuadrón de Mobil Exploración y Transporte)	Tapachula	Mi-8MTV, AS 555
QUINESCAMET (Quinto Escuadrón de Mobil Exploración y Transporte)	Campeche	Mi-8MTV

Shipborne

PRIESCEMB (Primer Escuadrón Embarcado)	Vera Cruz	BO105
SEGESCEMB (Segundo Escuadrón Embarcado)	La Paz	BO105

VIP transport

ESCAERTRANS (Escuadrón de Transporte)	Mexico City	Lear Jet 25/30/60, Sabreliner, DHC-8 Q202, Mi-8MTV

Training

Escuela de Aviación	La Paz	Baron, Maule MX-7, R22, R44, MD500E, UH-12

Additional bases include Acapulco, Tulum and Teacapan, which operate a handful of aircraft independent of any fixed unit. The new MD902 Explorers are thought to be operated by an Escuadrón Embarcado, although whether it is a new unit or an existing one is uncertain. They appear to be shore-based at Vera Cruz.

*A few Mi-8MTVs retain the old-style blue/yellow patrol colours (left), while one serves in a staff transport role with **ESCAERTRANS** (above). It carries a 'transport' serial, wears a VIP-style scheme and has a weather radar in the nose.*

Indian Air Force

The Indian Air Force is today the world's fourth largest air arm, and enjoys a formidable reputation. It also has a long and distinguished history. The IAF is well-equipped with a range of modern and upgraded aircraft from both Western and Russian sources, and embraces every air power role, from elementary training to nuclear strike.

Indian Air Force – Bharatiya Vayu Sena: History

The Indian Air Force has probably gained more operational experience since 1945 than any other air arm, and has done so against top-quality opposition. The Israeli Air Force has fought wars against its Arab neighbours in 1948, 1956, 1967, 1973 and 1982 but many of its operations have been very brief, and rather one-sided. India's Air Force has sometimes been more closely matched against its opposition, and has undertaken wars of attrition, in which small advantages have had to be fought for with grim determination.

The Indian Air Force has built upon firm foundations, having been established under RAF auspices on 8 October 1932. The new air arm formed a Flight-strength unit with four Westland Wapitis on 1 April 1933, with six RAF-trained officers and 19 sepoys. This unit, 'A' Flight of a planned No. 1 Squadron, went into action in 1937 in North Waziristan, supporting Indian Army operations against insurgent Bhittani tribesmen. By the time war broke out, No. 1 Squadron was up to full strength, with three Flights. An IAF Volunteer Reserve was then formed, parenting five Coastal Defence Flights, and these soon formed the basis of new squadrons, which were then thrown into action against the advancing Japanese.

By 1944, the IAF included squadrons equipped with modern Spitfires, Hurricanes and Vengeance dive-bombers, and played a major role in driving Japanese forces from Burma, while establishing a reputation for courage and efficiency. The re-naming of the force as the Royal Indian Air Force marked official British recognition of this contribution.

Spitfire squadrons

India re-equipped all of its front-line fighter squadrons with Spitfires by the middle of 1946, and in August 1945, one of these – No. 4 Squadron – was designated as a component part of the British Commonwealth Occupation Forces in Japan, deploying aboard HMS *Vengeance* and arriving in the enemy homeland on 23 April 1946, having just exchanged its Merlin-engined Spitfire Mk VIIIs for Griffon-powered Mk XIVs.

Plans were already underway to provide a more balanced force structure, with transport and bomber units. The IAF also planned to re-equip its Spitfire fighter units with Centaurus-engined Tempest Mk IIs, with No. 3 Squadron at Kolar becoming the first to re-equip in the autumn of 1946.

Britain had envisaged a post-war Royal Indian Air Force of almost twice its wartime size, with 20 squadrons of fighters, bombers and transports, though these plans were put on

hold as independence and partition approached. In the event, the Royal Indian Air Force actually shrank as it was divided to allow the formation of separate Indian and Pakistani air arms.

India was partitioned on 15 August 1947, and the two new states were soon in conflict. The RIAF was quickly in action. On 27 October 1947, No. 12 Squadron air-lifted Sikh troops from Palam into Srinagar to repulse insurgent forces pouring across the new border into Jammu and Kashmir, and India soon committed Advanced Flying School Spitfires and front-line Tempests to strafe the advancing enemy forces.

Fighting between India and Pakistan finally ended after some 15 months, and a ceasefire eventually took effect on 1 January 1949. Throughout the period, however, reorganisation and modernisation of the RIAF continued.

After the campaign, the RIAF acquired its first heavy bombers, simply refurbishing some of the redundant USAAF and RAF B-24 Liberators which were in storage at the immense Care and Maintenance Unit Depot at Kanpur and abandoned elsewhere. In the same month that the refurbished Liberators emerged from HAL, the RIAF also received its first jet fighters, in the shape of three de Havilland Vampires – the first of an eventual 400. These would even include Vampire NF.Mk 54 night-fighters which re-equipped No. 10 Sqn at Palam in May 1953, providing India with its first modern night-fighting capability.

RAF links maintained

The Air Force dropped its Royal prefix in January 1950, as India became a republic within the British Commonwealth, but links with the RAF remained strong. The IAF modelled itself on the RAF, and proudly retained the same rank and unit structure, while emulating RAF training methods and relying on the RAF's Central Flying School to train the first of its flying schools' instructors.

Expansion of the Air Force was accorded a high priority, and in 1955 an Auxiliary Air Force (previously raised at the beginning of World War II) was resurrected. Seven units (Nos 51 to 57 Squadrons) were formed, initially operating HT-2 trainers, but later converting to the Vampire jet fighter from 1959.

During the 1950s, India quite deliberately pursued a policy of arms diversification, reducing its reliance on the UK as its sole source of weapons. The first of over 100 Dassault Ouragans (known as Toofanis in IAF service), were delivered to Palam on 24 October 1953, and these were soon followed (in 1957) by 110 Mystère IVs. US equipment came with more

'strings' and although India did receive 79 Fairchild C-119G Flying Boxcars, acquiring front-line US types was never a real, meaningful possibility.

India turned to the USSR with similar caution, acquiring a small fleet of transport aircraft, initially purchasing eight Antonov An-12s, 24 Ilyushin Il-14s and 10 Mi-4 helicopters. The An-12 fleet was subsequently expanded through the purchase and loan of further aircraft, while the Mi-4 fleet eventually reached 120 helicopters. Both types quickly proved of enormous usefulness and value. There were also early successful examples of indigenisation, with the HT-2 primary trainer proving effective and popular in service.

While India diversified its sources of weapons, Britain remained vitally important, and 1957 saw the beginning of procurement of Canberra bombers and Hawker Hunter fighters, and these (together with licence-built Vampires) formed the backbone of the IAF's expansion from 15 to 33 squadrons, though six front-line units were formed on Vampires as interim equipment. 1957 also saw the retirement of No. 14 Squadron's last Spitfire F.Mk 18s, the last piston-engined fighters remaining in IAF service. From 1960, the new Hunters and Canberras were soon joined by large numbers of Folland Gnats, 23 of them built by the parent company, 20 supplied in kit form, and the remainder built under licence by HAL as the Ajeet.

The new Canberras were soon in action, with six B(I).Mk 58s of No. 5 Squadron being committed to support United Nations operations in the Congo (later Zaïre) during 1961-62. The aircraft were flown from Agra to central Africa, and operated from Leopoldville and Kamina, from where they provided the UN ground forces with their only long-range air support. The Canberras raided Katangan targets and soon destroyed the rebel air force.

The IAF's new helicopters and transports were also heavily used during the brief border war with China between 20 October and 20 November 1962, operating at very high altitudes. In the wake of this conflict, the Indian Government sanctioned a further expansion (to 45 squadrons), even though the 33-squadron target had not yet been achieved. India gained a further arms supplier immediately after the war, acquiring de Havilland Canada DHC-4 Caribous from the Canadian Government.

Having expected to face threats only from the west, the IAF had originally followed the RAF model of functional commands, but the realisation that the country faced similarly dangerous threats on both of its major borders led to a major restructuring, with three multi-function geographically-based operational commands, one in the east, facing China, one in the west facing Pakistan, and a central command providing defence in depth and capable of reinforcing either front if required. This followed the earlier establishment of a dedicated Maintenance Command in 1955.

The first 'Fishbeds'

In August 1962, India had decided to procure combat aircraft (and not just support types) from the USSR, and this decision resulted in the acquisition of an initial batch of 12 MiG-21F-13 interceptors and some SA-2 'Guideline' SAMs. The MiG-21s were still not operational when

This No. 24 Squadron Su-30M carries R-27 missiles under the wings and R-73s on the wingtips.

Above: The initial Russian-built batch of 18 Su-30Ms serves with No. 24 Sqn at Pune. HAL is building 140 more advanced Su-30MKIs under licence.

Right: The Jaguar IS provides the IAF with a potent strike/attack capability. This example is from No. 16 Squadron, based at Gorakhpur.

Above: The IAF received 15 Jaguar IT two-seaters initially, and is in the process of receiving 17 more. The latter batch is believed to have a combat role, with nuclear strike being one possible application.

Above: No. 6 Squadron at Pune flies the anti-shipping Jaguar IM version. The Agave nose radar is being replaced by HAL with an Elta EL/M-2022 radar.

Shown with acquisition rounds for Magic 2 and Super 530 missiles, this Mirage 2000H serves with No. 7 Squadron at Gwalior. The Mirage 2000 has proved to be arguably the most valuable type in IAF service, being used as a PGM platform, long-range air defence asset and, allegedly, as a vehicle for nuclear weapons delivery.

With R-27 missile under the wing, this MiG-29 wears the badge of No. 47 Squadron. The 'Fulcrum' fleet has an air defence commitment in the tense northern region of the Western Air Command area.

This MiG-29UB and MiG-29 serve with No. 28 Squadron. The MiG-29 has suffered from poor serviceability in IAF service, although the fleet is slated for a major upgrade programme to enhance its capabilities.

Pakistani armour pushed across the border in the Chhamb sector on 1 September 1965, and though they did fly local CAPs during the ensuing war, aircraft like the Hunter, Canberra, Gnat and Mystère bore the brunt of the fighting.

Pakistan proved far more adept in advertising its successes, and in releasing detailed records of at least some of its losses and kill claims, which India failed to challenge, leading many observers to assume that Pakistan had 'won' the air war. This was far from the truth, although Pakistan probably scored more air-to-air kills than India. At the same time, the IAF was beginning to demonstrate a degree of superiority by the time the war drew to a close,

and the PAF proved entirely unable to restrict Indian Canberra operations. Losses on both sides were similarly heavy, however, and each side suffered setbacks. While the IAF learned some painful lessons from the conflict, the air arm performed extremely well, especially in the offensive support and interdiction roles.

One of the lessons emerging from the war was 'the obsolescence of aircraft like the Vampire, and of the ageing Liberators still used in the maritime patrol role. The B-24s gave way to former Air India Constellations, while large numbers of MiG-21FLs were ordered to form the backbone of the IAF's fighter arm (nine squadrons re-equipped with the type between

1966 and 1969). The first of these were delivered in flyaway condition, but subsequent batches were delivered in kit form. With continuing delays to the indigenous HAL HF-24 Marut, the IAF ordered large numbers of Su-7BMs to re-equip a number of fighter-bomber squadrons.

Expansion of the force also continued apace, so that by the end of 1968, the IAF included 42 fixed-wing squadrons, including 23 fighter and fighter-bomber units and three Canberra bomber squadrons. The long-standing 45-squadron goal was achieved soon afterwards.

Great efforts were made to improve operational readiness, and training was refined to better reflect operational requirements. Front-line types like the MiG-21, Canberra and Gnat (which had been delivered in natural metal or silver finish) were increasingly camouflaged, and after encountering PAF Sabres armed with Sidewinders, the acquisition of AAMs became a priority.

The early days of 1971 saw attention switching from aircraft to infrastructure. The Air Force Academy at Dundigal (near Hyderabad) was inaugurated in January 1971, while planning for a new Air Defence Ground Environment System (ADGES) began in March.

1971 war with Pakistan

Another war between India and Pakistan broke out in late November 1971, after Pakistani forces put down an uprising in East Pakistan with great brutality, forcing India to intervene. Pakistan then mounted pre-emptive strikes against the IAF in the West, triggering a full-scale war, which India rapidly came to dominate. Pakistan was again better at advertising its claims (and at hiding many of its losses) than was the IAF, though it is now clear that air-to-air kills and losses were very closely balanced. More significantly, some 80 per cent of the IAF's sorties were close support or interdiction missions, and the IAF performed with distinction in the air-to-ground role. The PAF failed to conduct an effective offensive air campaign, or to prevent the IAF from achieving its objectives, and the war ended in a clear defeat for Pakistan.

Hunters and Canberras again played a vital role in India's air campaign, though newly acquired types also played their part. The IAF's six squadrons of MiG-21s performed with some success, in both the air-to-air and air-to-ground roles, while the new Su-7BMK flew large numbers of sorties, suffering heavy losses.

The 1971 war demonstrated the vital importance of the IAF, and maintaining the air arm's capability was accorded a high priority. Within a few years of the conflict, a range of requirements had been issued, and these would result in the induction of a whole new generation of combat aircraft, while the development and refinement of training and tactics continued apace.

After a period of relative calm during the 1970s and 1980s, the Indian Air Force found itself in action again at the end of the decade, in Operations Meghdoot, Pawan and Cactus. Operation Meghdoot was mounted from April 1984 in support of the Indian Army (and para-military forces) in Northern Ladakh, as they wrestled to secure control of the Siachen glacier, often dubbed the 'roof of the world'. IAF An-12s, An-32s and Il-76s transported men and material into the region, and air-dropped supplies to high-altitude airfields. Meanwhile, Chetak, Cheetah, Mi-8 and Mi-17 helicopters ferried troops and supplies to altitudes that were far above the limits set by the helicopter manufacturers.

From 1987, the IAF was heavily involved in Operation Pawan, supporting the Indian Peace Keeping Force (IPKF) in northern and eastern Sri Lanka. The IAF flew 70,000 sorties in the 30-month campaign, maintaining a continuous link between air bases in southern India and the various Indian Divisional headquarters at Palaly (Jaffna), Vavuniya, Trincomalee and Batticaloa. The main types involved were the An-32, the Mi-8 and Mi-17, while the Mi-25s of No. 125 HU were used to interdict clandestine coastal and riverine traffic, and to provide close air support when required.

Operation Cactus was the name given to the Indian special forces operation mounted in response to the Maldives Government's appeal for military help against a mercenary invasion. On 3 November 1988, Indian Air Force Il-76s airlifted a parachute battalion group from Agra to the Maldives, where they landed at 0030 hours on 4 November, securing the airfield and restoring government rule within hours. The IAF's transport force brought in more troops later that day, and Mirage 2000s made low-level passes over the scattered islands of the Maldives archipelago in an impressive show of force.

Indian Air Force – Bharatiya Vayu Sena: Equipment review

Between 1978 and 1988 India launched a major modernisation programme, and though aircraft like the Canberra and Hunter would continue in service in small numbers for another two decades or more, they would soon be replaced in front-line service by much newer aircraft types. Requirements issued in the wake of the 1971 war are largely responsible for the overall shape of the Indian Air Force today.

Air defence

Today, the IAF's front-line strength is increasingly composed of multi-role units, and relatively few squadrons have a pure air defence commitment. The introduction of the General Dynamics F-16 by the PAF in 1982 forced India to respond. In the interim, two squadrons of MiG-23MF air superiority fighters, equipped with R-23 (AA-7 'Apex') beyond-visual range missiles, were hurriedly rushed into service, equipping a single wing at Adampur in Western Air Command. However, the MiG-23MF was never considered as an adequate counterbalance to the PAF's F-16s and, in 1982, India also placed orders for two squadrons of Mirage 2000Hs. IAF pilots and technicians underwent conversion training at Mont de Marsan and ferried the first Mirage 2000s to India during the summer of 1985. This initial order comprised 36 single-seat Mirage 2000Hs and four two-seat Mirage 2000THs.

With their relatively long range, the two Mirage 2000 squadrons were stationed at Gwalior in Central Air Command. Soon afterwards, in 1984, India evaluated the new Russian MiG-29 (becoming the first export customer to do so) and finally ordered two squadron's worth (42 single-seaters and six or eight twin-stickers) in 1986, initially stationing these at Pune.

It soon became clear that the IAF was 'hedging its bets' and that these early batches of Mirage 2000s and MiG-29s were intended for prolonged operational evaluation, after which one type would be ordered in much larger quantities, probably with large-scale licence production.

Had the Mirage 2000 been selected, there were plans to order 110 further aircraft, including between 55 and 65 indigenously manufactured aircraft, while production of the MiG-29 was expected to be even larger. In the event, the Mirage 2000 proved expensive and maintenance-intensive, while the MiG-29's short range, lack of multi-role capability and poor spares support mitigated against increasing the 'Fulcrum' fleet.

Thus the IAF received a top-up attrition batch of six Mirage 2000Hs and three THs in 1987, while batches of 20 and 10 MiG-29s were delivered (allowing the formation of a third squadron) in 1989 and 1995. A further 10 Mirage 2000THs have been ordered as attrition/reserves, though the fact that all are two-seaters might indicate that they will be used to boost the Mirage 2000 fleet's air-to-ground capabilities. Delivery of the new batch might also be connected with reports that India has ordered, or is about to order, pod-mounted Elta EL/M-2060 synthetic aperture radars for the IAF's Mirage 2000H fighters, presumably for use in the air-to-ground and reconnaissance roles.

For a while, it seemed likely that the longer-range, multi-role MiG-29M would be acquired instead, but the failure of the Russian air forces to order this version sealed its fate, and India began to look elsewhere for its definitive air defence fighter. While further MiG-29 orders by the IAF are unlikely, the Indian Navy is showing interest in the type, and Air Force aircraft are expected to undergo an ambitious upgrade, with early aircraft being brought up to the same standard as India's newest 'Fulcrums' which, according to some reports, are MiG-29SEs capable of firing R-77 AAMs. The MiG-29s are also expected to gain an inflight-refuelling probe, and perhaps even an electronically scanned (phased-array) radar.

Multi-role 'Flanker'

Eventually, India settled on the two-seat Su-30, in its multi-role Su-30MKI form. An initial batch of 40 was ordered in November 1996, and orders for 10 more followed soon afterwards. India did not order an 'off-the-shelf' version of the aircraft, but instead required a new multi-role, PGM-capable aircraft with thrust-vectoring engine nozzles, canard foreplanes and an avionics suite blending Russian, French and Indian systems. The Su-30 programme was thus extremely ambitious and complex, and a complicated delivery schedule was drawn up, under which batches of aircraft would be delivered which would come progressively closer to the final standard. The final batch, to be delivered in 2003, would be full-standard Su-30MKIs, after which earlier aircraft would be retrofitted to the same standard.

Thanks to Russia's economic problems, development problems and delays in defining the avionics suite, the Su-30 programme was drastically delayed. The first eight aircraft (to little more than Su-27UB trainer standards, albeit fitted with retractable inflight-refuelling probes) were received 'on schedule', but only 10 more (to the same standard) have been delivered since then, while there have been suggestions that the thrust-vectoring system originally intended for the Indian aircraft has been cancelled, requiring the development and

Above: Procured as an interim air defender, the MiG-23MF remains in service with only one squadron – No. 224. Its roles include target-towing for other fighter units, having taken over the role from the Hunter.

Right: MiG-23UB and UM two-seat trainers were procured in sizeable numbers to train crews for the MiG-23BN/MiG-27 fleet. This aircraft is assigned to No. 29 Squadron, which flies the MiG-27 from Jodhpur.

Numerically the MiG-27M is the backbone of the attack fleet, equipping seven units. Another three squadrons fly the similar but less well equipped MiG-23BN. These aircraft serve with No. 222 Sqn (left, in the new grey colour scheme) and No. 18 Sqn (below).

Above: In the move to an all-grey fleet, many IAF aircraft have lost unit markings. This plain MiG-21U serves with No. 8 Squadron. Under the current training system, most IAF pilots begin their operational training in the MiG-21U and MiG-21FL.

Above: Remarkably, the first-generation MiG-21FL remains in service with three front-line air defence units, although they are located in the strategically less-sensitive eastern provinces. This example flies with No. 8 Squadron at Bagdogra.

Right: The MiG-21bis fleet bears the brunt of the air defence mission, and is consequently undergoing a major upgrade to add BVR capability, among other improvements. This specially marked MiG-21bis flies with No. 15 Squadron.

integration of an alternative system. These problems did not prevent India signing a contract on 28 December 2000 for the licence manufacture of some 140 further Su-30MKIs, the first of which is due to roll out in 2004. There have been reports that India might consider cutting back (or even cancelling) this order in favour of an order for one or two squadrons of Mirage 2000D fighter-bombers, and there have been reports that Dassault Aviation has offered India 18 Mirage 2000Ds. Although the D is the conventional version of the Armée de l'Air 2000N, Indian reports suggested that the new aircraft would be tandem-seat fighter-bombers 'hard-wired' for

the carriage of nuclear weapons.

At the time of writing, however, India has only a single Su-30 squadron, based at Pune within Southwestern Air Command. Meanwhile, the last MiG-23MF squadron moved from Adampur to Jamnagar in 1997, becoming a BVR training and trials and target-towing unit, replacing Kalaikunda's recently retired Hawker Hunters.

Mirages for attack

The 18 Su-30s of No. 24 Squadron do not yet have any meaningful air-to-ground capability, but the 40 remaining Mirage 2000Hs (and four THs) have proved to be useful PGM delivery

platforms, and the type's availability in the air defence role is thus constrained. Meanwhile, MiG-29 availability has been severely constrained by poor serviceability, inadequate factory support and spares shortages, and far fewer than the 67 survivors (plus about six UBs) nominally on charge are actually available for use. At one stage during the early 1990s, for example, MiG-29 availability was less than 55 per cent, by comparison with the 75 per cent achieved at the end of the 1980s. Altogether, these factors have left the bulk of the IAF's air defence commitments in the hands of 10 squadrons of about 200 MiG-21bis aircraft, and three squadrons with about 63 MiG-21FLs.

India received 75 Soviet-built 'third-generation' MiG-21bis aircraft during the 1970s and assembled or built 220 more, bringing total Indian MiG-21 procurement to more than 830 aircraft. Three Gnat air defence squadrons (Nos 15, 21 and 23) were re-equipped with the Russian-built MiG-21bis aircraft in 1976-1977, and further units converted as HAL-built aircraft became available. During the 1980s the IAF's MiG-21Ms and MiG-21bis were modified to carry the Matra R550 Magic II AAM, and later the Russian R-60 (AA-8 'Aphid') AAM. Some MiG-21bis were also retrofitted with a new head-up display (HUD), but the aircraft remained without BVR armament.

At one time it was assumed that the MiG-21 would be replaced by a mix of Mirage 2000s and MiG-29s, and by the new, indigenous LCA or Light Combat Aircraft, but as problems and delays made this progressively more distant, the decision was taken to upgrade some MiG-21bis aircraft for further service. Under this upgrade, some 125 MiG-21bis will be brought to MiG-21bis-UPG standards, with HAL incorporating the upgrade using kits supplied by the Sokol plant in Nizhny Novgorod, following the conversion of the first two aircraft in Russia.

The main aim of the upgrade was to make the MiG-21bis fleet BVR-capable, and to achieve this, the MiG-21bis was fitted with the lightweight Super Kopyo multi-mode radar, and gained provision for the Vympel R-77 AAM. The aircraft also gained a new IRST, a helmet-mounted sight, new avionics and a comprehensive defensive aids suite. The MiG-21bis-UPG will remain a multi-role aircraft, and has gained some PGM capabilities, using a range of TV- and laser-guided bombs, and the Kh-31 ASM. The first 36 upgrade kits arrived in India during the last quarter of 1999, delayed by two years due to centre of gravity problems and other minor difficulties.

Those MiG-21bis aircraft which are not upgraded (and the remaining ageing MiG-21FLs) are expected to be retired imminently, perhaps even without being replaced. In an effort to simplify logistics and improve operational flexibility, the IAF has successfully integrated Western air-to-air missiles with some of its Soviet-designed aircraft. MiG-21s were made compatible with the Magic II AAM some years ago, while some reports suggest that, more recently, the air force's MiG-29s have been adapted to carry the Super 530D.

There are also reports that the Mirage 2000 has been modified to carry Soviet missiles. There are suggestions that Mirage 2000s operating over Kargil in the summer of 1999 were armed with R-73 AAMs, and that the type may also now be compatible with the R-27 AAM, though the latter seems unlikely.

Ground-based air defence

Air defence is not only a matter for manned fighters, however, and the deployment of nuclear weapons by both India and Pakistan has placed greater emphasis on the importance of land-based air defence systems. The Air Defence Ground Environment System (ADGES) and the Base Air Defence Zones (BADZ) are two linked components of a well-integrated air defence network, controlling a variety of sensors and SAM units and manned fighters from both the Air Force and the Navy. The Air Defence Ground Environment System is based upon a chain of picket radars strung out along the western and northern borders, augmented by a series of mobile systems usually deployed to the northeast and south of the country, together with a string of visual Mobile Observation Posts, and a smaller chain of longer range radars controlled by Air Defence Control Centres. The ADGES network is responsible for overall airspace management and for the detection of intruders, and coordinates the air defence of large-area targets.

Base Air Defence Zones, by contrast, are tasked with the defence of individual high-value targets and are limited to an arc of 100 km (62 miles). These represent a dense, high-threat environment for attacking aircraft, with layers of long-range SAMs, L-40/70 radar-directed 40-mm anti-aircraft guns and man-portable Igla-1M SAMs.

The ADGES is currently being upgraded with new Indian and Israeli radar systems, and has been hardened and given multiple-redundant layers to enable it to survive an enemy first strike. Meanwhile, the IAF's SA-3 Pechora SAMs are also being fitted with more sensitive seeker heads, pending the service introduction of the next-generation Akash SAM and the associated Rajendra phased-array radar.

In order to protect against the threat of ballistic missile attack the current sensor network needs to be substantially upgraded in order to detect and track missile launches. Reports suggest that the Russian S-300 SAM is being considered to provide the cornerstone of the planned ATBM defences, and has undergone trials in India.

Offensive air operations

Compared to most air arms in the developing world, the Indian Air Force has always placed considerable emphasis on the air-to-ground role, acquiring and operating a succession of dedicated jet fighter-bombers, and a significant force of Canberra medium bombers.

In the future, the IAF's new Su-30s are expected to form the backbone of the force's strike/attack fleet. The public relations line was that a single Su-30 will be able to deliver as much ordnance as an entire squadron of MiG-21FLs, at greater ranges, and with greater accuracy.

The early aircraft delivered so far have only the most basic and reversionary air-to-ground capability, however, and the first Su-30 aircrew are understood to have trained only for air combat. India's Comptroller and Auditor General published a damning report condemning shortcomings in the Su-30K's electronic warfare and weapon delivery system, and criticising the delayed delivery schedule. The electronic warfare system was felt to be unable to cope with the Indian threat environment while radar performance was said to be 'below expectations', while the navigation system lacked accuracy and was thought to be insufficient to guarantee accurate weapon delivery. The aircraft's weapon systems controls were also said to be 'poorly integrated'. The Su-30K's generous payload was recognised and praised, but the Comptroller and Auditor General noted that it did not include any of the precision-guided munitions which had proved essential in the Kargil operations.

Unusually, the Mirage 2000, acquired as an interceptor, has been pressed into service in an offensive role. It is understood that the size and weight of the first Indian nuclear weapons was such that the Jaguar was seriously constrained when carrying them, and the Mirage 2000 was hastily pressed into service as the IAF's primary strike aircraft. When a second generation of nuclear weapons was produced, it could be carried by the Jaguar and MiG-27, although manned platforms have already lost their atomic monopoly. It is, however, believed that the IAF is entrusted with the operation of India's emerging force of nuclear armed missiles, which is understood to include unspecified numbers of Agni IRBMs and Prithvi SRBMs. Manned platforms have sufficient reach to deter Pakistan, but until long-range missiles are deployed, India's ability to inflict more than peripheral damage on China will be seriously restricted. The IAF is likely to remain India's sole strategic nuclear strike force for some years, although some believe that the Navy may assume a strike role when it deploys cruise missiles or when and if it leases or purchases Tu-22Ms.

The IAF's two Mirage 2000H squadrons have also increasingly been used in the air-to-ground role, thanks to their long range and to the easy integration of Litening laser designator pods and SAMP LGBs. The Indian deltas demonstrated this new PGM capability to devastating effect during the 1999 Kargil war, and this success resulted in the swift procurement of additional laser-guidance kits for standard 1,000-lb (454-kg) bombs from Elbit Computers of Israel. The Mirage 2000 has reportedly maintained the highest serviceability of any type in IAF service and is popular with air and ground-crew alike, though there are too few Mirages available, and other, older types still perform the bulk of the IAF's offensive missions.

Indian Jaguars

During the 1970s, the IAF's most urgent requirement (dating from 1966) was for a Deep Penetration Strike Aircraft (DPSA), to replace the ageing Canberras and Hunters still used in that role. Acquisition of 150 Jaguars was finally approved in 1978, with 40 BAC-built aircraft to be followed by 110 aircraft from HAL, 45 to be assembled from UK-supplied kits, and 65 which would incorporate progressively greater local content.

An interim batch of 18 ex-RAF Jaguars was provided on loan to equip the first IAF Jaguar unit, No. 14 Squadron, and these were ferried to India by their pilots in July 1979 after the latter completed their conversion training with the RAF and British Aerospace at Lossiemouth, Coltishall and Warton.

The 40 BAC-built Jaguars were delivered from March 1981, allowing No. 14 Squadron, to return its 'loan aircraft', and allowing No. 5 Squadron to retire its ancient Canberras. A proposal by India to purchase eight of the loan aircraft came to nothing, but one was retained for trials. The 35 single-seaters were similar to RAF GR.Mk 1s, with NAVWASS avionics, and were powered by Adour Mk 804E engines, similar to the Adour 104s introduced in RAF aircraft from 1978. Five of the aircraft were two-seaters equivalent to RAF T.Mk 2s. As Jaguar Internationals, the Indian single-seaters had overwing hardpoints for missile launch rails, and these were eventually activated to allow the carriage of MATRA R550 Magic AAMs.

Built under licence by HAL, the HS.748 continues to play an important role in the IAF, providing passenger and cargo transport. This example is assigned to Maintenance Command's HQ flight.

Above: Replacing the An-12 in the heavylift role, the Ilyushin Il-76MD 'Candid' serves with Nos 25 and 44 (illustrated) Squadrons. The type's good take-off performance is appreciated during resupply flights to forward bases situated high in the mountains along India's northern borders.

Below: AHQCS flies this Gulfstream III on VIP transport duties. Two further GIIIs are operated by the ARC, modified with a large side-facing camera window in the forward fuselage for long-range stand-off reconnaissance.

Backbone of the transport fleet is the Antonov An-32, which serves with six squadrons, including No. 12 Sqn 'Yaks' which flies this aircraft in the latest grey scheme. As well as front-line use, the An-32 also operates with the multi-engine training unit at Yelahanka and the parachute training school at Agra.

HAL's Kanpur plant produces the Dornier Do 228 under licence for light transport duties. A total of 80 has been built for all Indian armed forces, with another 22 on order. This grey-painted aircraft flies with Palam-based No. 41 Squadron 'Otters'.

The number of Jaguars assembled and built by HAL was reduced from 110 to 74 (but later increased again to a current total of 129, with deliveries continuing). The HAL-built Batch 3 aircraft were powered by Adour Mk 811 engines, and were fitted with a MIL STD 1553B digital databus, and a new locally integrated DARIN (Display Attack and Ranging Inertial Navigation) nav-attack system, incorporating a wide field-of-view Smiths HUDWAC, a GEC-Ferranti Combined Map & Electronic Display, and a SAGEM ULISS 82 INS. These HAL Jaguars re-equipped Nos 27 and 16 Squadrons at Gorakhpur, while 12 aircraft were delivered with nose-mounted Agave radar and compatibility with the Sea Eagle ASM, and these equipped No. 6 Squadron for anti-shipping.

Indian Jaguars are acknowledged by some India-based Internet sites as having a nuclear strike capability (with the introduction of a second-generation of lighter weapons). They also carry a wide range of conventional weapons, including Hunting BL755 CBUs, slick and retarded RAF-type 1,000-lb (454-kg) bombs, Matra 250- and 400-kg (551- and 882-lb) bombs, Matra Durandal anti-runway

No. 106 Squadron 'B' Flight at Agra operates the IAF's last Canberras for reconnaissance/survey/electronic warfare/target-towing tasks. Variants in use are the T.Mk 54 (two), PR.Mk 57 (two), B(I).Mk 58 (three, above), PR.Mk 67 (two, right) and TT.Mk 418 (three).

bombs, Lepus 8-in recce flares and Matra F1 and 155 (SNEB) rocket pods, and, in recent years, a range of LGBs. The BAC-built Jaguars used by the Ambala Wing (Nos 5 and 14 Squadrons) also have a recce commitment, using BAE recce pods and some of the 24 Vinten VICON 18 Srs 601 (GP) pods supplied more recently, and understood to be shared with the MiG-27 force.

Seventeen two-seat Jaguar IBs were ordered for delivery from 2001, and these will be fitted with the single-seater's Sextant INS-RLG system with embedded GPS, and may be intended to have an operational role as laser designators, using the Israeli Litening laser designator pod. Because Litening incorporates a FLIR and a CCD camera it confers improved night capability (which will be further enhanced by the addition of NVGs), and has a limited reconnaissance application. A further order for 20 more single-seat Jaguar ISs has since been placed and these

additional aircraft are expected to form a sixth front-line Jaguar squadron, perhaps with a 'target-marking' and designation role.

Upgrade programme

India's Jaguars are to receive an ambitious upgrade to allow them to remain viable until 2020 or 2032, with new avionics and displays, a new jammer, and other improvements. The anti-shipping Jaguars are to receive a new Israeli Elta radar.

While the Jaguar was selected to replace the longer-range Canberras (and the highly-prized, hard-hitting Hunters), shorter range air-to-ground fighter-bombers like the Su-7 and indigenous HAL HF-24 Marut also needed replacing, and the Air Staff drew up a Tactical

Air Strike Aircraft (TASA) requirement for such an aircraft. Having just ordered the Anglo-French Jaguar to fulfill the vital Deep Penetration Strike Aircraft requirement, it was always likely that 'non-aligned' India would select a Russian aircraft type to meet the less important TASA requirement, taking advantage of a lower unit price to augment the relatively small number of sophisticated but expensive long-range Jaguars with much larger numbers of shorter range Soviet-designed fighter-bombers. Some expected India to order the swing-wing Su-17, but instead, India turned to MiG to place an order for 95 MiG-23BNs, together with 15 (or 17?) MiG-23UB two-seat trainers. India's relationship with the MiG OKB was long-standing and happy, and its experience of the MiG-21 had been generally better than its experience with the Su-7.

Russian-built (but HAL-assembled) MiG-23BNs equipped four former Su-7 and HF-24 squadrons from 1981, while India selected the MiG-27 for licence production, to re-equip an eventual total of nine fighter-bomber squadrons, including one of the MiG-23BN units. 165 MiG-27s were ordered originally, and a four-phase production programme was drawn up. The first batch arrived 'crated', followed by aircraft in kit form, before HAL began actually manufacturing the type under licence, with an increasing proportion of local content, from 1988.

Licence-built 'Floggers'

Indian MiG-27s are known locally as MiG-27Ms, though they are actually something of a hybrid, with some simpler, 'export standard' systems and equipment, and are known to the MiG OKB as MiG-27Ls or MiG-27MLs. The total number of MiG-27s delivered is uncertain, because although most sources agree on a total of 165 aircraft, others give totals of 162, 167, or 210 aircraft.

India's MiG-23BNs, MiG-23MFs and MiG-27s were originally augmented by 15 MiG-23UBs, but these were insufficient to meet the training requirements of the combined fleet and perhaps as many as 26 MiG-23UMs were ordered to remedy this shortfall.

About 69 MiG-23BNs remain in service with three fighter-bomber units, and are expected to remain in service for at least another five years. The MiG-23BN was the first IAF fast-jet ground attack aircraft to be fitted with an automated EW/countermeasures system and, according to some local analysts, a number of MiG-23BN airframes (probably 16) have been modified for SEAD duties, with a new indigenous DRDO-developed Tranquil RWR, a new jammer, and armed with ARMAT, Kh-25MP and Kh-59 ARMs. It is not known which of the remaining MiG-23BN squadrons has taken on the 'Wild Weasel' role. Some local reports suggest that a recent defence review will result in the ageing fleet of MiG-23BNs being phased out. Presumably the SEAD aircraft would be retained, or replaced by the new batch of two-seat Jaguars or similarly equipped MiG-27Ls.

Some sources suggest that as many as 195 MiG-27Ms are in service, though this seems high. The type equips seven and a half squadrons, however, making it numerically the most important offensive aircraft in the inventory. One of these MiG-27 squadrons is reportedly being assigned to the maritime strike/attack role to cover the east coast, providing tactical air support to the navy in the same way in which the Jaguars of No. 6 Squadron cover India's west coast. Following combat experience in the Kargil operation, the IAF launched a limited upgrade of the MiG-27's EW and ECM systems, and Station Engineering Wings were quickly told to upgrade the aircraft's chaff/flare dispensers.

The IAF plans to keep the MiG-27 in service until around 2020, and although the IAF's MiG-27 airframes are relatively 'young', (the oldest being 18 years old and the youngest only six) and although the aircraft still enjoys excellent performance characteristics, its equipment and avionics were developed in the 1970s, and many of these systems are now obsolete. The aircraft is thus ideally suited to a more extensive and ambitious upgrade, which could produce an extremely effective strike/attack aircraft more cheaply than procuring a new-build replacement for the MiG-27.

The MiG OKB energetically marketed its MiG-27-98 upgrade configuration to the IAF, offering an inflight-refuelling probe, 'stealth coatings', a new digital databus, a new MVK digital computer and a redesigned cockpit with MFD LCD displays, HOTAS controls and an IN/GPS navigation system. Provision was also made for the installation of Western avionics systems, including state-of-the-art Israeli RWRs.

Local upgrade for the MiG-27

Surprisingly, India rejected the Russian proposal, preferring to opt for a local upgrade. India will therefore undertake much more of the MiG-27 upgrade in-country, with Hindustan Aeronautics Limited's (HAL) Nasik Division playing a dominant role and acting as the primary contractor. The Nasik Division's design department has been transformed into a fully-fledged Design Bureau, and is now known as the Aircraft Upgrade Research & Design Centre (AURDC). As well as leading the MiG-27 upgrade, the AURDC is expected to play a major role in India's planned Jaguar and Canberra upgrades. Work has already started on some of the structural modifications required as part of the planned upgrade and two prototype conversions are expected to fly within two years of contract approval, which was received in 2001.

Some confusion continues to surround the MiG-27 upgrade, with doubt as to whether it includes the entire active fleet, or only 40 aircraft. There may even be two separate programmes: one covering the 40 aircraft which will replace the soon-to-retire MiG-23BNs, and another covering the entire MiG-27 fleet. There may be a single upgrade configuration, to be applied to two consecutive groups of aircraft, or one upgrade may be more 'far-reaching', with speculation that the first phase of 40 aircraft will be modified to an advanced SEAD configuration, while the remainder will not have the same level of SEAD capability.

The main fleet upgrade is expected to feature a redesigned modern glass cockpit to reduce workload, a fixed inflight-refuelling probe to extend range and endurance, and a sophisticated new EW suite to enhance survivability in a hostile electronic environment. The aircraft's accuracy will be enhanced through the provision of more accurate navigation and mission planning systems, including IN/GPS and an Elbit digital moving map.

Stand-off capability will be enhanced by integrating new and more modern PGMs, and by providing a self-designation capability using the Litening laser designator pod, so that the MiG-27 will not simply be a 'bomb truck' hauling PGMs which will be guided by other platforms. Some of the upgraded MiG-27s may also carry a Russian Komar pod-mounted radar. Komar is a lightweight version of the multi-mode Super Kopyo now being installed in upgraded Indian MiG-21s, and would enhance the MiG-27's all-weather and anti-ship capabilities. The integration of Komar would also allow the MiG-27 to use BVR AAMs like the R-27 and R-77, allowing the aircraft to take on a limited BVR air defence role. This may be useful, since it is already expected that the MiG-27 will have to assume an air defence commitment mantle in the east, when Eastern Air Command's three MiG-21FL squadrons are withdrawn, in the absence of any other permanently based air defence units in the region.

Reconnaissance is also expected to assume an increased importance for the MiG-27 squadrons, and W. Vinten Ltd (now Thales) Vicon Series 18 Type 601 GP(1) reconnaissance pods (already in service on RAF and IAF Jaguars) will be integrated on the MiG-27. The upgraded aircraft, which may be designated MiG-27ML, will keep the MiG-27 force viable for some years, and will allow the MiG-27 squadrons to embrace new roles.

Fighter-bomber 'Fishbeds'

Jaguar, MiG-23BN and MiG-27 fighter-bombers are augmented by a dwindling number of MiG-21s. The MiG-21bis (in both upgraded and pre-upgrade forms) has a limited air-to-ground capability, though it is principally used in the air defence role. The older MiG-21M, by contrast, is regarded by the IAF as a fighter-bomber, and some 63 of these aircraft remain active, with three squadrons. About 12 more MiG-21Ms serve as ECM aircraft with No. 35 Squadron.

As older aircraft are retired, the Indian Air Force is likely to shrink slightly in size. The vastly increased cost of new generation combat aircraft makes it impossible to replace aircraft like the MiG-21 on a one-for-one basis, and in order to pay for its new generation of combat aircraft, the IAF will have to accept a reduction in force levels. If present procurement and modernisation programmes continue as planned, the number of fast-jet fighter squadrons could decline from the current peak of 39 to only 32 (or perhaps even 25) by 2005. The MiG-21FL, MiG-21M and MiG-23ML are already on the verge of retirement (accounting for seven units), and may be followed by the MiG-23BN (three more squadrons). Should the MiG-21bis-UPG programme remain set at only 125 aircraft, that fleet will reduce from 10 to only six squadrons, giving a further reduction in front-line strength.

With such a reduction in aircraft numbers, the IAF is making every effort to wring more from its reducing asset base, and this has driven the ongoing programme of modernisation and upgrade of existing aircraft types, in order to enhance their versatility and 'productivity', to improve survivability, and enhance lethality. The IAF has also embraced multi-role aircraft and units, for the same reasons.

Indian Air Force

India maintains tight security around its air force, and in particular the bases at which some of its units are located. While the equipment and locations of several 'high-profile' units are officially recognised, it should be noted that many others are not, and units frequently move between bases. Therefore, any order of battle is open to a degree of conjecture. The following incomplete and speculative order of battle has been compiled from a variety of open sources.

UNIT	NAME	TYPE	BASE
No. 1 Sqn	'Tigers'	Mirage 2000H, Mirage 2000TH	Gwalior
No. 2 Sqn	'Winged Arrows'	MiG-27M, MiG-23U	Kalaikunda
No. 3 Sqn	'Cobras'	MiG-21bis, MiG-21U	Pathankot
No. 4 Sqn	'Oorials'	MiG-21bis, MiG-21U	Jaisalmer
No. 5 Sqn	'Tuskers'	Jaguar IS, Jaguar T	Ambala
No. 6 Sqn	'Dragons'	Jaguar IS, Jaguar IM, Jaguar IT	Pune
No. 7 Sqn	'Battleaxes'	Mirage 2000H, Mirage 2000TH	Gwalior
No. 8 Sqn	'Eight Pursoot'	MiG-21FL, MiG-21U	Bagdogra
No. 9 Sqn	'Wolfpack'	MiG-27M, MiG-23U	Hindan
No. 10 Sqn	'Winged Daggers'	MiG-27M, MiG-23U	Jodhpur
No. 11 Sqn	'Rhinos'	HS.748	Gwalior
No. 12 Sqn	'Yaks'	An-32	Agra
No. 14 Sqn	'Bulls'	Jaguar IS, Jaguar IT	Ambala
No. 15 Sqn	'Flying Lances'	MiG-21bis, MiG-21U	Chandigarh
No. 16 Sqn	'Cobras'	Jaguar IS, Jaguar IT	Gorakhpur
No. 17 Sqn	'Golden Arrows'	MiG-21M, MiG-21U	Bhatinda
No. 18 Sqn	'Flying Bullets'	MiG-27M, MiG-23U	Kalaikunda
No. 20 Sqn	'Lightnings'	Su-30MKI	to form at Pune
No. 21 Sqn	'Ankush'	MiG-21bis, MiG-21U	Chandigarh
No. 22 Sqn	'Swifts'	MiG-27M, MiG-23U	Hasimara
No. 23 Sqn	'Panthers'	MiG-21bis, MiG-21U	Ambala
No. 24 Sqn	'Hawks'	Su-30K, Su-30MKI	Pune
No. 25 Sqn	'Himalayan Eagles'	An-32, Il-76MD	Chandigarh
No. 26 Sqn	'Warriors'	MiG-21bis, MiG-21U	Pathankot
No. 27 Sqn	'Flaming Arrows'	Jaguar IS, Jaguar IT	Gorakhpur
No. 28 Sqn	'First Supersonics'	MiG-29A, MiG-29UB	?
No. 29 Sqn	'Scorpios'	MiG-27M, MiG-23U	Jodhpur
No. 30 Sqn	'Charging Rhinos'	MiG-21FL, MiG-21U	Tezpur
No. 31 Sqn	'Lions'	MiG-23BN	Halwara
No. 32 Sqn	'Thunderbirds'	MiG-21bis, MiG-21U	Jodhpur
No. 33 Sqn	'Geese'	An-32	Gauhati
No. 35 Sqn	'Rapiers'	MiG-21M (ECM)	Bareilly
No. 37 Sqn	'Black Panthers'	MiG-21M, MiG-21U	?
No. 41 Sqn	'Otters'	Do 228, HS.748	Palam
No. 43 Sqn	'Ibexes'	An-32	Jorhat
No. 44 Sqn	'Mighty Jets'	Il-76MD	Agra
No. 45 Sqn	'Flying Daggers'	MiG-21bis, MiG-21U	?
No. 47 Sqn	'Black Archers'	MiG-29A, MiG-29UB	Adampur
No. 48 Sqn	'Camels'	An-32	Chandigarh
No. 49 Sqn	'Paraspears'	An-32	Jorhat
No. 51 Sqn	'Swordarms'	MiG-21bis, MiG-21U, MiG-27M	Jamnagar
No. 52 Sqn	'Sharks'	MiG-21FL	Chabua
No. 59 Sqn	'Hornbills'	Do 228	Gauhati
No. 101 Sqn	'Falcons'	MiG-21M, MiG-21U	Sirsa
No. 102 Sqn	'Trisonics'	MiG-25RBT, MiG-25RU	Bareilly
No. 104 Sqn	'Pioneer Rotarians'	Mi-25, Mi-35	Pathankot or Suratgarh
No. 105 HU	'Daring Eagles'	Mi-8	Gorakhpur
No. 106 Sqn	'Lynxes'	Canberra T.54/PR.57/B(I).58/ PR.67/TT.418, HS.748	Agra
No. 107 HU	'Desert Hawks'	Mi-8	Jodhpur
No. 108 Sqn	'Hawkeyes'	MiG-21MF, MiG-21U	?
No. 109 HU	'Knights'	Mi-8	?
No. 110 HU	'Vanguards'	Mi-8	?
No. 111 HU	'Snow Tigers'	Chetak (Mi-8)	?
No. 112 HU	'Thoroughbreds'	Mi-8	Yelahanka
No. 114 HU	'Siachen Pioneers'	Cheetah	Siachen Glacier
No. 115 HU	'Hovering Angels'	Chetak	?
No. 116 HU	'Tankbusters'	Chetak	Jodhpur
No. 118 HU	'The Challenger'	Mi-8	?
No. 121 HU	'Sea Eagles'	Mi-8, Chetak	?
No. 122 HU	'Flying Dolphins'	Mi-8	Port Blair
No. 125 HU	'Gladiators'	Mi-25, Mi-35	Pathankot
No. 126 HF	'Featherweights'	Mi-26	Pathankot
No. 128 HU	'Siachen Tigers'	Mi-17	?
No. 129 HU	'Nubra Warriors'	Mi-17	?
No. 130 HU	'Condors'	Mi-17-1V	Leh
No. 131 FAC Flt	'Airborne Pointers'	Chetak, Cheetah	Hindan

This No. 6 Squadron line-up at Pune includes all three versions of Jaguar in IAF service, although the majority are radar-equipped IMs. The Jaguar remains of great significance to the IAF, and the fleet is slated for a major upgrade programme. The type also remains in production in India, the last to be delivered in 2008.

No. 132 HU		Cheetah	?
No. 141 SSS Flt	'Flamingos'	Chetak	?
No. 142 SSS Flt	'Flying Amphibs'	Chetak, Cheetah	Bagdogra
No. 152 HU	'Mighty Armour'	Mi-17	?
No. 153 HU	'Daring Dragons'	Mi-17	Siachen region
No. 220 Sqn	'Desert Tigers'	MiG-23BN	?
No. 221 Sqn	'Valiants'	MiG-23BN	Halwara
No. 222 Sqn	'Tigersharks'	MiG-27M, MiG-23U	Hasimara
No. 223 Sqn	'Tridents'	MiG-29A, MiG-29UB	Adampur
No. 224 Sqn	'Warlords'	MiG-23MF, MiG-23U	Jamnagar

Training units

AFA		Kiran 1, Kiran 1A, HPT-32	Dundigal
FIS		HPT-32, Kiran 1	Tambaram
FTS		HPT-32	Allahabad
FTS		Kiran 1, Kiran 1A, Kiran 2	Bidar
(aerobatic team) 'Suryakirans'		Kiran 1	Bidar
FTS		Kiran 2, Iskra	Hakimpet
FWTF		An-32, Do 228	Yelahanka
HTS		Chetak	Hakimpet
MOFTU		MiG-21FL, MiG-21U	Tezpur
NAV TS		HS.748	Begumpet
OCU	'Young Ones'	MiG-21FL, MiG-21U	Chabua
PTS	'Skyhawks'	An-32	Agra
TTF	'Banners'	MiG-23MF	Jamnagar

Direct-reporting units

AHQCS	'Pegasus'	Boeing 737, Gulfstream III, HS.748, Mi-8S	Palam
ARC		Boeing 707, Il-76, An-32, Gulfstream III SRA, Astra SPX, Lear Jet 29	Palam
ADA		LCA, Mirage 2000TH	Bangalore
ASTE		various	Bangalore
DRDO		HS.748	Bangalore
HQ MC		HS.748	Nagpur
HQ TC		HS.748	Yelahanka
HQ SAC		HS.748	?
HQ CAC		HS.748	Allahabad
HQ SWAC		HS.748	Jodhpur
HQ EAC		An-32	Shillong
TACDE		MiG-21bis, MiG-27M	Jamnagar

No. 106 Sqn's mixed bag of Canberras includes three ex-bomber B(I).Mk 58s. Two of them, including this example, have been modified for unspecified electronic warfare duties, as have the MiG-21Ms of No. 35 Squadron.

The retirement of elderly aircraft types may reduce the IAF's infrastructure costs, although the IAF's desire to reduce the number of combat aircraft types in IAF service is unlikely to produce any significant results for some time. The retirement of MiG-21FL, MiG-21M and MiG-23ML, for example, will only reduce the number of fast-jet types in service from 10 to seven. The MiG-21's proposed replacement – the indigenous LCA – is now not expected to begin entering service until 2010, IOC having previously slipped from 2005 to 2008.

Reconnaissance

In addition to a relatively small force of dedicated reconnaissance aircraft, the IAF operates larger numbers of dual-role tactical reconnaissance/fighter-bombers. In the strategic reconnaissance role, the IAF operates a single squadron with the survivors of six MiG-25RBs and two MiG-25RUs delivered in 1981, and another squadron with about 12 Canberras (of various marks), which are currently being upgraded by HAL at a new maintenance, upgrade and overhaul facility established at the Canberra's base at Agra. Dedicated PR.Mk 57 and PR.Mk 67 reconnaissance aircraft (export versions of the PR.Mk 7) are augmented by converted bombers with newly-installed bomb-bay camera packs. These aircraft are expected to serve until 2010 and are augmented by a few Gulfstream III SRA-1s, which are believed to be fitted with oblique cameras, designed to look deep into enemy territory at stand-off ranges. In the longer term, the Canberras and Gulfstreams are likely to give way to satellites and high-altitude unmanned aerial vehicles (UAVs), the first of which have now been deployed, though the indigenous DRDO Nishant UAV reportedly suffers major problems, and has been unable to attain operational status. In particular, it has been alleged that the Indian drone cannot climb above 10,000 ft (3048 m).

Some reports suggest that the MiG-25RBs may give way to four Myasishchev M-55 high-altitude reconnaissance aircraft, and that India may acquire Elta EL/M-2060P synthetic aperture radars for these aircraft, with appropriate datalinks and ground imagery exploitation systems to allow real-time data collection and interpretation.

In the tactical reconnaissance role, the 12 MiG-21Ms and MiG-23BNs which previously formed the backbone of the force have largely given way to longer range Jaguars and MiG-27s. The Ambala Wing Jaguars, in particular, use BAe-supplied recce pods, and more recently supplied Vinten Vicon pods.

While India's photographic reconnaissance assets are operated directly by the IAF, Elint and survey assets (probably including three Antonov An-32, two Il-76, two Learjet 29, five HAL HS.748 and seven IAI Astra SPX aircraft) are operated under the auspices of the Aviation Research Centre (ARC) at Palam. Two Boeing 707s, and two Boeing 737s, are believed to operate as command posts, while two HAL HS.748s are used for AEW trials work.

AEW

India actually used MiG-21 fighters as airborne radar pickets during the 1971 war, but was unable to get funding for a dedicated AEW aircraft until much more recently. The Aviation Research Centre conducted trials with a number of HAL HS.748s, one of which was lost in a fatal accident, with most of the R&D team, before it became clear that producing an indigenous AEW aircraft from scratch was likely to be an over-ambitious project. The Soviet Beriev A-50 'Mainstay' was evaluated by the IAF in 1988 but was rejected as being incompatible with the existing air defence network, which relied heavily on Western systems.

The lack of an AWACS platform was noted during the Kargil war in 1999, and the IAF resumed its search for a suitable platform with renewed enthusiasm. A pair of Beriev A-50Us with the new Shmel-2 radar was leased from the Russian Air Force during May 2000 to help the IAF evaluate the AEW concept, although the Air Force remained reluctant to purchase the A-50 'as is' because it remained fundamentally incompatible with the Indian Air Defence network, and required too much reliance on ground stations for data processing.

The status of India's order for the A-50 remains unclear, though it has been suggested that three aircraft have been ordered, and that these will be A-50EhIs, produced through the conversion of new-build PS-90A-engined Il-76TD transports. These will be based on the cancelled A-50I development for China, with Israel's Phalcon AEW radar system. This will use three linked L-band active phased-array antennas in a non-rotating fairing above the fuselage. This may be a fixed version of the A-50's normal, flight-cleared disc-like rotodome, or could be replaced by a new triangular antenna fairing.

Tankers

With no inflight refuelling tankers, the IAF actually removed the inflight-refuelling probes from most of the Jaguars and Mirage 2000s which were delivered with them fitted, and omitted them from HAL-built aircraft. The prevailing view was that, in the event of a war with Pakistan, Indian attack aircraft could reach their targets without recourse to inflight refuelling, while no-one envisaged any more than tactical, border attacks against China. With the introduction of nuclear weapons (and a long-range strike role) and the adoption of loitering SEAD tactics, a need for inflight refuelling began to become apparent. The first inflight refuelling trials were conducted between IAF Jaguars and a loaned RAF VC10 tanker in 1996, and these trials proved the usefulness of the technique.

In the interim, the IAF acquired UPAZ buddy refuelling pods for its Su-30s, Mirages and Jaguars (all of which had their probes reinstated), and placed orders for a number of Il-78 'Midas' tankers (probably four surplus Russian aircraft), with an eventual requirement for eight of these tankers. The MiG-21 and MiG-27 fleets are being equipped with inflight refuelling capability as part of their ongoing upgrades.

Rotary-wing

India was quick to exploit the military potential of the helicopter, and the IAF's helicopter force has undergone rapid expansion, despite the loss of most of its HAL Chetaks and Cheetahs (HAL-built Alouette IIIs and Lamas) to the newly formed Army Aviation Corps on 1 November 1986. This left only three squadrons of about 40 Chetaks and Cheetahs under Air Force control with armed Chetaks operating in the anti-tank role, and unarmed versions performing casevac and liaison flying and with lighter Cheetahs mainly serving in the FAC role. The AOP role is now the province of the Army's 170 Chetak and Cheetah helicopters. These two types will eventually be replaced by the HAL Dhruv (Advanced Light Helicopter).

The formation of the Army Aviation Corps has, if anything, intensified a debate as to the appropriate command and control arrangements for the medium, heavy and attack helicopters which now come under Air Force command. The Army argues that, as the main 'customer' for these helicopters, it should 'own' them, while the Air Force points to the need for them to be integrated with transport and even offensive support aviation, and to its engineering and maintenance expertise.

The Mil Mi-8 'Hip' remains numerically dominant, with about 100 aircraft in 10 squadrons, though the more modern, more powerful Mi-17 is rapidly 'catching up' with 80 aircraft in eight squadrons, and with 40 further Mi-17Bs already on order. The Mi-8 and Mi-17 operate in the commando assault, support and utility helicopter, SAR and VIP transport roles, and are highly prized for their ruggedness and performance, having proved equally adept when operating from remote mountain helipads or restricted jungle clearings. The types can be armed to augment India's dedicated attack helicopters, and there are reports that some Mi-17s are to be upgraded with FLIR and other equipment to enhance their night capability.

With its superb hot-and-high performance, the Mi-17 has often been India's preferred attack helicopter in recent operations. Some IAF Mi-8 and Mi-17 aircraft are based in India, and in the island territories, and with the Indian permanent station in Antarctica.

The pride of the IAF's helicopter force is the Mi-26, 10 of which have been operated by No. 126 HU on special duties for many years. The Mi-26 is expensive to operate, but enjoys capabilities unmatched by any other type, and has achieved outstanding results in the mountains of northern India.

Arguably the most potent of the IAF's helicopters is the Mi-25 'Hind' – the export derivative of the Mi-24D 'Hind-D'. No. 125 Helicopter Unit was formed in May 1984 with Mi-25s, and used these in Sri Lanka. The improved Mi-35 'Hind-E' was received from April 1990, re-equipping No. 104 HU and, according to some sources, 116 HU as well. Recently, the IAF has commissioned an avionics and defensive aids suite upgrade worth $25 million for 20 Mi-35s and six Mi-25s helicopter gunships from IAI/Tamam. This Mission 24 upgrade will include a new stabilised FLIR and LLTV, a glass cockpit, helmet-mounted sights, and new navigation and ECM systems. These aircraft will also be compatible with the Rafael Spike ATGM. All surviving 'Hinds' and Mi-17s are expected to receive new chaff/flare countermeasures systems and missile approach warning systems from Elisra.

Transport aviation

Between them, the IAF's transport and helicopter elements account for more than half of the IAF's annual total flying hours. This is perhaps unsurprising, since India spreads over a huge area, and encompasses some of the world's most difficult and impassable terrain,

Above: Before going to the Air Force Academy, many prospective IAF pilots will have flown with the National Defence Academy, which has motor-gliders, or the National Cadet Force. Among the NCF fleet are several Zenair STOL CH 701s (illustrated)

Above: The TS-11 Iskra remains in service with the Flying Training School at Hakimpet, providing Stage 2A training alongside the Kiran 2, although students only fly one type. An advantage of the Iskra is that it has a similar cockpit layout to the MiG-21.

The HAL Kiran is available in unarmed Mk 1/1A and armed Mk 2 versions. Mk 1s and 1As are used for Stage 2 (basic) training at Bidar (left) and Dundigal, while the Mk 2 is used at Hakimpet (above) for Stage 2A. Kiran 1s also fly with the Flying Instructors School at Tambaram.

Primary training is undertaken on the HAL HPT-32, at either Dundigal or Allahabad. This example, however, flies with the instructor school at Tambaram.

The Fixed-Wing Training Faculty at Yelahanka trains transport pilots, who begin on the HAL/Dornier Do 228 (illustrated) before progressing to the An-32.

Stage 3 fast-jet (OCU) training is mostly accomplished at two units with MiG-21FL/U/UMs, one at Tezpur (MOFTU, illustrated) and another OCU at Chabua.

from dense jungle to barren deserts, and from frozen high mountains to baking plains. The IAF also has to support offshore oil rigs and a number of overseas territories, including the Laccadives and the Andaman and Nicobar islands, and India's Antarctic presence.

Air transport is the only practical way of deploying and supporting forces over these distances and across these types of terrain, and this becomes even more true when the armed forces are in action. During the Kargil operation, for example, transport aircraft and helicopters had to operate from some of the world's highest and least well prepared airstrips, and transport pilots became the unsung heroes of the war, flying day in and day out to support and supply the Army, maintaining an uninterrupted air bridge to the world's highest battlefield.

India has the largest air transport force in Asia, and is capable of airlifting the equivalent of a brigade plus its equipment in a single operation.

At much the same time as it drew up the DPSA and TASA requirements which resulted in the Jaguar and MiG-23BN, the Indian Air Force also formulated requirements for a pair of new transport aircraft. The METTAC (Medium

Tactical Transport Aircraft) requirement was intended to find a replacement for the IAF's ageing C-47 Dakotas and C-119 Flying Boxcars, and resulted in the procurement of 123 An-32s, all of which were built in the USSR, despite early intentions to build 45 in India. The An-32 was a dedicated high-altitude derivative of the An-26, and proved ideally suited to IAF use. The An-32 forms the backbone of the IAF's transport force, and today about 80 aircraft equip four operational squadrons. The An-32 has an unusual ability to be fitted with external bomb racks, and aircrew practise a medium-altitude level-bombing role, though this is not usually openly acknowledged by the Indian Air Force.

Serviceability of the An-32 was hit particularly hard by the disruption of spares supplies when the former USSR split apart. The IAF successfully fitted several An-32s with engines taken from redundant An-12s, and placed other An-32s into storage, forcing the disbandment of two of the IAF's original six operational An-32 squadrons. The An-32s will have to be replaced fairly soon, since their airframe life is finite, and is being consumed rapidly.

Augmenting the An-32 are some 32 HAL HS.748 military freighters in one and a half

squadrons. HAL built 89 748s (still colloquially known as Avros), 76 of them for the IAF, the last 28 fitted with wide freight doors. A total of 50 748s is being fitted with upgraded avionics for continued service with India's armed forces, although the airframes are now elderly. These aircraft are also augmented by about 40 HAL-built Do 228 light transports in two squadrons. The latter aircraft are relatively new, however, and the type is still in production by HAL, so its retirement remains some way off.

Heavy transport

The HETAC (Heavy Transport Aircraft) requirement was drawn up when fatigue cracks forced the premature retirement of the bulk of the Antonov An-12 fleet from 1981. Some 24 Il-76MDs were procured originally, and 28 of these versatile heavy lifters now equip two squadrons. Despite its size and payload, the Il-76 has good 'hot-and-high' performance, and regularly operated from airfields in Ladakh at over 10,000 ft elevation. Plans to add to the IAF's Il-76 flight remain active (quite apart from the Il-76-based A-50 'Mainstay' AEW aircraft and the Il-78 'Midas' tanker) and an upgrade is looking increasingly likely. Some reports suggest that the IAF is seriously considering re-engining the remaining aircraft with Western power-plants.

In addition to these more tactical transport aircraft, the IAF operates a sizeable VIP transport fleet, including a pair of Boeing 737-200s, some seven HAL HS.748s, and eight Mi-8 and six SA 365N Dauphin helicopters.

Training

With some 76 squadrons (including 40 front-line fast-jet units) and 835 combat aircraft (plus 154 combat-capable trainers, or combat aircraft serving with training units) as well as 60 armed helicopters and more than 230 transport aircraft, the IAF's requirement for pilots and aircrew is substantial. The Indian Air Force's 130,000 total manpower strength included 2,847 aircrew during March 1999, a figure which represented a shortfall of 500 pilots. That this shortfall is so low is perhaps remarkable, since the training system has had to deal with major problems, and it has also sometimes been quite fortuitous.

During the 1980s and early 1990s aircrew shortages were partly camouflaged by poor serviceability and funding shortfalls, which saw average flying hours per pilot tumble from 200 hours in 1988 to only 120 hours in 1992-1993. Had output from the flying training schools been higher, flying hours would have had to have been shared between a larger number of pilots.

As serviceability rates improved again in the mid-1990s, the pilot shortage again became more apparent. This was not merely a function of increased flying hours and increased tasking, however. During the 1990s, India experienced the highest levels of economic growth it has enjoyed since independence and there was an explosion in private sector salaries, which saw many aircrew being lured away to the airlines, or to other civilian careers. One response was to recruit women into the flying and technical branches of the Air Force, though recruitment of *ab initio* pilot trainees can never be a satisfactory answer to the problems of retaining fully trained, combat-capable personnel.

Flying training in the Indian Air Force still shows clear signs of RAF influence, following a similar pattern and structure. Since January 1988, cadet pilots begin their elementary flying training at the Basic Flying Training School at Allahabad, or at the Air Force Academy at Dundigal, flying the piston-engined HAL HPT-32 Deepak on a 40-hour course.

All then progress to a 90-hour Basic Flying Training phase on the HAL HJT-16 Kiran Mk I and Mk IA with the Air Force Academy at Bidar or Dundigal, after which pilots may be streamed for multi-engined or rotary-wing training, going to the transport training unit at Yelahanka for a 150-hour course on the Do 228 and An-32, or to the Helicopter Training School at Hakimpet for a 150-hour course on the Chetak. For fast-jet pilots, however, tactical and basic weapons training is then undertaken during a further 90-hour course, some pilots moving to the Kiran Mk II at the Air Force Academy at Hakimpet, where other pilots receive a similar course of instruction on the PZL Iskra.

In days gone by, pilots would then be commissioned, receive their Wings, and be posted either to the Hunter Operational Flying Training Unit (for those aircrew destined to fly Western aircraft types) or to the MiG Operational Flying Training Unit. The HOFTU closed its doors in 1995 and, since then, all fast-jet pilots have followed the same route through MOFTU, where applied operational and tactical flying training is carried out on the MiG-21FL and MiG-21UM.

When pilots could be streamed to fly either the delightful and rather benevolent Hunter or the hair-raising, high-performance MiG-21, it was possible to avoid sending the less confident students onto the more difficult aircraft, but that is no longer possible, and all potential fast-jet pilots now undergo 125 hours of training (expanded from a mere 75 hours) on the MiG-21FL and MiG-21UM before being assigned to their front-line aircraft type.

The MiG-21 has never been an ideal advanced trainer, and has been responsible for

an unacceptably high accident rate during training, and for high costs, poor morale and other problems. The conclusion that there has been an unacceptably high accident rate has been drawn by the IAF's Air Headquarters in successive reports, and is not 'mere sensationalism in the Indian media'. Some point out that the 1990s have been the safest decade in the IAF's history, and that in 1997-98 the IAF enjoyed its lowest ever accident rate, while also joyfully pointing out that the Pakistan Air Force has a much worse record. This, however, is in spite of the MiG-21, and certainly not because of it.

Advanced Jet Trainer

The IAF has a long-standing requirement for a new Advanced Jet Trainer, for which the BAE Systems Hawk was provisionally selected during 1999, though negotiations then stalled over the Hawk's $21m price tag. Some 66 Hawks were to be acquired for the IAF, with 11 more for the Navy, and this remains the IAF's preferred option, if agreement can be reached on pricing. Meanwhile, HAL is aiming for a December 2002 first flight for the HJT-36 Intermediate Jet Trainer (IJT), of which 200 are required to replace the Kiran and Iskra.

From advanced flying training trainee pilots go to front-line squadrons for conversion training, with one squadron in each 'fleet' sometimes operating in the OCU role, in addition to its front-line commitments. Some of these units have simulators, and most have extra two-seat trainer aircraft.

Like the best of the world's other professional air arms, the IAF provides a degree of 'post-graduate' training for its front-line aircrew. The Tactics & Air Combat Development Establishment (TACDE) at Jamnagar, for example, offers courses in fighter, ground-attack and helicopter combat tactics, and also provides Dissimilar Air Combat Training (DACT) facilities. India makes every effort to allow its pilots to practise live firing, and there are a number of dedicated target-towing aircraft, including Canberras at Agra.

Indian Air Force – Bharatiya Vayu Sena: Facing the future

Tested in war on several occasions, most recently in 1999, the Indian Air Force has won itself an enviable reputation, though it still faces a number of problems, with several key programmes delayed by technical difficulties, funding problems and delayed decisions. An Air Force which was largely composed of 'cutting-edge', 'state-of-the-art' aircraft in the early and mid-1980s has become one which relies on largely elderly aircraft types. Now, some older aircraft types will be retired without being directly replaced, forcing the IAF to accept a reduced force structure, and to compensate for this 'weakening' it will have to improve capability through the adoption of advanced technologies in avionics, systems and weapons, and through pursuing ambitious upgrade programmes on those aircraft that are being retained.

Fortunately, India's most frequent enemy, Pakistan, has also failed to adequately modernise its air force. Pakistan has been constrained by a long-standing US arms embargo, which has forced Pakistan to concentrate on acquiring second-hand Mirages from

Australia and France, and newly-built but obsolescent MiG-21 and MiG-19 derivatives from China.

However, no-one would pretend that India could or should tailor its armed forces to meet only one threat, or to keep pace with only one potential enemy. India is a major regional power and a leader of the non-aligned world, and has always followed a philosophy of providing its forces with the best possible equipment in order to guarantee decisive and swift victories at minimum cost. Pakistan's relative weakness (which could easily be temporary) is no excuse for India to field 'bargain basement' air power.

And while Pakistan's air force has not been greatly modernised, China's air force remains dominated by old-fashioned Chinese-built fighters and bombers, but has undergone major modernisation in certain key areas, including the procurement of Sukhoi Su-27 fighters, Su-30 multi-role strike attack aircraft, and indigenous Xian JH-7 fighter-bombers.

While the IAF has plentiful combat experience of its own from which to learn, the Air

Force has always made great efforts to study and learn from other air campaigns. Operation Desert Storm, the Allied air campaign over Iraq, for example, was closely studied by the IAF and many lessons from it were incorporated into India's own operational doctrines and procurement plans. Subsequent Allied operations in the Balkans and the Middle East have been studied with equal interest, as have US operations over Afghanistan.

A formal Air Power Doctrine was published in 1997, recording and disseminating the results of intensive study into the changing nature of air power, and drawing up doctrine that would underpin the IAF's new-found nuclear role and would be sufficient to deal with the changing threat to India.

The new Air Power Doctrine marked a shift away from the pre-eminence of defensive fighter aviation, and recognised the need to accord at least equal priority to offensive air operations. It also marked the IAF's formal recognition of the inevitable reduction in force levels, and the need to compensate for these through the adoption of new technology and

Above: Now named the Dhruv, HAL's ALH is being procured for all Indian armed forces. The first two were handed over to the air force in March 2002.

For utility transport the Mi-8 serves with many Helicopter Units, including 109 HU (above). Type conversion and multi-engined helicopter training is undertaken by 112 HU (right) at Yelahanka. The more powerful Mi-17 is in use mainly in the mountainous regions.

Above: This Mi-8 has a smart scheme and VIP interior. It flies with the AHQCS from Palam, near New Delhi.

The HAL Chetak – a licence-built Alouette III – has been around for many years in the FAC and utility transport roles. BZ798 (left) of 121 HU is the oldest Chetak still flying in IAF service. The camouflaged aircraft above is operated by 116 HU at Jodhpur.

the acquisition of force multipliers like tankers and AEW platforms, as well as through improved command, control and communications and air defence systems.

The IAF is in the middle of a major and quite controversial shift in offensive air power operations. Operations in Kargil (and combat operations by other air forces) clearly showed that air power is most effective when carefully applied against key targets, destroying an enemy's war-making potential, preventing his freedom of movement, closing enemy supply lines and taking out C^{3}I nodes rather than simply attacking enemy forces in the field, in direct support of company- and battalion-sized units. There is, of course, still a place for such direct close air support operations, (especially in halting enemy armour in open country), but making precision attacks against other targets in a semi-strategic or interdiction campaign is rightly being accorded a much higher priority.

National pride, a fierce independence and strategic necessity have always combined to make India keen to stand on its own two feet as much as is possible, especially in equipping its armed forces. Over the years, India has built up an enviable indigenous aircraft industry, and

this has produced vast numbers of aircraft for the Indian Air Force, including the majority of the MiG-21s and Jaguars now in service, as well as all of the MiG-27s, Chetaks and Cheetahs.

But licence manufacture of aircraft types designed elsewhere does not provide the level of autonomy that many Indians aspire to, and there has been a drive towards greater indigenous design input into aircraft upgrades, as well as a continuation of the local design and manufacture of new aircraft types.

In previous decades, India successfully designed and built the HAL HT-2 elementary trainer, the HAL Krishak AOP aircraft, and even the HAL HF-24 Marut, which suffered some problems and shortcomings but which was nonetheless a great achievement – an indigenous supersonic multi-role fighter which saw genuine front-line service. The HAL Kiran jet trainer was less glamorous, but even more successful.

Today, the need for self sufficiency is arguably even greater, since the growth of 'ethical foreign policies' makes it even harder for any country to rely on the uninterrupted flow of arms, equipment and even spares from abroad. The crisis in the supply of spares from Russia

and the Ukraine during the early 1990s demonstrated India's vulnerability, and this was further highlighted by Britain's temporary inability to complete the overhaul of Sea Harriers and Sea Kings due to US pressure resulting from India's nuclear tests.

India has tried to reduce its reliance on other nations for overhaul and maintenance facilities, and for the supply of spares and weapons, and is building up sufficient and substantial stocks of critical spares and ordnance, and wherever possible is co-producing those weapons and systems which cannot be produced entirely autonomously.

Two completely indigenous aircraft programmes are currently underway, these being the HAL ALH (Advanced Light Helicopter) and the LCA (Light Combat Aircraft) though the latter relies on a US powerplant and US flight control system.

Both have had a prolonged development programme, and are likely to be useful rather than spectacular performers, but the next generation of Indian combat aircraft may be worth waiting for, and may further enhance the reputation of the Indian Air Force.

Jon Lake

Indian Air Force organisation

Air Headquarters

The Indian Air Force Air Headquarters is located in New Delhi, where the Chief of Air Staff, currently Air Chief Marshal Anil Yashwant Tipnis, has his offices. Reporting to Air HQ are the IAF's five regional operational commands, and two functional commands. The commands are Western, South Western, Central, Southern and Eastern Air Commands, Training and Maintenance Commands. Each of these is commanded by an AOC-in-C (Air Officer Commander-in-Chief) with the rank of Air Marshal.

There are a number of units which report directly to Air HQ in New Delhi, including a VIP transport unit and the Air Research Centre and Analysis Wing, concerned mainly with survey, reconnaissance and Sigint. Both of these units are based at Palam, near Delhi.

Palam: Air HQ Communications Squadron (transports), Air Research Centre and Analysis Wing (special mission and reconnaissance types), Indian Air Force Museum Historic Flight (Tiger Moth, Harvard, Spitfire LF.Mk VIIIC, Vampire T.Mk 11, HT-2, HAL Ajeet, Hunter)

Western Air Command

Western Air Command is also headquartered at Palam, near New Delhi. Unusually, Western Air Command also has a forward headquarters located near the Army's Western Command at Chandigarh. Because it is the closest to Pakistan, Western Air Command is said to be the most important of the five regional commands. The Command controls air operations north of Jaipur, from Kashmir south to Rajasthan, including the capital and the Punjab. An Air Operations Group was formed at Udhampur AFB in 1982, dedicated to the defence of Jammu and Kashmir and also including Ladakh.

Western Air Command has permanent airfields at Adampur, Ambala, Avantipur, Chandigarh, Halwara, Hindan, Leh, Palam, Pathankot and Srinagar, and also has several Forward Base Support Units. They accommodate around nine air defence squadrons equipped with the MiG-21bis and MiG-29 and about seven ground attack squadrons equipped with MiG-21M/MFs, MiG-23BNs, MiG-27Ms and the Direct Supply BAC-built Jaguars, the latter also having a tactical reconnaissance commitment.

Providing an accurate air order of battle for Western Air Command is even more difficult than it is for the rest of the air force. The Indian obsession with secrecy and security is exacerbated by the Command's proximity to Pakistan, while the Command is frequently reinforced by units from other commands, which may be based at the Western FBSUs, or at the permanent airfields, while the Command's own units frequently deploy away from their nominal homes to bases closer to their operational areas. The MiG-29s at Adampur, for example, have recently spent a much time operating from Leh.

It is most likely that most of the front-line units without a known location are based in the WAC region. Leh and Srinagar (classed as permanent airfields on an official IAF administrative chart) are believed to be used mainly as forward deployment airfields, without many permanently based units. The WAC area also covers the hotly contested border region with Pakistan in the Karakoram range, including the Kargil area, Siachen Glacier and Nubra valley. This region is thought to be base (on at least a temporary basis) to several Helicopter Units, especially those equipped with the Mi-17.

WAC airfields, assigned units and mission equipment

Adampur: No. 47 (MiG-29), No. 223 (MiG-29)
No. 28 Sqn (MiG-29) has also been reported here, but is believed to be at another, unknown, location
Ambala: No. 5 (Jaguar IS), No, 14 (Jaguar IS), No. 23 (MiG-21bis)
Bhatinda: No. 17 (MiG-21M)
Chandigarh: No. 15 (MiG-21bis), No. 21 (MiG-21bis), No. 25 (An-32/Il-76), No. 48 (An-32)
Halwara: No. 31 (MiG-23BN), No. 221 (MiG-23BN)
No. 220 Sqn (MiG-23BN) has also been reported here, but is believed to be at another, unknown, location
Hindan: No. 9 (MiG-27), No. 131 FAC Flt (Chetak/Cheetah)
Leh: No. 130 HU (Mi-17)
Palam: No. 41 (Do 228/HS748)
No. 109 HU has also been reported here
Pathankot: No. 3 (MiG-21bis), No. 26 (MiG-21bis), No. 125 (Mi-25/35), No. 126 HF (Mi-26)
Sirsa: No. 101 (MiG-21M)

WAC air bases with no confirmed assigned units: Amritsar, Avantipur, Udhampur, Srinagar
No. 37 Sqn (MiG-21bis) has been associated with Srinagar, but this is unlikely. Units in the Siachen region are believed to include No. 114 HU (Cheetah), No. 128 HU (Mi-17), No. 129 HU (Mi-17) and No. 153 HU (Mi-17)

South Western Air Command

South Western Air Command is the newest and the smallest of the IAF's operational commands. It was formed in 1980 at Jodhpur, from units and airfields which had previously come under the operational control of WAC.

The Command is today headquartered at Ghandinagar (in Gujarat), and controls air operations in the south western sector, which includes most of Rajasthan, and south to Saurashtra, and from Kutch to Pune. South Western Air Command controls Permanent Airfields at Bhuj, Jaiselmer, Jamnagar, Jodhpur, Nalia and Pune, with FBSUs at Ahmadabad, Barmer, Nal, Suratgarh and Uttarlai.

With no immediately adjacent threat, South Western Air Command has a few air defence units equipped with the multi-role MiG-21bis, and has tended to accommodate ground attack and offensive support units which may deploy forward to Western Air Command bases, and has also frequently hosted units while they work up with a new aircraft type. Thus the MiG-29 was originally based at Pune, before the front-line squadrons moved to Adampur, closer to the Pakistani border. No. 24 Squadron's Su-30s are now based at Pune, in just the same way.

No. 224 Squadron is the only MiG-23MF squadron in the IAF, and began to adopt a secondary target-towing commitment while still flying in the air defence role. It may now be entirely dedicated to target facilities, having split into a number of detachments, some aircraft having moved to Kalaikunda and perhaps Gwalior. The remaining aircraft at Jamnagar may still be referred to under the No. 224 Squadron designation, or may now simply be a Target Towing Flight.

South Western Air Command does have a vital maritime defence role, and Pune accommodates a single maritime attack squadron, equipped with the radar-equipped and Sea Eagle-armed Jaguar IM. Other units in the Command include four squadrons equipped with the MiG-23BN and MiG-27M.

SWAC airfields, assigned units and mission equipment

Jaisalmer: No. 4 (MiG-21bis)
Jamnagar: No. 51 (A Flt - MiG-21bis/B Flt - MiG-27), No. 224 (MiG-23MF), TTF (MiG-23MF), TACDE (MiG-21bis/MiG-27)
Jodhpur: No. 10 (MiG-27), No. 29 (MiG-27), No. 32 (MiG-21bis), No. 107 HU (Mi-8), No. 116 HU (Chetak), HQ SWAC (HS748)
Pune: No. 6 (Jaguar IM), No. 20 (Su-30, to form), No. 24 (Su-30)
Suratgarh: No. 104 (Mi-25/35)
No. 104 Sqn has also been reported as based at Pathankot

SWAC air bases with no confirmed assigned units: Ahmadabad, Barmer, Bhuj, Nal, Nalia, Uttarlai
Before the major earthquakes in the area, Bhuj is believed to have accommodated two MiG-21 squadrons. No. 45 Sqn (MiG-21bis) has formerly been mentioned in connection with Uttarlai but is no longer thought to be at the base

Central Air Command

When India first switched from functionally-based to regional Commands the priority was to provide forces in the West, facing Pakistan and in the East, facing China. The need for an operational Command in the Central sector soon became apparent, in order to protect the Indo-Nepalese and Indo-Tibetan borders. Central Air Command's Headquarters was established in March 1962 at Ranikutir in Calcutta, moving to Bamrauli, a former civil airfield just outside Allahabad in the state of Uttar Pradesh.

Central Air Command is the IAF's largest Command in terms of area, and the most geographically diverse, stretching from the snow-capped Himalayas in the north to the Gangetic plains and the Central Highlands. The Command's responsibilities include the states of Bengal and Assam, and the eastern states of Arunachal Pradesh, Meghalaya, Mizoram and others bordering Tibet, Bangladesh and Burma.

Central Air Command has no forward-based support units, but has permanent Air Force stations under it at Agra, Allahabad (Bamraulli), Bareilly, Gorakhpur, Gwalior, Kanpur and Nagpur, and has Care and Maintenance units at Bihta, Dharbhanga, Bakshi-ka-Talab and Nagpur, with unspecified 'other units' at Nainital, Tiwari-ka-Talab and Memaura. The Air Force Selection Board is located at Varanasi.

Central Air Command is principally a support organisation, with the bulk of the IAF's transport, ECM and reconnaissance assets, which are not duplicated in the east and west. It does also have two Mirage 2000 squadrons (which have added nuclear strike and PGM delivery to their original air defence commitment) and two Jaguar Offensive Support squadrons. The latter include the Indian-built, DARIN-equipped aircraft, and are assumed to be the Jaguars which are also tasked with the nuclear strike role, since they have no reconnaissance commitment.

CAC airfields, assigned units and mission equipment

Agra: No. 12 (An-32), No. 44 (Il-76), No. 106 (Canberra), PTS (An-32)
Allahabad: FTS (HPT-32), HQ CAC (HS748)
Bareilly: No. 35 (MiG-21M ECM), No. 102 (MiG-25R)
Gorakhpur: No. 16 (Jaguar), No. 27 (Jaguar), No. 105 HU (Mi-8)
Gwalior: No. 1 (Mirage 2000), No. 7 (Mirage 2000), No. 11 (HS748)
A MiG-23MF-equipped target-towing flight, detached from Jamnagar, is also believed to be based here
Nagpur: HQ MC (HS748)

CAC air base with no confirmed assigned units: Kanpur

Eastern Air Command

After Western Air Command, Eastern Air Command is the most important of the IAF's operational formations, facing, as it does, the threats from China, and bordering both Bangladesh and Myanmar. Headquartered at Shillong (Meghalaya), Eastern Air Command controls air operations in the eastern sector, encompassing the states of West Bengal, Assam, Mizoram and the states bordering Bangladesh, Burma and Tibet.

Eastern Air Command has permanent air bases at Bagdogra, Barrackpore, Chabua, Gauhati, Hashimara, Jorhat, Kalaikunda and Tezpur, with FBSUs at Agartala, Calcutta, Panagarh and Shillong.

Though a front-line command, Eastern Air Command is relatively modestly equipped. Air defence rests in the hands of three squadrons of ageing MiG-21FLs, and four ground attack squadrons with the MiG-27M. With China's recent acquisition of Su-27SK fighters, Su-30 and Xian JH-7 fighter-bombers, many commentators expect a major modernisation of Eastern Air Command, but others merely expect the command's rapid reinforcement arrangements to be streamlined.

EAC airfields, assigned units and mission equipment

Bagdogra: No. 8 (MiG-21FL), No. 142 SSS Flt (Chetak/Cheetah)
Chabua: No. 52 (MiG-21FL)
Gauhati: No. 33 (An-32), No. 59 (Do 228)
Hashimara: No. 22 (MiG-27), No. 222 (MiG-27)
Jorhat: No. 43 (An-32), No. 49 (An-32)
Kalaikunda: No. 2 (MiG-27), No. 18 (MiG-27)
A MiG-23MF-equipped target-towing flight, detached from Jamnagar, is also believed to be based here
Shillong: HQ EAC (An-32)
Tezpur: No. 30 (MiG-21FL), MOFTU (MiG-21FL)

WAC air bases with no confirmed assigned units: Agartala, Barrackpore, Calcutta, Panagarh

Southern Air Command

Southern Air Command was formed in July 1984, by splitting the southern part of Central Air Command away under a new headquarters at Trivandrum (Kerala). Southern Air Command controls air operations in the southern sector, which includes all the southern states, the Bay of Bengal, the Andaman and Nicobar Islands, and Lakshwadeep.

The Command has no front-line squadrons of its own, but does host units from other commands when required. During the Maldives operation in 1987, for example, Mirage 2000s from Gwalior were deployed to a Southern Air Command base.

Southern Air Command includes permanent airfields at Bangalore, Begumpet, Bidar, Dundigal, Hakimpet and Tambaram, and at Port Blair and Car Nicobar, and also has FBSUs at Madurai and Salur. Most of these house Training Command units, and the Command's sole front-line transport unit, No. 19 Squadron at Tambaram, disbanded in 1997, one of two An-32 units forced to go through spares shortages.

SAC airfields, assigned units and mission equipment

Bangalore: ADA (LCA, Mirage 2000), ASTE (various trials), DRDO (HS748)
Begumpet: NAVTS (HS748)
Bidar: FTS (Kiran 1/2)
Hakimpet: AFA (HPT-32/Kiran 1/2)
Port Blair: No. 122 HU (Mi-8)
Tambaram: FIS (HPT-32/Kiran 1)
Yelahanka: No. 112 HU (Mi-8), FWTF (An-32/ Do 228), HQ TC (HS748)

SAC air bases with no confirmed assigned units: Car Nicobar, Madurai, Salur

Training Command

Training Command has its headquarters at JC Nagar, Bangalore (Karnataka), and the majority of flying training schools and ground training establishments are located within the Southern Air Command area.

All Indian Air Force pilots begin their careers by undertaking an Ab Initio Pilots Course conducted under the auspices of the Air Force Academy at Dundigal, though flying is conducted from a number of other nearby airfields.

The new Air Force Academy, Hyderabad, was formally inaugurated on 16 January 1971, and flying training began immediately. The Ground Duty Officers' course moved from Air Force Administrative College at Coimbatore in early 1975 and the Air Traffic Control Officers' Training Establishment moved in from Air Force Station Hakimpet in July 1977, bringing most initial officer training and all basic flying and air traffic control training 'under one roof'.

Pilot trainees arrive at the Air Force Academy from the National Defence Academy, or from the National Cadets Corps or from the universities. Non-NDA cadets first undergo a six-month Pre-Flight Training School course at Air Force Station Begumpet. All flying trainees then go to the Technical Type Training (TETTRA) School for Technical Training on either the Kiran or HPT-32.

The Air Force Academy's Faculty of Flying is split into four squadrons, two each for each training stage. Stage I (the 22-week basic stage) introduces ab initio pupils to flying, and teaches basic flying skills while also being used to grade and systematically filter out those unlikely to complete training. The basic stage is conducted at Hakimpet on the HPT-32 and HJT-16 Kiran I, while the HPT-32 phase is also conducted at Allahabad.

The Stage II flying training course (the 24-week advanced phase) is conducted at Air Force Station Bidar and Air Force Station Hakimpet on the Kiran Mk II and the PZL Iskra aircraft. This phase teaches students to fly a modern jet aircraft to its maximum performance limits by day with skill and confidence, while handling more complex systems and higher workloads, and introducing basic low-level, formation and night flying proficiency. After completion of Stage II training pilots are commissioned and awarded their wings at one of two annual Combined Graduation Parades.

Since July 1995, the Stage I training of Army Pilots has also been carried out at the Air Force Academy, on a separate 17-week course on the HPT-32. Stage II training of Army trainees is carried out on air force helicopters (Chetaks) at Air Force Station Hakimpet. The Air Force Academy also operates a Detachment at Bihta in the North Indian State of Bihar. Newly commissioned pilots are then streamed for Stage III (Applied) flying training on fighter, helicopter and transport aircraft.

Engineering officers are trained at AFTC, Bangalore, while ATC officers are trained at AFA, Hyderabad, Intelligence and Security officers at AFIS, Pune, and Fighter Control branch training is conducted at ADC, Lucknow. After initial training, Admin officers are trained at AFAC, Coimbatore, while all aeromedical training is provided by the IAM at Bangalore. The AFTC, Jalahalli teaches Management Techniques to senior technical officers, and provides training for prospective officer instructors, quality control inspectors and for Warrant Officers who are about to be commissioned. Senior officers undergo Staff training at the Defence Services Staff College at Wellington, while a Joint Air Warfare Course is provided by the College of Air Warfare at Secunderabad.

Additional training units include the instructor's school at Tambaram, parachute training school at Agra, and transport aircraft training unit at Yelahanka. There are thought to be two OCUs operating the MiG-21FL, one being known as MOFTU and based at Tezpur. A tactical evaluation/ development unit is based at Jamnagar with various MiGs.

Maintenance Command

Headquartered at Nagpur in Madhya Pradesh, Maintenance Command was formed on 26 January 1955 at Chakeri, Kanpur, moving to Nagpur in 1963. It is responsible for the maintenance, overhaul and repair of all IAF equipment, including aircraft and helicopters, undertaking these tasks through a network of Base Repair Depots, including units at Bangalore, Bhatinda, Calcutta, Chandigarh, Coimbatore, Kanpur, Nagpur, Tughlakabad and Salur. It is also increasingly responsible for spares procurement and indigenisation.

Indigenisation promises to offer reduced costs and reduce reliance on foreign suppliers, and is therefore accorded a high priority, with three main areas of effort. Group 1 looks after aircraft, helicopters and aero-engines, Group 2 radar, navigational aids and missiles, and Group 3 aircraft support vehicles, arrester barriers and fabrics.

All Maintenance Command units are based at airfields and stations within the areas of responsibility of the operational commands, sometimes sharing these bases with front-line units.

Maintenance Command is also responsible for storage using a network of Equipment Depots (EDs) and Air Storage Parks (ASPs) and today Maintenance Command is responsible for 46 units including nine Base Repair Depots, eight Equipment Depots, three Air Storage Parks and 27 miscellaneous units and detachments.

Maintenance Command sites and activities

Avadi (Chennai): 8 BRD – aircraft support vehicles, arrester barriers and fabric used in parachutes
Chandigarh: 3 BRD – aero engines and aircraft specific materials
Coimbatore (Sulur): 5 BRD – modification kits and testing equipment
Kanpur (Chakeri): 1 BRD – transport aircraft of Russian origin, 4 BRD – aero engines of Russian and French origin, and for British transport aircraft
Nasik (Ozhar): 11 BRD – MiG overhauls
Palam: AMSE (ADGES Maintenance and Servicing Establishment) – servicing, repair and overhaul of communication, radar and avionics systems, PRD – Parachute Repair Depot
Pune: 9 BRD – Electronic ground equipment
Tughlakabad (New Delhi): 7 BRD – ground-based radar, nav and comms equipment

Other sites: Bangalore, Bhatinda, Calcutta, Nagpur, Salur

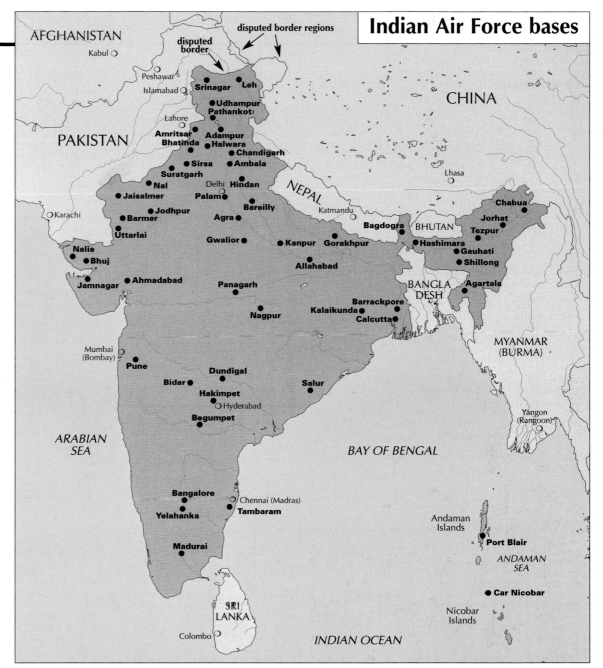

Indian Air Force bases

USAF gunships in Vietnam

Part Two: Gunship II (AC-130A/E/H Spectre) and Gunship III (AC-119G Shadow/AC-119K Stinger)

Having proved the fixed-wing, sideways-firing gunship concept with the AC-47, Captain Ron Terry was able to produce a set of requirements for a follow-on aircraft. These centred around the need for greater endurance, increased firepower and improved armour protection and added up to a much larger machine than the AC-47. Gunship II and Gunship III were the result.

When Captain Terry, who had headed the team that had brought the sideways-firing gunship idea to fruition, returned from South Vietnam in early 1965 with the initial results of the FC-47 Gunship I programme, he brought with him the basic requirements for a follow-on aircraft – initially known as Gunboat. Terry and the Air Material Command personnel were excited about the early success posted by the FC-47D 'Dragonship' crews, but those early successes also pointed up the limitations of the aircraft.

The FC-47 was in need of many things that could make it a much better aircraft for the mission – such as a greater fuel capacity to increase both range and loiter time over the target area, and much heavier armament.

Missions against targets along the Ho Chi Minh Trail in early 1965 not only highlighted the fuel capacity problem, but also the need for heavier armour plating. These requirements led to a necessary increase in payload capacity – something that could not be matched by an aircraft the size of the C-47.

Project Gunboat

Project Gunboat would thus need a much larger aircraft to meet the new requirements. Three types were considered for Gunboat – the Boeing C-97 Stratofreighter, the Lockheed C-130 Hercules and the Fairchild C-119 Flying Boxcar. Air Staff recommended a mix of types for a limited war such as that in Southeast Asia, suggesting an establishment comprising two

squadrons of FC/AC-47s, one heavy squadron (26 aircraft) of AC-119Gs, another heavy squadron of improved capability AC-119Ks, one squadron of AC-130s, and one squadron of AC-97(X)s.

The projected AC-97(X) was one of the answers to the requirement for a heavy weapons gunship type. It had the payload capacity necessary for the heavier-calibre guns that would be used in the Gunboat programme, and could handle the greatly increased and much heavier ammunition load. With four turboprop and two jet engines, it had the required performance and airframes were readily available for conversion, since many could be drawn from existing stocks of C/KC-97 aircraft that were being phased out in favour of Boeing C/KC-135s. However, problems with the conversion, plus higher projected 'in the field' maintenance costs, made the AC-97(X) less desirable than either of the other types under consideration. Thus, the Air Force chose the C-119 and C-130 as the basis for its follow-on gunships, selecting the Lockheed C-130A for Project Gunboat.

Above: The AC-130s were, by far, the most successful of the USAF's fixed-wing gunships in Southeast Asia. Surviving examples remained in USAF service into the mid-1990s. This 1st SOW machine fires its forward pair of the 20-mm rotary cannon over the Eglin Water Range in the Gulf of Mexico during 1972.

Left: An AC-119K assigned to the 18th SOS opens fire with one of its four MXU-470 Miniguns on a target during the North Vietnamese Easter Offensive of spring 1972. All AC-119Gs and Ks were involved in fire suppression and armed reconnaissance missions in an attempt to stem the advance of the North Vietnamese armoured columns.

Terry, now promoted to Major for his work in the gunship programmes, chose JC-130A 54-1626 as his Gunboat prototype. An armament upgrade was the first thing Terry addressed when deciding how to equip the new aircraft; he chose the General Electric MXU-470 Minigun system for the secondary armament, four of which were to be mounted in the upper part of the left side of the fuselage.

Primary armament would be the GE M61 Vulcan 20-mm rotary cannon, on a stationary mount assembly – the same basic weapon that was fitted to aircraft like the F-104 Starfighter and F-105 Thunderchief. Four M61s would be mounted on the left side of the fuselage, below the MXU-470 Miniguns – two in front of the main

landing gear, with the other pair mounted aft.

To alleviate identification problems during night missions, a Night Observation Device (NOD), or Starlight Scope, was fitted in the forward fuselage and an AN/AVQ-8 1.5 million-candle power search light was mounted on the rear ramp (lowered in flight). To co-ordinate these systems, Gunboat had a computerised fire control system that used both the NASARR F-151-A radar (mounted in the rear paratroop doorway) and the AN/AAD-4 side-looking infra-red (SLIR) system (in an aperture situated aft of the forward pair of M61 cannon), though the latter was not part of the original equipment fit.

Much of the equipment installed in the Gunboat prototype was virtually 'off-the-shelf',

drawn from the inventories at Wright-Patterson AFB (the FCS radar, for example, was adapted from that fitted to the F-104). The new systems were literally bolted together on wooden boards and installed in the aircraft, giving rise to the term 'bread-board computer'. In addition to the new weapons and sensors, Terry and his crew installed a new Bell optical gunsight, an AN/ALE-20 semi-automatic flare dispenser, and armour plating in the floors and around the ammunition and flare boxes.

Communications and navigation equipment changes included the fitting of a beacon tracker, DF homing instruments and an FM radio transceive, the latter to allow communication with troops on the ground and the Airborne Battlefield Command and Control Center (ABCCC) aircraft that were rapidly coming into their own in the Southeast Asian theatre. This equipment was installed in a small cubicle in the forward part of the fuselage, commonly known the 'war room' or 'booth'.

Gunboat becomes Gunship II

In the spring of 1967, the Gunboat programme underwent a name change. Due to the naval connotations associated with the term 'gunboat', the Air Force renamed the programme Gunship II.

Gunboat/Gunship II prototype JC-130A 54-1626

Left and above: These views of the AC-130A prototype show the aircraft around the time of its conversion, still in the colour scheme worn by many JC-130 test machines. The various new apertures in the side of the aircraft (left) for the eight rotary guns, the NOD and the SLIR are evident. The SLIR was fitted in the rear paratroop door opening (above); the searchlight was mounted on the rear ramp, which was usually opened on an operational sortie to allow for AAA/SAM 'spotting'.

As the 13th production C-130A, 54-1626 was a so-called 'Roman-nosed' aircraft, with the original AN/APS-42 radar (a feature of the first 28 C-130As); all later AC-130A conversions were of later machines. 54-1626, and 'production' AC-130A 54-1630, were both preserved and are displayed at the USAF Museum at Wright-Patterson AFB.

Right and far right: Later views of the aircraft (at Nha Trang in October 1967) show various modifications carried out in the light of experience in the field. Aft of the forward pair of M61 cannon a new aperture has been provided for the AN/AAD-4 SLIR system. The air scoops protruding from the side of the fuselage (feeding the gun gas extraction system) have also been modified.

Above: Test firing of the Gunship II prototype's M61 cannon took place off the Florida coast during the summer of 1967. Here the aircraft's pilot 'looses off' a burst of cannon fire at 2,500 rounds per minute.

On 6 June 1967, the Gunship II JC-130A was flown to Eglin AFB, Florida, for flight testing of its weapons and associated systems. Throughout the summer of 1967, Major Terry's crews worked the 'bugs' out of the new weapons system. The FLIR system was the most impressive component of the new package, being able to pick up the heat generated by the engine and exhaust of a vehicle at night, in the densest jungle environment. Once the computerised fire control system had an accurate 'fix' on the target, it put the aircraft into a so-called 'search or attack orbit' and waited for the pilot to press the trigger.

The resultant display was the most spectacular anyone had seen. An AC-47D with three Miniguns could fire at a maximum combined rate of 18,000 rpm (though the use of all three guns at once was a rarity in combat); the Gunship II had four Miniguns, for a maximum combined rate of fire of 24,000 rpm. In addition, the M61 20-mm rotary cannon were capable of firing another 24,000 rpm, though these were much heavier high-explosive 20-mm rounds. A single 20-mm round could usually stop a truck dead; even at its lower 3,000 rpm rate of fire the M61 could destroy soft-skinned vehicles and most light armoured vehicles, including those of the Soviet PT-76 type. The M61s installed in the AC-130 were almost always locked at the slower speed of 3,000 rpm.

The Gunship II prototype undertook combat trials in Southeast Asia between September and December 1967, assigned to the 14th ACW at Nha Trang. In this February 1968 view, 54-1626 is seen leaving Wright-Patterson AFB, Ohio, following its hastily completed minor overhaul. By this time many of its systems had been 'standardised' to allow the conversion of further aircraft.

The Gunship II Combat Test Team poses for a group portrait at Ubon in 1968. Major Ron Terry is second from the left kneeling. Early AC-130As flew with a crew of 11, comprising two pilots, flight engineer, navigator, fire direction officer, NOD operator, radar/sensor operator, loadmaster, master armourer, 7.62-mm Minigun gunner/armourer and 20-mm cannon gunner/armourer.

The 16th Special Operations Squadron was activated at Nha Trang AB at the end of October 1968, shortly before the first production AC-130As were due to arrive in Southeast Asia. Initially 54-1626 was the unit's only aircraft; in this view it sports the unit's 'EA' tailcode, used until April 1969.

24 May 1969: first Spectre combat loss

Aircraft commander Colonel William Schwehm, USAF, describes the mission on which his aircraft – AC-130A 54-1629 (callsign CARTER) of the 16th Special Operations Squadron at Ubon RTAFB – was lost on 24 May 1969.

"The pilot/navigator briefing was held at Ubon at around 16.00, with the crew briefing at about 17.30. It's interesting to note that intelligence had advised us that Russian advisors were now assisting the North Vietnamese gunners in an effort to bring down one of the elusive gunships.

"We took off from Ubon at 18.00 and the weather was very good. There was no moon that night, so we should have been pretty safe, at least from small arms fire and non-radar-directed anti-aircraft fire. We arrived over our assigned area, which was a section of road just south of Tchepone, Laos, at about 18.30. Almost immediately, an Army FAC called us to advise that a number of trucks were in our area. At the same time, our F-4 escort from the 497th 'Night Owls' Squadron arrived.

"About five minutes later we located the trucks and rolled into our firing orbit, making a single pass. We then continued on to the next road intersection and made a 120° turn to come back. Suddenly, Staff Sergeant Jack Troglen, our illuminator operator, came over the radio – "Triple A at 6 o'clock and accurate!" The NVA fired 10 37-mm rounds at us. Four [passed] on each side of the aircraft but missed. The remaining two rounds hit us, one in the tail and the other hit the right wing just inboard of No. 3 engine.

"I immediately turned the aircraft westward and started for Ubon, calling out the "May Day". Suddenly, the hydraulic system warning lights came on, followed immediately by the back-up system lights. This meant the loss of the flight controls as well as landing gear, flaps and nose wheel steering. The aircraft started into a turning descent to the right, which would soon become an uncontrollable spiral if I didn't do something immediately.

"I had no flight controls at all, no rudder, no elevators, no flaps. But I did have aileron trim. With the use of engine power and aileron trim, I was able to regain control slightly, and the aircraft started to fly straight and level. Then it started to climb and we were unable to stop it. My co-pilot, Major Gerald Piehl, and I fought the controls, and finally jammed the control columns full forward. Still the aircraft continued to climb. I ordered all members of the crew who were able (Sergeant Troglen was badly injured and could not be moved) to come forward onto the flight deck, which changed the aircraft centre-of-gravity.

"Slowly but surely, we regained control of the aircraft and continued toward Ubon. Because of the wounded personnel and the fact that we were still over Laos (the Pathet Lao Communists were known to kill American prisoners), I decided to try to make it to Ubon. We had two wounded members of the crew. Sergeant Troglen had been hanging over the open rear ramp watching for 'Triple-A' and SAMs when we got hit; he sustained massive head injuries and was dying. And the photographer who was aboard, using a starlight scope to confirm our truck kills, had been hit in the legs.

"We managed to get the aircraft to within 19 miles [30 km] of Ubon and I ordered most of the crew to bail out – everyone except Major Piehl, flight engineer Staff Sergeant Cecil Taylor, myself, and Sergeant Troglen. Major Piehl and I kept the aircraft under control while Sergeant Taylor manually cranked down the landing gear. We set up for an emergency landing at Ubon. Ubon tower notified us that the runway had already been foamed for us.

"We only had 10° of flap since they were already cranked in during the attack orbit, and that was all we were left with due to the loss of hydraulic pressure. I managed to get the aircraft lined up with the runway and we touched down reasonably smooth, but before I could get excited about a 'good landing', things started going wrong. As soon as I pulled the power off, the nose tucked and dropped very hard onto the runway. The aircraft bounced once and hit heavily on the main landing gear.

"We had taken a hit about a month earlier and the right main gear had been badly damaged. That same right landing gear failed when the aircraft bounced back onto the runway and started sliding. An attempt to reverse engines was futile, and some 2,000 ft [609 m] down the runway the right main gear collapsed and the aircraft slid off the runway to the right.

"The right wing hit the barrier shack and broke off with a big explosion and fireball. We continued skidding along the ground until we hit the second barrier head-on. We came to an abrupt stop and I told Major Piehl to get out. He started to leave but then stopped very suddenly. He had forgotten to release his seat belt. After punching the release, he was gone, and I was right behind him. My legs felt like rubber.

"The aircraft was engulfed in fire by now. Since we had lost the right wing, that side was clear of fire, and that's where Major Piehl and I went out. As I entered the escape hatch, I looked out and saw my two crew members running away from the aircraft. However, when I joined them a minute or so later, the two I thought to be my co-pilot and flight engineer were actually Major Piehl and one of the sensor operators, who I didn't know had refused to bail out when I ordered the rest of the crew to do so. Sergeant Taylor was still on-board the aircraft. We didn't find him until the next day when the crash crew brought his and Sergeant Troglen's bodies out of the burned-out wreckage.

"One of the problems the crash crews had was that one of their trucks got too close on the initial [approach]. When the crash truck engine stalled and quit, it had to be abandoned and was burned up with our aircraft. After that, the Fire Marshall ordered all the fire-fighting equipment to be pulled back to a safe distance, and the aircraft and both men inside were left to burn. I was extremely disturbed about this and threatened bodily harm to the Fire Marshall, but it was already too late. All I could do is what I did, and then file a report with 7th Air Force about the fire equipment being removed. The rest of the crew, who had bailed out, were rescued that afternoon.

"I flew 126 AC-130 missions during my tour, and was credited with 228 trucks destroyed, 67 trucks damaged, three anti-aircraft guns destroyed, and one Soviet helicopter destroyed on the ground. But all of this success does not make up for the loss of the two fine men who died aboard my aircraft that fateful day in May 1969."

Inside the AC-130A

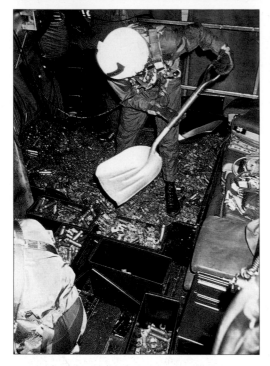

Above: This 16th SOS armourer uses a snow shovel to empty 7.62-mm and 20-mm shells and ammunition links that have accumulated beneath the forward M61 battery.

Below: This view shows the original lightweight mounts used to carry the forward M61 battery in the AC-130A prototype. Though capable of a 6,000 rpm rate-of-fire, the M61s were generally set at 3,000 rpm to avoid overheating and prolong barrel life.

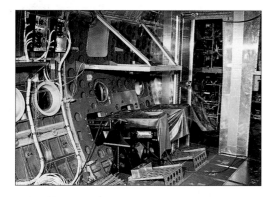

Below: One of the most hazardous duties was that of AAA/SAM 'spotter', who had the unenviable task of hanging over the edge of the aircraft's open rear ramp to watch for AAA fire and/or SAM launches. As violent manoeuvring to avoid enemy fire occasionally threw the 'spotter' out of the aircraft, he needed to be securely tethered so that fellow crew members could haul him back inside. Later AC-130s had an observation bubble installed in the rear ramp.

Above: Mounted in the front crew entry door on the port side of the aircraft, the AN/ASQ-24A Stabilised Tracking Set replaced the NOD in later AC-130As. In aircraft equipped to Pave Pronto standard, this system included an AN/ASQ-145 LLLTV camera, an AN/AVQ-18 laser designator and a BDA camera.

Above: Surprise Package aircraft introduced a new AAD-7 FLIR in place of the earlier AAD-4 SLIR; the FLIR Operator's station is pictured here. Just visible to the right is part of the ECM Operator's position, including part of the APR-25/26 RHAWS scope.

Above: The Fire Control Officer's station, dominated by a large CRT display and including an altimeter, air speed indicator and sensor angle display, was situated in the air-conditioned Battlefield Control Center (usually nicknamed the 'booth' or 'war room').

Below: The AN/ASQ-145 Low Light Level TV camera was controlled from this station by the NOD Operator.

In late September 1967, Major Terry and the Gunship II task force were to test their new mount in combat. With Major Jack Kalow at the controls, the Gunship II arrived at Nha Trang, home of the 14th ACW, on 21 September 1967. The Gunship II test programme had three phases, the first involving tests against troops in contact in the IV Corps area of the Mekong Delta. The second phase included tests against traffic and their defences on the Ho Chi Minh Trail in the Tiger Hound area of Laos, while the third part of the test programme covered use on armed reconnaissance and ground support missions in the II Corps area of the Vietnamese Highlands. The Gunship II, commonly known as the 'Vulcan Express' or 'Super Spooky', flew its first test mission three days after its arrival at Nha Trang.

By 8 December 1967 the combat test and evaluation programme for Gunship II had been completed and the AC-130A (as it was now designated) had returned to the United States to have the temporary equipment, i.e. the 'bread-board computers', standardised and permanently mounted. Test results were analysed; the aircraft had met or exceeded expectations. During the three-month combat test and evaluation, the AC-130A expended almost 223,000 rounds of 7.62-mm and 88,000 rounds of 20-mm ammunition, and dropped 310 flares, resulting in the destruction of at least 38 enemy trucks along the Ho Chi Minh Trail and major damage to another 50. Enemy troops killed in action numbered in the hundreds during the TIC and ground support mission portions of the test.

Although the AC-130A had been remarkable against enemy troops, it was its record against vehicular targets on the Trail that caught the eyes of 7th Air Force officers. They indicated that additional AC-47s could handle the TIC and ground support missions within the borders of South Vietnam, and that the AC-130A was needed in the hostile environment over the Trail in Laos. General William Momyer, commander of 7th Air Force, ordered that a minimum overhaul of the AC-130A be accomplished in the shortest amount of time so that the aircraft could return to Southeast Asia and be available for combat before the current monsoon season ended. On 12 February 1968, the 'Vulcan Express' returned to the theatre, where its mission would be almost strictly interdiction of the Trail.

Nightly operations

The single AC-130A operated on an almost-nightly basis throughout the spring and summer of 1968, equipment breakdowns and very bad weather being the only inhibiting factors. (The original JC-130A was not a true all-weather aircraft.) On 18 November 1968, following 151 missions and countless equipment breakdowns that had halted operations several times, the AC-130A prototype was ferried back to Wright-Patterson AFB for a long-overdue major overhaul. The record it left in Southeast Asia was formidable – 228 trucks destroyed, 133 trucks heavily damaged, 17 sampans either sunk or heavily damaged, and some 240 enemy troops killed during actions supporting TIC operations in South Vietnam.

So successful were the initial combat tests flown in late 1967 that the Air Force authorised an additional seven JC-130As to be converted to Gunship IIs as early as 20 December 1967.

The AN/ASD-5 Black Crow installation was tested in combat aboard C-130A 56-0471, a Blindbat flareship used to illuminate targets, under the direction of a FAC aircraft, for night attack by jet fighter-bombers. Here the aircraft is seen here with apparently hastily applied black underside paint work.

C-130A 55-0011 (above) was the first of 10 C-130As converted to Pave Pronto AC-130A standard, a conversion based on Surprise Package aircraft 56-0490. A pair of Bofors 40-mm cannon replaced the Miniguns and M61s in the rear of the aircraft; some Pave Pronto AC-130As (including 56-0469, below) also had their forward Miniguns deleted, though these are clearly visible on 55-0011. Pave Pronto aircraft were equipped with the new AN/ASD-5 Black Crow truck ignition sensor (immediately behind the nose radome), but did not receive some items of Surprise Package equipment, including an INS, laser range-finder and a digital FCS computer.

These new aircraft would be modified by LTV's Electrosystems plant at Greenville, Texas. E-Systems had been providing support for the AC-130A prototype from 12 February 1968 at Ubon RTAFB, which was the home of the Gunship II for all operations against the Ho Chi Minh Trail. On 6 August 1968, the first production AC-130A came off the E-Systems line; the seventh aircraft was finished four months later, on 9 December. These machines differed from the prototype principally in being converted from later production C-130A aircraft (with improved weather/navigation radar) and being equipped with a new FCS radar (AN/APQ-133).

On 31 October 1968 – with the imminent delivery of the additional aircraft – 7th Air Force activated the 16th Special Operations Squadron with the callsign SPECTRE, at Ubon. All subsequent AC-130As were assigned to the new squadron at Ubon, under the control of the 8th Tactical Fighter Wing. Four of the seven production aircraft were sent directly to Ubon, the remaining three going to Lockbourne AFB, Ohio, and Eglin AFB, Florida, to train crews for combat in Southeast Asia.

The mission of the 16th SOS was multi-faceted and made good use of the AC-130A's capabilities – night interdiction, close fire support of TICs, search and rescue fire support, daylight escort of road and water convoys, and harassment of enemy troop concentrations. The 'Fabulous Four Engine Fighter' scored its first aircraft kill on 8 May 1969 when BENNET 01, AC-130A 56-1629 with Lieutenant Colonel William Schwehm as aircraft commander, destroyed a NVAF helicopter on the ground in Laos.

The ongoing duel between AC-130 and North Vietnamese AA artillery was one the Spectre could handle reasonably well. Though flying low and slow, the AC-130's sophisticated sensors were often able to locate an AA battery before the gunship came under fire.

On 30 December 1968, 7th AF introduced a new and very potent night attack duo against NVA traffic along the Trail – a hunter-killer team of AC-130As and 497th TFS 'Night Owls' F-4D Phantoms. Acting as forward air controller and bait, the AC-130A would mark the target with

With the advent of Pave Pronto, the original AC-130As became known as 'Plain Janes'. In the spring of 1970 all were returned to the US and fitted with the 40-mm Bofors cannon and AN/ASD-5 Black Crow, thereby becoming AC-130A 'Updates'. The following year the remainder of the Pave Pronto equipment was fitted, with the exception of the LLLTV system. 54-1623 Ghost Rider is seen after modification to 'Update' standard.

flares and tracer rounds, then direct the F-4D to the threat area. In the first four months of 1969, the AC-130/F-4D hunter-killer teams destroyed or silenced over 120 37-mm guns, and destroyed more than 25 enemy trucks during the same actions.

As successful as the gunship forces were – in spite of a total of just four aircraft – they began losing aircraft to the enemy defences. On 24 May 1969, Colonel Bill Schwehm was aircraft commander of AC-130A 56-1629 (the aircraft that had destroyed the NVAF helicopter earlier in the month) when NVA 37-mm AA guns badly damaged the aircraft over Laos. The AC-130 was destroyed in the subsequent crash-landing at Ubon, thereby reducing the 16th SOS's operating strength by 25 per cent, though shortly afterwards the remaining three AC-130As, which had been employed stateside to train new Spectre crews, arrived in Southeast Asia.

Coronet Surprise/Surprise Package

As early as June 1968, 7th Air Force began calling for major improvements in the AC-130.

One of the first and most frequent requests was for larger-calibre weapons for use against the hardened targets that were increasingly showing up along the Trail. Targets such as T-34 tanks, concrete bunkers and bridges could not be destroyed using the 20-mm guns. A further request was for upgraded sensors and fire control computers. On 18 July 1969 the Air Force approved a programme known officially as Coronet Surprise, though more commonly as Surprise Package. Under this programme, the aft 7.62-mm Miniguns were removed, and the aft pair of M61 20-mm Gatling cannon was replaced by a pair of M2A1 40-mm Bofors automatic cannon, with a rate of fire of up to 100 rpm. A total of 256 rounds of 40-mm ammunition was carried. The accurately fired 40-mm Bofors round from above was enough to knock out an armoured vehicle like the Soviet T-34.

Along with the 40-mm gun installation came a variety of new sensors, including the AN/ASQ-24A Stabilised Tracking Set, which included the General Electric ASQ-145 low light level television (LLLTV) system, a Korad

Above: From 1972 AC-130A/Es often carried as many as four AN/ALQ-87 ECM pods, in pairs (mounted tail to tail) on modified triple-ejector racks.

AVQ-18 laser designator and the AN/ASD-5 Black Crow system. The Black Crow sensor allowed the AC-130A crew to locate an enemy vehicle by picking up radio frequency radiation emitted by unshielded ignition systems, as fitted to Soviet ZIL truck engines.

The first Surprise Package AC-130A was ready for testing at Eglin AFB on 27 October 1969. After a month of successful testing, the Surprise Package aircraft left for Southeast Asia, arriving at Ubon on 5 December 1969.

Combat evaluation flights took place between 12 December 1969 and 18 January 1970. Despite numerous equipment failures, mostly in the new sensors (which had not been tested in combat, much less in a humid climate) the results were deemed satisfactory. Of the 313 trucks spotted by its crew, the Surprise Package aircraft destroyed 178, damaged 63, and officially recorded 'no results seen' (even though crew members were sure they had hit something) against another 37.

Over the next two months, the Surprise Package AC-130A participated in Operation Commando Hunt III. On 14 February 1970 the AC-130A set a new record for a single mission – 43 trucks destroyed and two damaged. On 21 February 1970 the 16th SOS destroyed its 5,000th truck. By the end of Commando Hunt III, the Surprise Package gunship had destroyed 604 trucks, while damaging an additional 218.

By late spring 1970, the Surprise Package prototype had seen considerable service in the combat zone. As with the original prototype Gunship II, the 'bread-board computer' assemblies began breaking down on a regular basis,

Pictured over Laos in June 1972, AC-130A 55-0011 is finished in the, by then, standard AC-130 colour scheme of overall black with camouflaged upper surfaces and red serials and codes. To counter the SA-2 missile's 'Fan Song' radar the aircraft carries a pair of AN/ALQ-87 ECM pods on each outer wing pylon. Aircraft of the 16th SOS were recoded 'FT' (from 'EA') in April 1969.

greatly hindering combat operations. On 4 June 1970, the aircraft was returned to Wright-Patterson AFB for overhaul and standardisation of the on-board equipment.

During a January 1970 fact-finding tour of the combat area in Southeast Asia, Secretary of the Air Force Robert Seamans made a special point of reviewing the gunship fleet at Ubon, with particular emphasis on Surprise Package and the results of the limited combat that the aircraft had seen. Sufficiently impressed, he directed that all the AC-130As be brought up to a limited Surprise Package configuration. At the same time, the Air Staff recommended that the C-130E variant be used as the basis of the next gunship model. Perhaps unsurprisingly, this was opposed by Tactical Air Command Headquarters, given the requirement for airlift capability in Southeast Asia. By late January 1970, the Air Force had ordered the upgrading of all the original AC-130As to full Surprise Package specification be initiated. These aircraft were designated AC-130A Pave Pronto.

AC-130E Pave Spectre

At the same time, arguments regarding the use of the C-130E airframe carried on in the corridors of the Pentagon. The Air Force, ASD

Above: Chaff and flare dispensers were also carried, the latter to counter the SA-7 threat; an SUU-42A/A flare dispenser is pictured.

Below: So-called 'sugar scoop' exhaust shrouds were fitted to the AC-130 fleet as a means of reducing the IR signature and vulnerability of these aircraft to the SA-7.

Below left: One of the first Pave Spectre AC-130Es shares the ramp at Ubon with an AC-130A. The AC-130E, with uprated T56-A-15 engines, was distinguished by its four-bladed Hamilton Standard propellers. As well as additional ammunition and armour plating, the AC-130E introduced a new digital fire control system, based on that fitted to the Vought A-7 attack aircraft.

Below: Pave Aegis introduced the M102 105-mm howitzer for use against armoured targets, replacing one of the 40-mm cannon. All 11 AC-130Es were fitted with the M102 during 1972, most at Ubon RTAFB using kits shipped from the US.

Above: 1973 saw the 10 surviving AC-130Es re-engined with T56-A-15 engines and redesignated AC-130H. New defensive systems were also introduced and an overall black finish adopted.

Below: The 16th SOS moved to Korat RTAFB in mid-1974, in the process coming under the control of the 388th TFW. The following April the squadron covered the final withdrawal of US personnel from both Phnom Penh (Cambodia) and Saigon; the squadron returned to the US in December 1975, joining the 1st SOW at Hurlburt Field.

and PACAF all wanted to use the C-130E as the basis for the next AC-130 upgrade: it was more reliable, could carry a heavier payload (30,000 lb/13608 kg more than the AC-130A), had a loiter time two hours longer than the AC-130A, and the aircraft were some 15 years younger. On 7 May 1970 the Air Force finally approved the use of the C-130E as the next-generation gunship. It was known as Pave Spectre.

Above: In spring 1973 a new compass grey finish replaced overall black. The 16th SOS adopted the 8th TFW's 'WP' tailcode in April 1974; this aircraft (69-6576, pictured at Ubon in April 1974) served for another 20 years, only to be lost as a result of structural failure off the Kenya coast during Operation Continue Hope on 15 March 1994. Sadly, eight of the 14 crew were lost.

While work was progressing on the initial AC-130E, 69-6567, modifications began to several of the original AC-130As to bring them to Pave Pronto standard. The dual 40-mm guns were installed, as were the AN/APN-59B (with its moving target indicator), battle damage assessment equipment, and the new Hayes International 2-kW illuminator. Much-needed IRAN (Inspection and Repair As Necessary) work was also carried out on the airframe. Lockheed then installed the AN/ASD-5 Black Crow package. By late 1970, the five AC-130As with Pave Pronto equipment were back on the ramp at Ubon.

On 6 October 1970, the first C-130E was delivered to Warner-Robins Air Material Area for conversion to AC-130E, followed by a second in early January 1971. Both aircraft received the Pave Pronto equipment, including the dual 40-mm Bofors cannon. The sensor systems were upgraded with a digital fire control computer for increased target acquisition capability, and an AN/APR-25/26 radar homing and warning system (RHAWS) was installed. In light of an increase in the surface-to-air missile threat along the Trail, the AC-130E had a modified flare launcher – AN/ALE-20 – and SUU-42A/A flare pods under the wings that dispensed chaff for self-protection.

Whereas the the AC-130A had flown with a crew of 11, the AC-130E generally had a crew of 14, comprising two pilots, navigator, flight engineer, fire control officer, LLLTV operator, FLIR operator, illuminator operator, EWO, loadmaster, master armourer, two 40-mm cannon gunner/armourers and two 20-mm cannon gunner/armourers. All sensor operators were enclosed in a small compartment in the forward portion of the modified cargo bay; officially known as the Battlefield Management Center.

The AC-130Es deployed to Ubon in autumn 1971, the first aircraft arriving on 25 October.

Pave Aegis

By the late summer of 1971, the Air Force was again considering an upgrade of the weapons system. Tests in South Vietnam using a truck loaded with rice – the primary staple of the VC and North Vietnamese Army – indicated that the 40-mm round might not be enough if it impacted in the rice cargo. The next step was to incorporate the US Army M102 105-mm

AC-130 losses in SEA

After the loss of AC-130A 54-1629 on 24 May 1969, five other AC-130s were downed during the fighting in Southeast Asia – all during night missions over southern Laos. Brief details of the circumstances of each are given here.

■ **22 April 1970: AC-130A 54-1625** *War Lord*
Hit by 37-mm AAA during a Commando Hunt mission with two fighter-bombers over Route 96A, 54-1625 (callsign ADLIB 1), commanded by Major William Brooks, caught fire near its port wing root. Attempts by some of the crew to control the fire failed due to the intensity of the blaze. Groping his way from the rear of the aircraft, SSgt Eugene Fields found the gunner's position unmanned and a hatch open so decided to bail out. With burns to his face and hands he landed in a tree, but managed to climb down and hide until the following morning, when a SAR squad located and rescued him. No other members of the 11-strong crew were able to escape before the aircraft crashed.

■ **28/29 March 1972: AC-130A 55-0044** *Prometheus*
55-0044 was hit by an SA-2 fired from a newly-established SAM site about 35 miles (56 km) west of Khe Sanh. The aircraft, commanded by Major Irving Ramsower, burst into flames and crashed. The Pathet Lao later claimed to have shot the aircraft down. A SAR team found no trace of survivors. 55-0044 had suffered major battle damage before; during December 1971 it had lost both starboard propellers to AAA fire.

■ **30 March 1972: AC-130E 69-6571**
With the callsign SPECTRE 22, 69-6571 (the first Pave Aegis aircraft) was attacking a convoy of trucks 35 miles (56 km) north of Muang Fangdeng when it

was hit by 57-mm AA fire while at an altitude of 7,500 ft (2286 m). The starboard wing and fuselage were hit, leaking fuel from the pylon tank starting a fire. Aircraft commander Captain Waylon Fulk turned northwest towards Thailand, but all 15 crew were forced to abandon the aircraft over Laos. A massive SAR operation was launched that night, in which other Spectre aircraft and FACs conducted visual and radio searches in order to locate the downed crew. The following day 13 crew members were picked up by four HH-53s; the remaining pair (who had actually bailed out of the AC-130 shortly after it was hit and were some 40 miles/64 km from the main group) were rescued by Air America UH-34s.

■ **18 June 1972: AC-130A 55-0043**
Flying over border country 25 miles (40 km) southwest of Hué, 55-0043 was hit by an SA-7, the missile striking the starboard inner engine and starting a fire which soon reached fuel tanks, causing an explosion which blew the aircraft's wing off. Three crew bailed out and were rescued by SAR helicopters the following day; the other 12 personnel aboard the aircraft, including its commander, Captain Paul Gilbert, were lost in the ensuing crash. This was the first AC-130 loss to an SA-7; the first recorded SA-7 launch against an AC-130 had taken place on 5 May and there had been a number of near misses recorded over the following six weeks.

■ **21 December 1972: AC-130A 56-0490** *Thor*
Piloted by Captain Harry Lagerwall, 56-0490 (which had been the Surprise Package prototype) had found three trucks 25 miles (40 km) west of Saravan and was firing on the target from 7,800 ft (2377 m) when it was hit by 37-mm AAA fire. The aircraft exploded and crashed in flames, though not before two members of the 16-man crew parachuted to safety. Located using the LLLTV equipment aboard another AC-130, both were picked up that night by an HH-53 of the 40th ARRS.

Above: AC-119G 52-5927 of the 71st SOS pauses at Elmendorf AFB, Alaska, en route to Southeast Asia in December 1968. The 71st had been flying the 'Shadow' since June from Lockbourne and Eglin AFBs and began deploying to Vietnam during early December.

Left: The first four AC-119Gs were in place at Nha Trang AB by 27 December 1968, when this aircraft arrived at the base. Aircraft commander Major B.R. McPherson and his crew of four are seen here enjoying a beer with base personnel upon the arrival of their 'very special delivery'. Once celebrations were over, armament was hastily installed so that the 71st SOS was able to fly its first mission just over a week later.

howitzer. Its installation in the AC-130E was conducted under Project Pave Aegis, which was approved on 18 November 1971.

Pave Aegis was the result of several test flights conducted during the summer of 1971. A number of new weapons were tested, including a 57-mm anti-tank gun, a 106-mm recoilless rifle, and a 105-mm howitzer. The howitzer was chosen for Pave Aegis. The original howitzer installation was mounted directly to the rear floor of the aircraft, taking the place of the aft 40-mm gun. Later, the gun was placed on a trainable mount controlled via the aircraft's fire control computer. In addition, the Pave Aegis AC-130E had the APQ-150 beacon-tracking radar system installed.

The FCS operator could move the gun vertically and horizontally to keep the weapon trained on the target. Fired by the pilot, the gun unleashed a 44-lb (20-kg) shell. On 17 February 1972, the first Pave Aegis 105-mm gun was installed in an AC-130E at Ubon. Eventually, all AC-130Es had the Pave Aegis howitzer installed; AC-130As were not so-equipped.

Unfortunately, the Pave Aegis test aircraft, AC-130 69-6571, was shot down over Tchepone, Laos, on the night of 31 March 1972.

The entire crew was plucked from the Laotian jungle the following morning. It was the second AC-130 to be shot down over Laos in two days: AC-130A 55-0044 had been hit by an SA-2 SAM on 29 March, with the loss of all 15 crew.

In late 1972, the Air Force decided to re-engine the C-130E fleet with new Allison T56-A-15 engines, as fitted to the later C-130H. As the AC-130Es were rotated through the IRAN facility to receive the engines, any old or missing equipment was either replaced or updated. These 'new' AC-130Es were then redesignated AC-130H. The first AC-130Hs began arriving at Ubon in March 1973.

Easter Offensive

The arrival of the Pave Spectre AC-130Es – especially the Pave Aegis – came just in time to counter the Easter Offensive launched by North Vietnamese troops on 30 March 1972. Pave Aegis was the only type able to destroy the North Vietnamese tanks that began pouring across the borders that fateful Sunday. VNAF AC-47s and AC-119Gs, plus USAF AC-119Ks, could handle the infantry and truck traffic, but it took the 'big gun' of the Pave Aegis AC-130E to stop the tank forces.

The Air Force rushed several Pave Aegis kits to Ubon and installed the 105-mm gun in several more AC-130Es. Numerous incidents attest to the effectiveness of the 105-mm gun. On 5 May, an AC-130E tackled an NVA regiment attacking the camp at Polei Kleng, killing well over 350 enemy troops. At Kontum City, an AC-130E helped the ARVN troops beat off a combined attack by VC and NVA troops and tanks. Between 14 May and 6 June, ARVN troops, supported by AC-130Es, destroyed 38 tanks and killed over 5,600 enemy troops.

However, the increase in combat sorties naturally increased the risk of losing an aircraft; the NVA had recently unveiled a new threat to American air power – the SA-7 Strela shoulder-fired, heat-seeking SAM. On the night of 28/29 March 1972, just prior to the launch of the NVA's Easter Offensive, AC-130A 55-0044 was shot down by an SA-2 SAM over Laos. On 5 May the NVA launched the first SA-7s against an AC-130A operating in support of the troops near An Loc. The aircraft was hit in the tail section, but the pilot managed to limp to Tan Son Nhut where he made a successful landing.

The Air Force countered the new SAM threat by installing upgraded RHAW and jamming systems, including ALQ-87 ECM pods to jam the Fan Song radars of the SA-2, and underwing flare launchers like the SUU-25 and SUU-42 to counter the SA-7 threat. In addition, metal shrouds were installed over the AC-130's exhausts, lowering their IR signature. Many of the aircraft began appearing in a new overall grey paint scheme, called Gunship Grey, which reduced their visibility during night operations.

The last gunship mission over South Vietnam was flown on 27 January 1973; the last over Laos followed on 22 February. Operations over Cambodia continued until 15 August.

53-7852 Midnight Special of B Flight, 17th SOS rests between missions in a revetment at Phan Rang AB during 1969. Visible in the open door is the aircraft's Night Observation Device; immediately aft of the door are retractable air scoops used to ventilate the fuselage of gun gases during firing.

In this rear view of a B Flight, 17th SOS aircraft (52-5892) at Phan Rang in 1970, the LAU-74/A flare launcher is visible in the rear door. Up to 60 Mk 24 flares, used to illuminate a target, were carried on a typical night mission. The aircraft was also equipped with an AN/AVQ-8 1.5 million-candlepower illuminator.

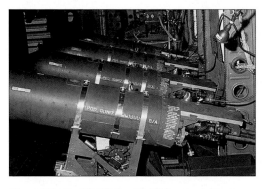

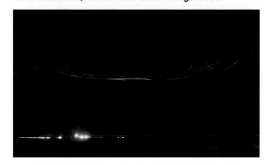

Above: Like the AC-47D, the AC-119G was equipped with modified SUU-11B/A Miniguns (pictured) until the modular MXU-470/A variant became available. On daytime missions aircraft are known to have carried as many as 50,000 rounds of 7.62-mm ammunition; only 35,000 rounds were carried at night as a supply of flares was also required after dark.

Below: An AC-119G from B Flight, 71st SOS, fires on Viet Cong mortar teams near Phan Rang AB on 20 April 1969. The AC-119G's fire control system had fully automatic, semi-automatic, manual and offset firing modes.

Gunship III – AC-119G/K

The AC-119 was the follow-on aircraft to both the AC-47D and the AC-130A. It was known as Gunship III, and evolved into two distinctly different aircraft and missions. Both types started out as Fairchild C-119G Flying Boxcars and, although both were developed under the same programme, one was developed to replace the AC-47D 'Spooky' gunship that was so successful in defending villages, hamlets and forts throughout South Vietnam, while the other was intended to take some of the pressure off the AC-130A crews flying the truck-busting mission along the Ho Chi Minh Trail.

The first variant developed was the AC-119G. On 17 February 1968 the Air Force awarded a letter contract to Fairchild-Hiller Corporation to modify 52 C-119Gs – 26 as AC-119Gs and 26 as AC-119Ks. Equipment installed in the AC-119G Gunship III prototype included four SUU-11A/A 7.62-mm Minigun pods (as installed in the early AC-47D), an AN/AVQ-8 airborne illuminator light set with 20-kW xenon illuminator in the aft paratroop door on the left side, a night observation device (NOD) in the left forward door, an LAU-74/A flare launcher in the aft paratroop door on the right side, a DPN-34 Doppler radar, a General Precision Systems analog fire control computer; and racks for a total of 31,500 rounds of 7.62-mm ammunition.

Later in the production run, the SUU-11A/A Minigun pods were replaced with, first, the GAU-2B/A, then the MXU-470A Minigun modular weapons system. The AN/APR-25/26 RHAWS was installed as both Gunship III types were slated for combat in the hostile air space of Laos, which would put them under threat from both radar-directed AAA and SAMs.

Both C-119G aircraft and their crews were taken from Air Force Reserve inventories, the majority from the 71st Tactical Airlift Squadron, 930th Tactical Airlift Group, based at Bakalar AFB, Indiana. On 13 May 1968 the 71st TAS was called to active duty for 24 months. Squadron personnel were told to report to Clinton County AFB, Ohio, where they would undergo C-119 training in anticipation of the gunship mission. Actual gunship training in the AC-119

Intended to replace the AC-47D, the AC-119G was a hastily conceived and implemented programme. Though the resultant aircraft was better armed and equipped, it was seriously underpowered. Losing an engine on take-off was especially serious; two fatal crashes in such circumstances led to the reduction of maximum permissible take-off weights.

comprised Phase II, and took place at Lockbourne AFB, Ohio. At the same time, the squadron was redesignated the 71st Special Operations Squadron and was assigned to the 1st SOW at Eglin AFB. The first AC-119G was delivered by Fairchild-Hiller on 21 May 1968, with the last of 26 AC-119G aircraft being delivered on 11 October 1968.

Training continued for the crews of the 71st SOS throughout the summer and autumn of 1968. On 5 December the first AC-119Gs deployed from Lockbourne AFB to Nha Trang AB, Republic of Vietnam; they arrived on 27 December after the crews had gone through survival training, otherwise known as 'snake school', at Clark AB, Philippines. The mission of the 71st SOS was quite similar to that of both

AC-119 and AC-130 training was assigned to the 4413th CCTS, 4410th SOTW at Lockbourne AFB, Ohio. The 4413th CCTS was subsequently redesignated the 415th Special Operations Training Squadron in 1970. Aircraft assigned to the 415th SOTS were under the control of the 1st SOW at Eglin AFB. This aircraft appears to carry 'IH' tailcodes, indicating assignment to the 4413th CCTS.

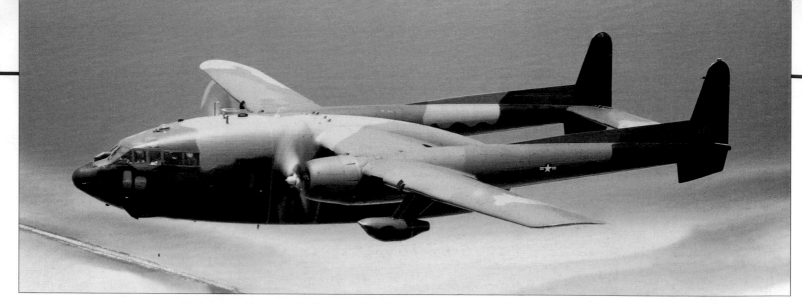

Above: The AC-119K was initially engaged in CAS provision for TICs and armed reconnaissance duties, though from 1970 the variant operated alongside AC-130s flying night interdiction missions over the Ho Chi Minh Trail.

Left: Pictured at Lockbourne AFB in December 1968, this AC-119K carries the 'IH' tailcode of the 4413th CCTS, until 1970 the gunship crew training squadron. The first examples of the 'Stinger' had been delivered to the squadron in September.

Below: The 18th SOS deployed the first AC-119Ks to Southeast Asia in late October 1969, with the first aircraft (pictured) landing at Phan Rang AB on 3 November; their first combat mission was flown 10 days later.

of the AC-47D squadrons operating in South Vietnam: close air support of troops in contact, air support of hamlets, forts and villages under attack, armed reconnaissance and interdiction missions, SAR support, night escort of road and water convoys, illumination and harassment of enemy troops during the night.

Headquarters for the 71st SOS was at Nha Trang with the 14th SOW Headquarters. A Flight with five AC-119Gs was stationed at Nha Trang initially, B Flight with six aircraft went to Phan Rang, and C Flight had five aircraft at Tan Son Nhut. The first combat mission was flown on 5 January 1968. The radio callsign for the AC-119G squadron was the name that became associated with the aircraft – SHADOW. It was not until 11 March 1969 that the 71st SOS was up to full combat strength and considered fully operational.

AC-119K Stinger

The second variant of the Gunship III series was the AC-119K. The K-model differed from the AC-119G in so many ways that it should have been considered an entirely different project type. Starting with a C-119G, Fairchild-Hiller added a pair of GE J85-17 jet engines, for additional power, in pods beneath the outer wings. The AC-119K used a Hamilton Standard three-bladed propeller, whereas all other C-119s, including the AC-119Gs, used an Aeroproducts four-bladed version. In addition to all the equipment installed in the AC-119G, the K had the Motorola AN/APQ-33 beacon tracking radar installed just aft of the 20-kW xenon illuminator. Under the cockpit was a small fairing with a retractable Texas Instruments AN/AAD-4 FLIR, and a Texas Instruments AN/APQ-136 search radar with a moving target indicator mode.

The most important difference, however, was in its armament. The K-model had four MXU-470/A Minigun modules, the same weapon as found on the AC-119G, but with a 2,000-round ammunition storage drum. On either side of the Miniguns was a GE M61 20-mm rotary cannon. As with the AC-130A, this could be fired at a rate up to 6,000 rpm, but was generally fixed at 2,500 rpm. The ammunition storage drum for the M61 held a total of 1,200 rounds. The AC-119K carried 31,500 rounds of 7.62-mm ammunition, plus an additional 4,500 rounds of 20-mm.

Problems with supply of the Texas Instruments AAD-4 FLIR unit delayed delivery of the first AC-119Ks; the first AC-119K was completed and delivered to the Air Force, minus its FLIR, on 24 September 1968. The first FLIR units were not available until 3 May 1969.

The AC-119K gunships were assigned to the 18th Special Operations Squadron at Clinton County AFB, Ohio, which was activated on 1 February 1969. The first operational aircraft arrived the following month. Training contin-

This AC-119K of the 1st SOW, on the Fairchild-Republic ramp at Dulles International Airport during an air show in August 1969, carries the Fairchild company logo on its tail. To the left is a similarly adorned AU-23 Peacemaker I (licence-built Pilatus Turbo-Porter).

Inside the AC-119K

Above: The reprofiled nose of the AC-119K contained an AN/ALQ-136 radar set, while the AN/AAD-4 FLIR was housed in a fairing to the left of the nose gear.

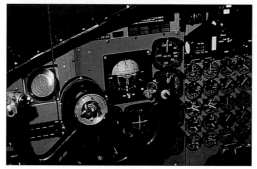

Left and above: This HUD-equipped gun sight was common to both AC-119 variants; the guns were fired by the pilot using the red button on the right side of the control wheel. The pilot also has an AN/APR-25/26 RHAWS scope on the far left of the control panel.

Right: Like the AC-119G, the K was equipped with a Night Observation Device (NOD) mounted in the forward door immediately behind the cockpit.

Lower left: From the outset the AC-119K was equipped with four MXU-470/A Miniguns and a pair of M61A1 20-mm rotary cannon. This view shows the forward pair of Miniguns and the ammunition drum of the forward M61.

Right: The 'black boxes' to the left and centre of this view are components of the AC-119K's fire control system, while the small scope (lower left) is an AN/APR-25/26 RHAWS display. To the right is the AN/AAD-4 FLIR scope.

Above: The positioning of the AC-119K's armament is clear in this view. Note also the AN/ALQ-133 Beacon-Tracking Radar (BTR) mounted in the rear fuselage. For a time ground beacons were used by TICs to guide aircraft to their target, a ground beacon signal plus range and bearing information allowing the AC-119 to remain above cloud and still direct its fire on to an enemy position. However, use of the BTR in angular tracking mode required the aircraft to fly round its target on a predictable left-turning track, leaving the slow-moving aircraft vulnerable to attack from radar-guided AAA fire. Use of the BTR set ceased in December 1970.

ued at Clinton County AFB until FLIR units were installed and the aircraft were ready for combat operations. On 21 October 1969 the first six AC-119Ks departed Lockbourne AFB for combat in Southeast Asia, arriving at Phan Rang AB, Vietnam on 3 November 1969. Like the 71st SOS AC-119Gs, the 18th SOS AC-119Ks came under the command of the 14th SOW.

The 18th SOS flew its first mission on 13 November 1969, but was not completely combat-ready with a full complement of AC-119Ks until 4 February 1970. In the middle of February, the 18th SOS received its callsign – STINGER. (This was an unused callsign from the 366th Tactical Fighter Wing at Da Nang.)

Overall gunship strength in Southeast Asia was five squadrons – two of AC-47Ds, two of AC-119s and one of AC-130A. However, as the AC-119s were phased in, both squadrons of AC-47Ds were phased out of service and their aircraft assigned to units in the Vietnamese and Laotian air forces.

The AC-119s played a major role during the incursion into Cambodia conducted by US and ARVN troops from 1 May 1970. Flying both armed reconnaissance and convoy escort sorties supporting the invasion force, AC-119Gs and Ks flew a total of 178 missions during the Cambodian incursion between 1 May and 30 June 1970. Although ground operations ceased in July, the AC-119s flew continuously against Cambodian targets well into 1971. Between 1 July 1970 and 31 March 1971, AC-119s destroyed some 609 trucks and 237 sampans, while damaging another 494. Enemy troops killed numbered well over 3,100.

The 7th AF ordered the AC-119Ks into Udorn RTAFB, Thailand, on 15 February 1970 for operations against traffic along the Ho Chi Minh Trail. They flew against the Trail for the first time on 17 February. By June, the 18th SOS 'Stingers' accounted for 70 per cent of all truck kills in the Barrel Roll area of Laos.

As 1970 ended, the AC-119s had destroyed some 312 trucks and damaged another 196. That figure jumped significantly in the first months of 1971: by the end of March, AC-119 crews accounted for 1,845 vehicles destroyed or damaged. Included among those totals were eight North Vietnamese PT-76 armoured vehicles destroyed by 18th SOS 'Stingers' during

when both aircraft were fired upon by a 37-mm AA gun. This immediately became the priority target but, before long, the AC-119 was hit, its wing catching fire. Its starboard piston engine soon stopped and, with the adjacent jet engine having been blown off and with the right undercarriage hanging down, the aircraft became uncontrollable. Seven crew bailed out of the aircraft as its pilot, Captain Terrance Courtney, tried to maintain control, but it crashed five miles (8 km) from An Loc, killing the three crew who remained aboard. Over the next four hours the survivors were rescued by two HH-53s and a US Army UH-1, supported by AC-130s, fighter-bombers and helicopter gunships. Following this episode no further daylight 'Stinger' missions were flown.

Whereas both the AC-47 and AC-119G were turned over to the VNAF forces during 'Vietnamisation', none of the AC-119K 'Stingers' was transferred. The 18th SOS did move to Da Nang in November 1972 and painted Vietnamese air force markings on the AC-119Ks, but USAF crews continued to fly the missions through December. The 18th SOS was deactivated after the ceasefire, but the crews remained in Vietnam to train Vietnamese aircrews in the operation of the AC-119K.

Larry Davis

Operation Lam Son 719 from 8 February to 24 March 1971, when ARVN troops crossed into Laos in an attempt to cut the Ho Chi Minh Trail near Tchepone.

At the end of 1970, the AC-119 force had been divided between five bases in South Vietnam and Thailand – the 17th SOS had seven Gs at Phan Rang, and another nine aircraft at Tan Son Nhut; the 18th SOS had three 'Stingers' at Phan Rang, seven at Da Nang, and six more at Nakhon Phanom, Thailand. However, the process of 'Vietnamisation' transferred the entire 17th SOS AC-119G force into the VNAF on 1 September 1971.

When the NVA launched its Easter Offensive in spring 1972, most of the gunship forces were used for support of the troops attempting to

stop the NVA assault on South Vietnam. AC-119Ks were credited with helping stop the NVA forces attacking at Kontum City, An Loc, and dozens of forts and fire bases throughout the I Corp region of Vietnam along the DMZ.

During operations in Southeast Asia, only one AC-119 was lost to enemy fire, on 2 May 1972. AC-119K 53-7826, flying as STINGER 41, was among a detachment of 18th SOS aircraft operating from Bien Hoa to assist in the defence of An Loc. Ammunition dropped to ground forces by C-130 had fallen in enemy-held territory and, to prevent it from falling into enemy hands, AC-119s were required to fly daylight missions in an attempt to destroy it. Accompanied by a FAC aircraft, STINGER 41 was circling the target at 4,700 ft (1433 m)

AC-119 losses in SEA

Only one AC-119 was lost in combat in Southeast Asia, though a further four were written off on operational missions to other causes. The most serious losses were those of a pair of AC-119Gs where engine failure on take-off led to fatal crashes; it soon became clear that the AC-119G was underpowered for the gunship role and that action would need to be taken to improve take-off safety.

■ 11 October 1969: AC-119G 52-5907
SHADOW 76 of Det 1, 17th SOS crashed when an engine failed and caught fire on take-off from Tan Son Nhut AB. With a full load of fuel and ammunition, the AC-119G was unable to maintain height on one engine. Six of the aircraft's 10 crew, including aircraft commander Lt Col Bernard Knapic, were killed.

■ 19 February 1970: AC-119K 53-3156
This 'Stinger' of the 18th SOS's Da Nang detachment was returning from an armed night reconnaissance of

the Ho Chi Minh Trail when it landed short at Da Nang. Fuel starvation resulted in both engines on the port wing losing power; the pilot was unable to maintain control of the aircraft. Though the 'Stinger' was written off in the crash, all 10 crew escaped with only minor injuries.

■ 27 April 1970: AC-119G 53-8155
In similar circumstances to those on 11 October 1969, SHADOW 78 of the 17th SOS was just 100 ft (30.5 m) AGL when it suffered an engine failure after taking-off from Tan Son Nhut AB on a mission over the Trail. The crew struggled to keep the aircraft airborne, and it crashed two miles (3.2 km) from the end of the runway, killing six of the eight crew aboard. After this accident the USAF limited the AC-119G's MTOW to ensure that it could, if necessary, maintain a safe climb rate on one engine.

■ 6 June 1970: AC-119K 52-5935
This detached 18th SOS aircraft suffered a runaway propeller on its number one engine on take-off from Da Nang AB. The pilot attempted to return to the airfield and make an emergency landing, but lost control of the aircraft. The crew bailed out over the

While only one AC-119 was lost in combat, numerous others suffered damage like this; despite being hit by 57-mm AAA fire, which took 15 ft (4.6 m) off its starboard wing, this AC-119K limped back to Udorn and was able to make a successful landing. The incident took place during June 1970.

sea, but the aircraft continued flying and, briefly, appeared to be heading for Chinese airspace over Hainan Island, though it eventually crashed in the sea well short of Hainan. Nine of the 10 crew were rescued.

14TH AIR COMMANDO WING (8TH TFW, 388TH TFW)
16TH AIR COMMANDO/SPECIAL OPERATIONS SQUADRON

The 16th was activated at Nha Trang in August 1968 with one AC-130A aircraft and was assigned to the 14th ACW with the tailcode 'EA'. On 31 October 1968 the squadron moved to Ubon AB, Thailand, was assigned to the 8th Tactical Fighter Wing (with the tailcode 'FT') and was redesignated 16th Special Operations Squadron .

Operations initially used the single AC-130A Gunship II prototype aircraft, then AC-130A Pave Pronto production gunships. Equipment was upgraded to AC-130E Pave Spectre aircraft in 1972, including addition of Pave Aegis AC-130E aircraft in February 1972, and AC-130H aircraft in March 1973, with the tailcode 'WP'.

The 16th SOS moved to Korat RTAFB, Thailand, on 19 April 1974 after the closure of Ubon RTAFB, and was attached to the 388th TFW; tailcodes were removed. It returned to Eglin AFB, Florida, on 12 December 1975 and was assigned to the 1st Special Operations Wing at Hurlburt Field, with the tailcode 'AH'. All AC-130As were transferred to the 919th SOS/AFRes. The 16th SOS was equipped with only AC-130E and AC-130H aircraft.

14TH AIR COMMANDO WING (14TH SPECIAL OPS WING)
71ST SPECIAL OPERATIONS SQUADRON

Originally designated the 71st Tactical Airlift Squadron, AFRes, the 71st TAS was called to active duty on 13 May 1968 at Bakalar AFB, Indiana. The squadron moved to Lockbourne AFB, Ohio, and was redesignated the 71st Special Operations Squadron on 1 June 1968 and equipped with Fairchild AC-119G 'Shadow' aircraft between 1 June and 15 June 1968. The 71st trained in the AC-119G mission at Lockbourne AFB in the summer and autumn of 1968. It deployed to Southeast Asia on 5 December 1968 and had four aircraft in place at Nha Trang AB, RVN, on 27 December 1968. The 71st flew the first AC-119G combat sortie on 5 January 1969.

Three flights of six aircraft each were assigned at Nha Trang AB, which was Headquarters for the 71st SOS, Phan Rang AB, and Tan Son Nhut AB. 71st SOS Headquarters moved to Phan Rang AB with the 14th SOW. The 71st SOS was inactivated 18 June 1969.

17TH SPECIAL OPERATIONS SQUADRON

The 17th Special Operations Squadron was activated at Phan Rang AB, RVN, on 1 June 1969, absorbing most of the active-duty US Air Force personnel who remained from the 71st SOS. The 17th SOS, with Headquarters at Phan Rang AB, had three flights: A Flight (four AC-119Gs) was initially based at Tuy Hoa AB, RVN, then moved to Phu Cat AB, RVN; B Flight (seven AC-119Gs) was based at Phan Rang; and C Flight (five AC-119Gs) was based at Tan Son Nhut AB, RVN. By 1971, A Flight had been consolidated with B Flight at Phan Rang AB, with seven AC-119Gs, while C Flight at Tan Son Nhut had the remaining nine AC-119Gs within the 17th SOS.

On 30 September 1971, the 17th SOS was inactivated and all remaining AC-119G assets were turned over to the Vietnamese air force and formed into the 819th Combat Squadron. Two aircraft crashed and were written off – 52-5907 on 11 October 1969, and 53-8155 on 27 April 1970.

18TH SPECIAL OPERATIONS SQUADRON

The 18th Special Operations Squadron was organised at Lockbourne AFB, Ohio, on 1 February 1969 and equipped with Fairchild AC-119K 'Stinger' gunships. During training, the 18th SOS was assigned to the 1st SOW between 25 January 1969 and 15 July 1969. The 18th SOS deployed to South Vietnam on 21 October

1969, being assigned to the 14th SOW and based at Phan Rang AB, RVN. The first aircraft arrived on 3 November 1969, and flew their first mission, from Da Nang AB, RVN, on 13 November 1969. On 25 August 1971, with the inactivation of the 14th SOW, the 18th SOS was transferred to control of the 56th SOW, before being inactivated on 31 December 1972. Its aircraft were passed to the VNAF and equipped the 821st Combat Squadron.

Headquartered at Phan Rang, the 18th SOS initially had three flights: A Flight (six AC-119Ks) was based

AC-130A 55-0029 of the 16th SOS rests at Ubon AB during 1972. The aircraft carries AN/ALQ-87 ECM pods.

first at Phan Rang, then moved to Da Nang AB, RVN; B Flight (three AC-119Ks) was based first at Phu Cat AB, RVN, before moving to Udorn AB, Thailand; and C Flight (three AC-119Ks) was based at Phan Rang AB, RVN. On 17 February 1970, D Flight (three AC-119Ks) was activated at Udorn RTAFB, with aircraft and crews taken from B Flight at Phu Cat AB. By 1971, the inventory of the 18th SOS was distributed as follows: A Flight at Da Nang with seven AC-119Ks, B Flight at Phan Rang AB with three AC-119Ks, and the remaining six AC-119Ks assigned to C Flight at Nakhon Phanom RTAB, Thailand.

The 18th SOS lost three AC-119Ks during operations in Southeast Asia – AC-119K 53-7826 was shot down by 37-mm gun fire over South Vietnam on 2 May 1972, and aircraft 53-3156 and 52-5935 were lost in crashes after mechanical failure.

Left: The badge of the 18th SOS adorned an outside of the operations building at Phan Rang AB.

Below: Colonel Frank Eaton, commander of 17th SOS, flies Oklahoma Representative over Tan Son Nhut AB (home of C Flight, 17th SOS) during 1969. As with other gunship units in South Vietnam the AC-119Gs of the 17th were deployed in six-ship flights at various other bases, including Da Nang, Tuy Hoa, Nha Trang and Phu Cat.

Left: The 17th SOS 'business card' of late 1969 neatly encapsulated the role of the AC-119G.

Below: On 30 September 1971 the 17th SOS turned its aircraft over to Vietnamese Air Force crews assigned to the newly-created 819th Combat Squadron. This AC-119G in VNAF markings is pictured on the ramp at Da Nang AB in October 1971.

Dornier Do 217
The 'Baedeker' bomber

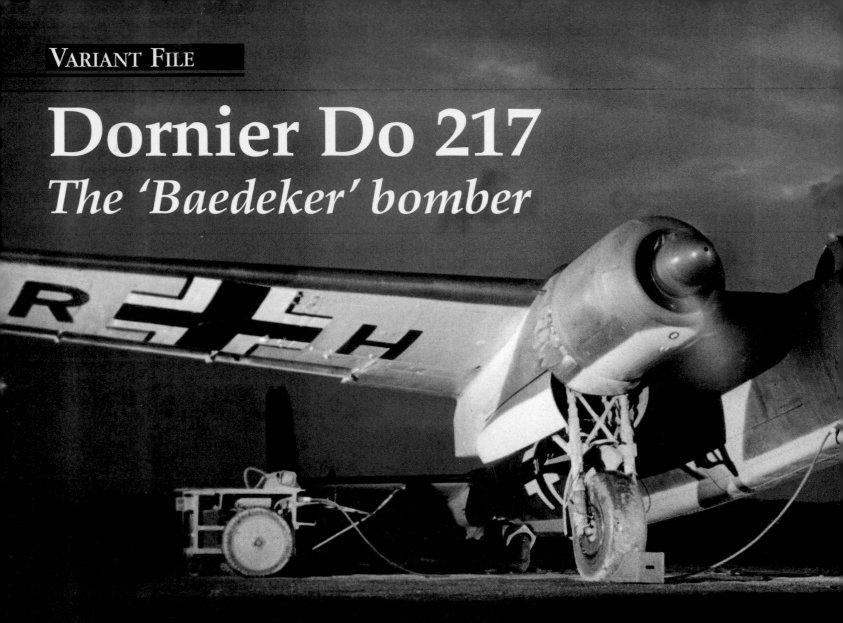

Developed by Dornier-Werke GmbH from its successful Do 17/215 family of bomber aircraft, the Do 217 was the first new bomber design to enter large-scale service in the Luftwaffe following the outbreak of World War II, and was to be the last Dornier bomber to see front-line service with the Luftwaffe.

The Dornier Do 217 first saw service in 1940, when aircraft of the pre-production batch flew clandestine reconnaissance missions over the Soviet Union towards the end of that year, preparing the way for the attack on that country. During 1942 and 1943 the Do 217 units bore the brunt of the bombing operations against Great Britain, and suffered heavy losses in the process. A few Do 217s also performed night reconnaissance missions on the Eastern front. More than three hundred Do 217s were modified to serve as night-fighters, but the type proved too unwieldy for this role and was soon replaced. In the late summer of 1943 the Do 217 began a new lease of life, when it went into action carrying the newly developed air-launched guided missiles. Until the Allies took the measure of the new threat, it scored some major successes against warships. Production of the Do 217 ended in May 1944, after 1,887 had been built.

The initial pre-production batch of eight Do 217A reconnaissance aircraft went to the Aufklärungsgruppe des Oberbefehlshabers der Luftwaffe, the special reconnaissance Gruppe operating under the direct control of the Luftwaffe High Command. Based at Cracow in Poland and Bucharest in Romania, these aircraft flew clandestine reconnaissance missions over Soviet-occupied Poland and the Soviet Union during the final months of 1940 and the first half of 1941. The photographs they took would assist the planning of the German offensive against the Soviet Union.

At the end of 1940, the first Do 217E-0 and E-1 aircraft were delivered to regular Luftwaffe units. Several of these aircraft, fitted with recon-

Pictured on an airfield at Bordeaux-Merignac in 1941, Do 217E-1 'F8+GN' carries the well-known badge of II./KG 40. Formed in March 1941 and based at Soesterberg, Holland, and Bordeaux, this unit was engaged in anti-shipping operations over the Atlantic, often with an escort of Junkers Ju 88C fighters provided by V./KG 40.

Dornier Do 217 production

...able shows annual Dornier Do 217 production by each of ...ree factories which built the type between 1939 and 1944. ...of 1,887 aircraft is accounted for; some sources give this ...acceptance figure, actual production totalling some 1,966 ...es (including the six Do 317A aircraft). Whatever the true ...it may be seen that over half of all aircraft were built at ...'s Munich plant.

	1939	1940	1941	1942	1943	1944
...n	5	15	130	235	552	50
...shafen	–	–	6	300	296	–
	–	–	130	168	–	–
...tals	5	15	266	703	848	50

...n on the night of 4/5 May, RAF intrud-
...ked a couple of Dorniers as they were
...o land at their base at Eindhoven in
..., and shot down both. Piloting one of
...rcraft was the commander of KG 2,
...alter Bradel, who was killed.
...n 1942, increasing severity of the RAF
...acks on Germany, coupled with short-
...Messerschmitt Bf 110s and Junkers
...d to consideration of the Do 217 for
...-fighter role. The Do 217J was a night-
...nversion of the Do 217E bomber, with
...ned 'solid' nose housing a fixed arma-
...our MG FF 20-mm cannon and four
...9-mm machine-guns. Later, some of
...raft were fitted with the FuG 202
...in BC airborne intercept radar with a
...range of about 4 miles (6.4 km). Yet
...t's performance was insufficient for
...nd its size and weight made it too

...ffort to improve matters the next
...r sub-type, the Do 217N, was fitted
...ore powerful DB603 engine. While
...a marked improvement in perfor-
...mpared with the J sub-type, the
...was still unpopular with front-line
...r Wilhelm Herget, commander of
...f NJG 4 who ended the war cred-
...night victories, told this writer:
...943 my Gruppe received sufficient
...'N aircraft to equip one Staffel,
...he time the Messerschmitt 110 was
...pply and the High Command
...the converted bomber's extra
...ight be useful. We found the 217
...stable, excellent for instrument
...viously a very nice bomber; but it
...y on the controls, and it climbed
...be much good as a fighter. I flew
..., just to try it. But after that I
...it on operations and reverted to
...usted [Bf 110], which was greatly
...ight interceptor."
...summer of 1942 half a dozen
...aft were delivered to the Italian
...e they formed the equipment of
...riglia based initially at Treviso
...ate Pozzolo. Early in the follow-
...were replaced with radar-fitted
...lian Dorniers saw little action,
...fter the armistice in September
...aft were either destroyed by the
...sessed by the Luftwaffe.
...ing of 1944, the Do 217 had
...n front-line night-fighter units
...he Luftwaffe received a total of
...versions of the Do 217, but
...type achieved little. Of all the
...by this versatile aircraft, this

...three Gruppen of
...operating in the West,
...7Es. Initially the new
...ed on anti-shipping
...issions and minelaying
...und Great Britain. From
...however, the strength of
...and fighter defences
...oastal convoys began to

make these attacks unprofitable. During a five-week period in February and March 1942, KG 2 lost 13 crews engaged in such operations.

The 'Baedeker' raids

Due to the demands of the Eastern front, during this period the Luftwaffe flew relatively few bombing sorties against targets in Britain. That quiescent phase came to an abrupt end on 28 March 1942, however, following the particularly destructive RAF attack on Lübeck. In reply, Adolf Hitler demanded retaliation from his bomber units in the west. On 23 April a force of 45 German bombers – for the most part Do 217s of KG 2 – attacked Exeter. On the following night, 60 aircraft repeated the assault on that city. During the following two nights the target was Bath, which suffered severe damage in raids totaling 250 sorties.

Yet, even as German aircraft pounded Bath, RAF bombers wrecked the German town of Rostock with a series of four destructive fire raids. Hitler was beside himself with rage at this development. At a Foreign Office briefing in Berlin, a government press officer spoke of taking a copy of Baedeker's guide-book and marking-off each British city as it was destroyed. Because of this, the series of attacks that followed became known in both Great Britain and Germany as the 'Baedeker Raids'.

Bath again, then Norwich and York, all suffered heavily. Then, on the night of 3 May, came the most devastating of the reprisal attacks. Again the target was Exeter, but on this occasion the target marking was accurate and the bombing concentrated. Fires quickly took hold amongst the heavily timbered medieval buildings. Unhindered by the narrow streets, they raged unchecked until a large part of the city had been gutted.

During the remainder of May 1942 German bombers, for the most part the Do 217s of KG 2, struck at Cowes, Hull, Poole and Grimsby and, on the final day of the month, Canterbury. Throughout these attacks, losses had mounted steadily, and in June and July there was a marked reduction in the number of attacks on Britain. The dying spasm at the end of July – three raids on Birmingham and one on Hull – cost the Luftwaffe 27 bombers and caused little damage.

Ice covers standing water either side of the taxiway of this airfield, probably in Holland, on a clear winter's day during 1941/42. Do 217E-4s of III./KG 2 are pictured taxiing towards the runway at the beginning of another anti-shipping sortie.

Following this painful mauling the Baedeker units began a period of recuperation, but this was not allowed to take place without interruption. On 19 August Allied forces launched a large-scale seaborne raid on the port of Dieppe in northern France. To meet it, virtually every operational Luftwaffe unit in France and

Front-line Do 217 units, 17 May 1943

By 17 May 1943, about 1,360 Do 217s of all versions had been delivered to the Luftwaffe. Production was then running at about 80 aircraft per month. On this date the deployment of Do 217s was almost at its peak. The Luftwaffe Quartermaster General's census of aircraft in service with combat units listed 278 Do 217s, of which 159 were combat-ready. Many more were held in aircraft parks ready for delivery to combat units.

The two Do 217 Gruppen of KG 100 were preparing to go into action with the new guided missile systems. Many of their aircraft were being modified for the new role, hence the low proportion of combat-ready aircraft reported by these units.

The number of aircraft on strength is given, with the number in serviceable condition given in brackets.

Luftflotte 3 (units based in France, Holland and Belgium)

Kampfgeschwader 2	Stab	2	(2)
	I. Gruppe	21	(8)
	II. Gruppe	26	(26)
	III. Gruppe	18	(17)
Kampfgeschwader 40	II. Gruppe	21	(19)
Kampfgeschwader 66[1]	I. Gruppe	23	(7)

Luftwaffenbefehlshaber Mitte (Central Air Command)

Nachtjagdgeschwader 1	II. Gruppe	6	(3)
Nachtjagdgeschwader 3[2]	I. Gruppe	11	(9)
	II. Gruppe	29	(20)
Nachtjagdgeschwader 4[2]	I. Gruppe	11	(8)
	II. Gruppe	11	(11)
	III. Gruppe	6	(5)
	IV. Gruppe	3	(3)
Nachtjagdgeschwader 5[2]	II. Gruppe	2	(1)
Kampfgeschwader 100	II. Gruppe[3]	37	(0)
	III. Gruppe[4]	35	(11)

Eastern Front

Nachtaufklärungsgruppe[5]		16	(9)

Notes:
1. Pathfinder unit
2. Reich Air Defence night-fighter units
3. Unit preparing to go into action with Hs 293 glided bombs
4. Unit preparing to go into action with Fritz-X guided bombs
5. Night reconnaissance gruppe; also operated Do 17s and He 111s

Belgium went into action. KG 2 launched nearly its entire strength of about 80 aircraft against the concentration of Allied shipping, but found the way barred by numerous Allied fighter patrols. That day the Geschwader lost about 20 bombers – about a quarter of those sent – in the fierce air battles over and around Dieppe.

Again KG 2 underwent a quiet period as it gradually rebuilt its strength. Although there was no shortage of replacement aircraft, the provision of trained replacement crews fell far short of that needed. At the close of the year the unit resumed sending Dorniers in ones and twos to deliver daylight nuisance attacks on peripheral targets in Britain. Typical of these was that against Eastbourne on 18 December. A solitary Do 217 swept in low over the sea, flying under low cloud and below the British radar cover. It dropped its stick of four 500-kg bombs across the town centre, then escaped out to sea. At the time the streets were crowded with Christmas shoppers and there was scarcely any warning. Eighteen people were killed and 37 more suffered injuries.

New sub-types

At the close of 1942 two important new sub-types of the Do 217 entered service; the K and the M. Both featured a redesigned forward section, with a rounded and more streamlined nose profile, and were about 28 mph (45 km/h) faster than the earlier E model. They also carried a couple of new systems to improve their ability to penetrate the defences. The FuG 101 radio altimeter enabled these aircraft to make low-altitude approaches to targets at night or in poor visibility. The FuG 217 tail warning radar provided warning of enemy fighters approaching from astern.

Strengthened by these improvements, the revitalised KG 2 resumed its attacks on Britain with a heavy raid on the capital on the night of 17/18 January 1943. In the meantime, the defences had also improved, however, and during this and later raids the Geschwader suffered heavy cumulative losses. During March 1943 KG 2 lost 26 complete crews, and that steady drain in men and machines would continue throughout the spring.

Even after they had regained 'friendly' territory, German bomber crews could not consider themselves safe. Following the attack on

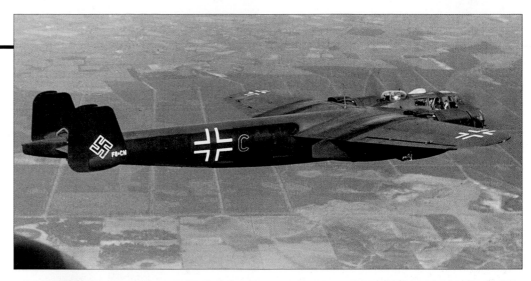

'F8+CN' (right) is a II./KG 40 Do 217E-2 finished in typical night camouflage of the 1942/43 period. Aircraft markings were progressively toned-down, fuselage codes being applied instead to tailfins as all fuselage and wing markings were oversprayed black (as shown on the 5./KG 2 aircraft, below right). Even swastikas were eventually deleted to reduce conspicuity.

was its least effective.

From the early part of 1943 until the end of the war, a dozen or so Do 217s operated on the eastern front as part of the Nachtaufklärungs-gruppe (night reconnaissance Gruppe). Before they launched attacks, Soviet troops often moved into position under cover of darkness, so the unit's reports were a valuable part of the overall reconnaissance effort. Moreover, as the conflict wore on, the strength of the Soviet air defences reached the point where daylight photography became a hazardous business.

The procedure for night photography was as follows. The aircraft approached the target flying at maximum speed at altitudes around 19,700 ft (6,000 m). Initially the camera shutters were locked open. Over the target, the Do 217 released a series of 66-lb (30-kg) photoflash bombs at about 10-second intervals. After a pre-set time delay, the first flash bomb ignited at a point about 4,000 ft (1200 m) above the ground. On ignition the bomb produced a flash of 6 million-candle power lasting for one-third of a second, which lit up the ground and produced an image on the film. A photoelectric cell then closed the cameras' shutters, wound on the film then re-opened the shutters ready for the next picture. When the second flash bomb ignited, the process was repeated. Usually four or five pictures were taken, by which time the defences were thoroughly alerted and the crew had to beat a hasty retreat.

In action with guided weapons

From the beginning of the war, the Luftwaffe had devoted much thought to increasing the effectiveness of its attacks on enemy shipping. One answer was to deliver the weapons in dive attacks, but as already noted the Do 217 was too large and too heavy to make these without risk of overstressing the airframe. Horizontal bombing attacks from high altitude were notoriously ineffective against ships manoeuvring in open water. Low-altitude attacks took the aircraft within range of the ships' gun defences, and were in any case ineffective against armoured vessels since regular bombs dropped from low altitude lacked penetrative power.

The ideal weapon against a moving ship was an air-launched guided missile, the course of which could be corrected in flight to compensate for the movement of the target. That would give a good chance of hitting a manoeuvring ship, while the launching aircraft stayed beyond reach of its guns.

To meet this requirement two German firms each produced a radio command-guided anti-shipping weapon: the Henschel Hs 293 glider-bomb and the Ruhrstahl Fritz-X guided bomb. Early in 1943 two Gruppen from KG 100 re-equipped with Do 217s modified to carry the new missiles. II. Gruppe operated Do 217E-5s with racks to carry one Hs 293 glider bomb under each outer wing section while III. Gruppe operated Do 217K-2s with racks to

carry two Fritz-X guided bombs under the wing centre-section, between engine and fuselage.

After some unsuccessful missions, the first partially successful attack using air-launched radio-guided missiles took place on 25 August 1943. Hauptmann Molinus led 12 Dorniers of II./KG 100 in an attack on a Royal Navy escort group off the northwestern tip of Spain. During this action the sloop HMS *Landguard* suffered damage from four near misses.

Two days later, the missile-carrying Dorniers struck at another Royal Navy escort group in the same area. Several missiles were aimed at the sloop HMS *Egret*, five of which exploded in the water either short or over the ship. *Egret's* 20-mm Oerlikon cannon shot down another of the missiles. Then a glider bomb came straight at the sloop and hit its starboard side near the after magazine. This detonated the ammunition and depth charges stored there, causing a large explosion. The vessel broke up and sank rapidly and from a complement of 250, only 28 survived. During the same engagement the destroyer HMCS *Athabaskan* suffered damage.

At the end of 1943, the Luftwaffe began assembling units in the West for what was to be its final manned bomber offensive against Britain: Operation Steinbock. By now the Do 217 was nearing the end of its career as a regular bomber, and higher performance types like the Heinkel He 177, the Junkers Ju 88S, the Ju 188 and the Messerschmitt Me 410 made up

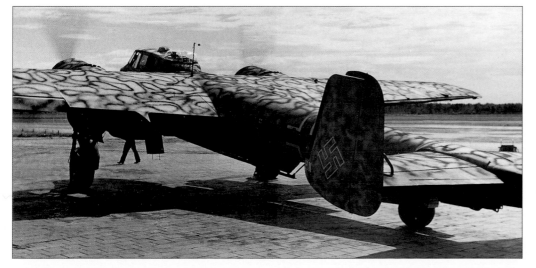

So-called 'wave-mirror' camouflage was applied to a limited number of aircraft engaged in night operations. Patterns varied and depended on the mode of application; the scheme applied to this 5./KG 6 Do 217E-2 was clearly applied using a spraygun.

Attack on the battleship Roma

While II./KG 100 operated with varying degrees of success over the Bay of Biscay, its sister unit III./KG 100 stood by at its base at Istres in southern France awaiting the appearance of targets suitable for its armour-piercing Fritz-X bombs. This unit operated the Do 217K-2 variant, specially modified as a high-altitude bomber with the wingspan increased by 19 ft (5.8 m) to 81 ft 4 in (24.8 m). Like the Hs 293-carrying Do 217E-5, the K-2 could carry two missiles but only at some expense in range. On operational missions only one Fritz-X was carried, on the rack between the starboard engine nacelle and fuselage.

On 9 September 1943, the Italian government capitulated. Under the terms of the armistice the main body of the Italian battle fleet, comprising three battleships, six cruisers and eight destroyers, left La Spezia to sail to Malta to surrender. German aircraft shadowed the warships, and early that afternoon Major Bernhard Jope led a striking force of 12 missile-carrying Do 217Ks to attack them. It was a beautiful summer's day with almost unlimited visibility, and Jope's crews had little difficulty in finding their quarry.

Above: Do 217K-2s of III./KG 100 are pictured at Istres near Marseilles in the summer of 1943. This variant was modified to carry two Fritz-X bombs on racks between the engines and the fuselage. To improve altitude performance (allowing the guided bombs to achieve maximum penetration against armoured warships), the K-2's span was extended by 19 ft (5.8 m).

Left: Major Bernhard Jope, commander of III./KG 100, led the attack on Roma and his bomb was the first to strike the battleship.

The Dorniers attacked from altitudes around 23,000 ft (7000 m), where the warships' gunfire was inaccurate and ineffective. Far below, the Italian ships twisted and turned in their efforts to throw the German crews off their aim. Had these aircraft carried regular bombs, that tactic might have been successful. But Jope's aircraft were loaded with radio-controlled bombs.

The first hit was on the Italian flagship, the battleship *Roma*. A Fritz-X struck it near the after mast, punched its way clean through her hull and exploded immediately underneath. Seriously damaged, *Roma*'s starboard steam turbines ground to a halt and the ship lost speed. A few minutes later *Roma* took a second hit, this time in front of the bridge. This knocked out the steam turbines on its port side and started a fierce fire below decks. Raging out of control, the fire reached one of the magazines. There was a huge explosion, and the ship broke into two and sank with heavy loss of life.

Roma's sister ship, the *Italia*, suffered a Fritz-X strike on its bow. The vessel took on some 800 tons of water and its speed was reduced to 24 knots, but it was able to reach Malta unaided.

On the same day Allied troops landed at Salerno to the south of Naples and established a beachhead. Jope's unit went into action against the concentration of Allied shipping off the coast and scored hits on the battleship HMS *Warspite* and the cruisers HMS *Uganda* and USS *Savannah*; all three ships sustained heavy damage but survived.

The time of the easy pickings came to an end as soon as Allied fighters began operating from an airstrip in the beachhead on 12 September. Between then and 18 September KG 100 lost 10 aircraft and crews, before it was forced to cease its attacks.

The next Allied landing operation in the Mediterranean was at Anzio north of Rome, in January 1944. By now, however, the threat of the German missile-carrying aircraft was fully appreciated and the ships enjoyed lavish fighter protection. The only major German success was the sinking of the cruiser HMS *Spartan*, by a glider bomb.

Below: These attack photographs, taken from a Do 217 of III./KG 100 on 9 September 1943, show the attack on the Italian battleship Roma. The warship was in a sharp turn to port. The Fritz-X missile tracking flare (marked with a semi-circle and dashed line) was being steered towards the Roma. The warship took hits from two of these weapons, which started uncontrollable fires below decks and set off ammunition in the forward magazine. The battleship then exploded and sank.

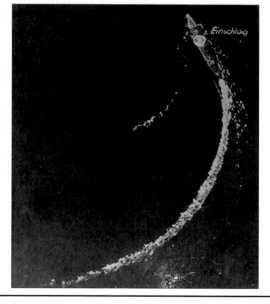

the bulk of the attacking force. Of more than 500 bombers assembled for the operation only 76 were Do 217s, all from Kampfgeschwader 2.

Operation Steinbock

Operation Steinbock opened on the night of 21 January 1944 with two large-scale attacks on London. During the month that followed the capital was raided in force on five more occasions. Typical of these attacks was that on the night of 23/24 February. A total of 161 German bombers took part, their aiming point being the Millwall dock area. That night I./KG 2 sent 15 Do 217s. A typical incendiary load carried by the Dorniers was one AB 1000 container filled with 590 stick incendiary bombs each weighing 2½ lb (1 kg), and two AB 500 containers each with a further 140 stick incendiaries. The aircraft took off from Melun-Villaroche, south of Paris, and flew via Evreux, aiming to cross the French coast at St Valery-en-Caux at 16,500 ft (5000 m). From a position 25 miles (40 km) from the coast of England on the outbound flight, to a similar position on the return flight, each bomber

released one bundle of *Dueppel* radar reflective foil every 30 seconds. After crossing the English coast near Eastbourne, each bomber commenced a gentle descent to bring it over the target at 13,000 ft (4,000 m), where they delivered a concentrated attack between 22.30 and 22.42 hours. Beforehand, pathfinder Ju 88s and Ju 188s had laid out yellow sky markers which ignited at 10,000 feet (3048 m) over the target, to serve as aiming points for the main force. While in the target area, each aircraft increased its rate of releasing *Düppel* to one bundle every 4 seconds. After dropping their bombs, each raider turned through a semicircle and withdrew flying parallel to its inbound track. During their withdrawal, bombers flew at maximum speed in a steady descent, aiming to regain the French coast at about 650 ft (200 m). This high-speed descending withdrawal, combined with the *Düppel* released during both the inbound and outbound flights, made interception difficult for the RAF night-fighters. Although these claimed four raiders shot down that night, none was from I./KG 2.

I./KG 2 lost one aircraft that night, but not to a fighter. Over the target an anti-aircraft shell detonated close to one of its Dorniers. Shell fragments damaged the starboard engine and wing, and struck the cabin dousing the cockpit lights. Seeing little chance of regaining his base at Melun near Paris, the pilot engaged the autopilot and ordered the crew to bail out. Abandoned, the aircraft headed north some 50 miles (80 km) in a shallow descent. Finally, it made a near perfect belly landing on allotments near Cambridge.

Although the raiders got off lightly on the night of 23/24 February, that was not normally the case during attacks on London. During January and February, Steinbock cost the Luftwaffe 129 bombers destroyed or damaged beyond repair.

The attacks on Britain continued in a similar vein through March and most of April 1944, though in the latter part of March they broadened in scope to include Hull, Bristol and Portsmouth. On the night of 29/30 April there was an unusual departure when about 10

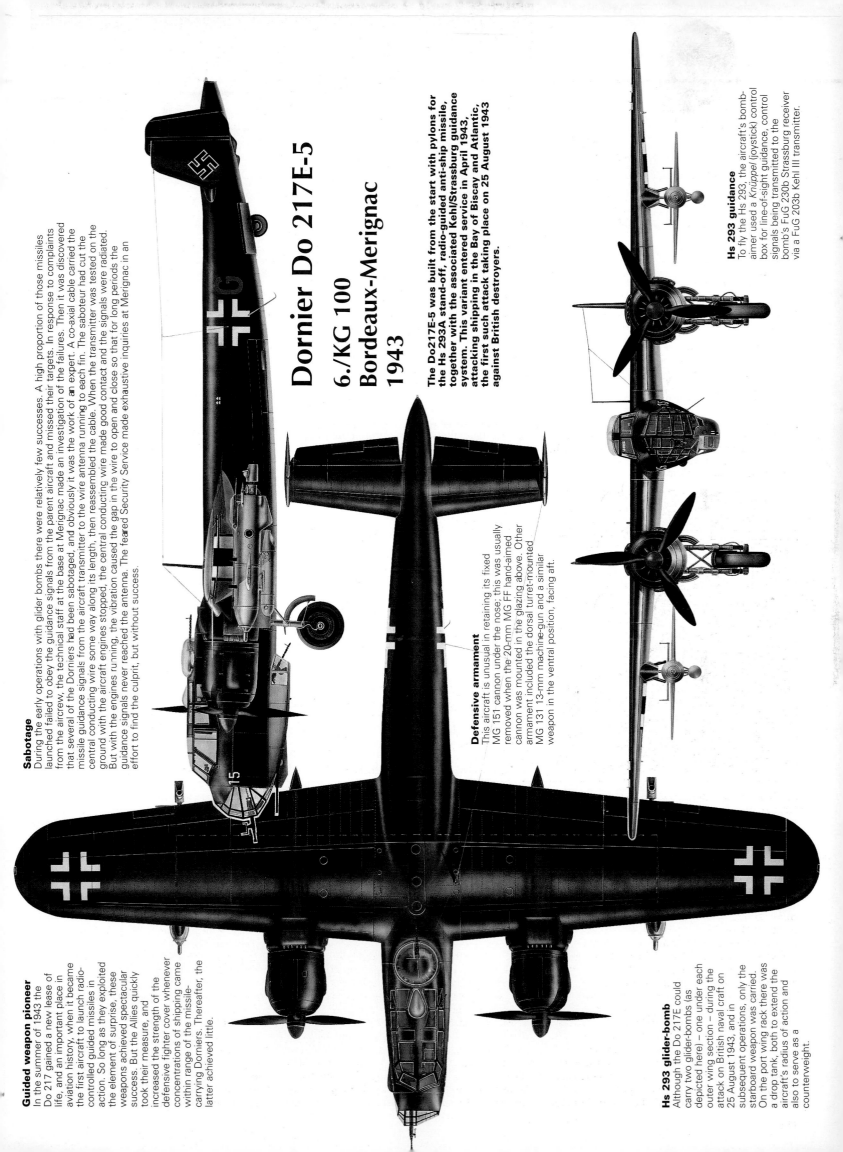

Guided weapon pioneer

In the summer of 1943 the Do 217 gained a new lease of life, and an important place in aviation history, when it became the first aircraft to launch radio-controlled guided missiles in action. So long as they exploited the element of surprise, these weapons achieved spectacular success. But the Allies quickly took their measure, and increased the strength of the defensive fighter cover whenever concentrations of shipping came within range of the missile-carrying Dorniers. Thereafter, the latter achieved little.

Sabotage

During the early operations with glider bombs there were relatively few successes. A high proportion of those missiles launched failed to obey the guidance signals from the parent aircraft and missed their targets. In response to complaints from the aircrew, the technical staff at the base at Merignac made an investigation of the failures. Then it was discovered that several of the Dorniers had been sabotaged, and obviously it was the work of an expert. A co-axial cable carried the missile guidance signals from the aircraft transmitter to the wire antenna running to each fin. The saboteur had cut the central conducting wire some way along its length, then reassembled the cable. When the transmitter was tested on the ground with the aircraft engines stopped, the central conducting wire made good contact and the signals were radiated. But with the engines running, the vibration caused the gap in the wire to open and close so that for long periods the guidance signals never reached the antenna. The feared Security Service made exhaustive inquiries at Merignac in an effort to find the culprit, but without success.

Dornier Do 217E-5

6./KG 100 Bordeaux-Merignac 1943

The Do217E-5 was built from the start with pylons for the Hs 293A stand-off, radio-guided anti-ship missile, together with the associated Kehl/Strassburg guidance system. This variant entered service in April 1943, attacking shipping in the Bay of Biscay and Atlantic, the first such attack taking place on 25 August 1943 against British destroyers.

Defensive armament

This aircraft is unusual in retaining its fixed MG 151 cannon under the nose; this was usually removed when the 20-mm MG FF hand-aimed cannon was mounted in the glazing above. Other armament included the dorsal turret-mounted MG 131 13-mm machine-gun and a similar weapon in the ventral position, facing aft.

Hs 293 guidance

To fly the Hs 293, the aircraft's bomb-aimer used a *Knüppel* (joystick) control box for line-of-sight guidance, control signals being transmitted to the bomb's FuG 230b Strassburg receiver via a FuG 203b Kehl III transmitter.

Hs 293 glider-bomb

Although the Do 217E could carry two glider-bombs (as depicted here) – one under each outer wing section – during the attack on British naval craft on 25 August 1943, and in subsequent operations, only the starboard weapon was carried. On the port wing rack there was a drop tank, both to extend the aircraft's radius of action and also to serve as a counterweight.

Twin-engined bombers were often useful as testbed aircraft, given their weight-lifting capability and range performance. The Do 217 was no exception, several examples being used as testbeds for new weapons and powerplants. Do 217E-2 'RE+CD' was employed during 1942 to test fly 500-mm (above) and 1000-mm (above right) versions of the Sänger ramjet, under test as a potential powerplant for a new type of fighter. Results were unimpressive and though interest was revived in 1944, no German ramjet-powered aircraft ever flew.

Do 217Ks of III./KG l00 attacked warships in Plymouth harbour using Fritz-X guided bombs. The missiles scored no hits, but two Dorniers fell to anti-aircraft fire. During May the Steinbock attacks petered out, as German bomber units licked their wounds as they prepared for what was to be their greatest trial of all: the long-heralded invasion of France.

Invasion of France

On 6 June 1944, Allied troops punched through the German coastal defences at Normandy and established a beachhead. Among the bomber units launched that night against the massive concentration of shipping off the coast, the only Do 217 unit was III./KG 100. Over the invasion area the attackers encountered a thicket of fighter patrols. Also, the adroit use of smokescreens laid by the warships prevented the effective use of guided missiles. During the 10 days following the invasion, and despite crippling losses, the German bomber force sank only five small ships through direct air action. Thereafter the Luftwaffe shifted to a large-scale minelaying

Though this Do 217M-1, pictured at Copenhagen in 1945, appears to be a standard bomber variant, its 'K7' codes betray it as a Do 217M-1/R6 night photographic reconnaissance aircraft of the Aufklärungsgruppe Nacht.

effort off the coast but this, too, failed to achieve more than a nuisance effect.

At the end of July 1944, American forces pushed westwards out of their beachhead and advanced rapidly down the western edge of the Cherbourg Peninsula. On the final day of the month they seized intact the bridges over the See and Sélune Rivers at the southern end of the peninsula. The move outflanked the main German defensive line in Normandy, for ahead lay an area that was almost undefended. General George Patton, commander of the US 3rd Army, realised his heaven-sent opportunity and poured troops across the bridges and into the open countryside beyond.

In a desperate attempt to halt or at least hinder the advance, Do 217s of III./KG 100 attacked the bridges with Henschel Hs 293 glider bombs on the nights of 2/3, 4/5, 5/6 and 6/7 August. For the first time, the Hs 293 was used against land targets. During the last of those attacks six Dorniers took off from Toulouse/Blagnac, each carrying a single glider-bomb. Their crews had orders to hit the bridges, or crater the approaches to them. Shortly after 03.00 hours, Leutnant Hans Kieffer arrived in the area in his Do 217 only to find the target shrouded in a thick haze. That was not his only misfortune, for attentive British eyes had followed his progress. Flight Lieutenant J. Surman, piloting a Mosquito of No. 604 Squadron, afterwards reported:

"At 02.45 hours when being vectored home the controller informed me that an aircraft was 2 miles ahead and on my starboard, and asked whether I would like to investigate. I turned starboard and my navigator obtained a contact at 2 miles range, well to starboard. I closed in slowly on a vector of 280º and then 240º, and from 800 feet obtained a visual and identified it

as a Do 217. I drew up and gave a short burst from 600 feet; the port engine exploded but did not catch fire, and I overshot."

The Dornier's ventral gunner, Feldwebel Karl Salzer, failed to see the Mosquito until the cannon flashes revealed its position. Kieffer then threw the bomber into a violent corkscrew manoeuvre, which caused Surman to overshoot. The Mosquito pilot realigned himself and made two further firing runs, the last of which caused an explosion in the bomber's fuselage fuel tank. The Dornier went down out of control and Kieffer ordered his crew to bail out. Soon after landing by parachute, the German pilot was captured by American troops.

During the series of attacks one missile struck the bridge at Pontaubault, but the damage was not serious and the structure continued in use. The actions cost the Gruppe seven Do 217s destroyed.

The last Do 217 was delivered to the Luftwaffe in May 1944, and thereafter the number of these aircraft in front-line service dwindled rapidly. Soon after the Pontaubault operation III./KG 100 was disbanded, and with that the bomber versions of the Do 217 virtually passed out of service. The Do 217 continued in service to the end of the war in the night reconnaissance role, however.

In April 1945 there was a brief resurrection of the Do 217 in the bomber role. A dozen of these aircraft, operated by the Versuchs-kommando (test unit) of KG 200, delivered attacks on bridges over the Oder River with Hs 293 glider bombs. Although the raiders claimed some hits, the action appears to have done little to slow the Red Army's steamroller advance on Berlin.

Dr Alfred Price

Do 217 V prototypes

The Dornier Do 217 was designed to meet a 1937 specification from the German Air Ministry, for a larger and more capable aircraft to replace the twin-engined Dornier Do 17 in the bomber and reconnaissance roles. The requirement called for an aircraft with improved load-carrying and performance characteristics, capable of delivering bombs in both the horizontal and the dive attack modes.

At that time the German Air Ministry demanded that all its new bomber types be able to make steep diving attacks, at angles of 60° or greater. This caused enormous problems for the aircraft's designers, yet this was no mere whim. Until the advent of the guided missile, the steep diving attack was by far the most accurate method for attacking defended targets. This would be borne out by the formidable reputation gained by the Junkers Ju 87 Stuka and the Ju 88 during the early war years. That formula worked well enough for the Ju 87 with a loaded weight of 9,500 lb (4300 kg) or the Ju 88 with a loaded weight of 27,000 lb (12250 kg). It remained to be seen whether it would work when applied to the Do 217 with a loaded weight of 33,000 lb (15000 kg).

Detailed design work on the Do 217 began in mid-1937. The aircraft that emerged was essentially a larger and much heavier version of the Do 17, with a redesigned internal structure and more powerful engines. The bomb load was increased from the Do 17's 2,200 lb (1000 kg) to 6,600 lb (3000 kg), the wingspan was extended by about 3 ft 4 in (1 m) and the fuselage was lengthened by nearly 8 ft (2.40 m). As with the Do 17, the crew of four was accommodated close together in the nose section. To satisfy the dive-bombing requirement, the extreme rear section of the fuselage opened out to produce a cruciform air brake to slow the aircraft in during its dive.

Do 217 V1
The first prototype, the Do 217 V1 (V for *Versuchs* – test) first flew in August 1938, powered by two 1,075-hp (802-kW) Daimler-Benz DB 601A liquid-cooled inline engines. Preliminary flight tests revealed poor directional stability and sluggish control response. In the following month, while flying at low altitude on one engine, the prototype went out of control and crashed. Both crewmembers were killed.

Do 217 V2 and V3
By the time of the crash of the V1, the Do 217 programme was considered too important to allow the accident to hold up the development of the new bomber for long. Before the end of the year two

Do 217 V1

- D/F loop
- DB 601A liquid-cooled in-line engines
- three-bladed propellers
- tail-mounted air brake (stowed)

Do 217 V7

- BMW 139 air-cooled radial engines
- provision for defensive armament
- four-bladed propellers
- air brake deleted

further Do 217 prototypes, the V2 and V3, were flying. In place of the DB 601 engines of the V1, these aircraft were powered by 950-hp (708-kW) Junkers Jumo 211A engines. One of these prototypes was employed in tests with the novel tail air-brake system, but these ended prematurely after the discovery that the aircraft had been overstressed during the pull-out manoeuvre. The stringers of the rear fuselage were distorted and the stressed skin was buckled in places.

Do 217 V4
Engineers at Dornier worked to cure the faults revealed during the test programme. One problem concerned the dangerous tendency of the tail fins to stall if the aircraft was yawed at low speed – a condition likely to occur during a landing approach on one engine. A fixed slot built into the leading edge of each fin cured the problem. This and other changes were incorporated in the fourth prototype (Do 217 V4).

This aircraft first flew early in 1939. Otherwise similar to the V2 and the V3, it was the first variant to be armed, carrying an MG 15 7.9-mm machine-gun in a flexible mounting firing forwards, and similar weapons in the ventral and dorsal gun positions firing rearwards.

Do 217 V5, V6 and V1E (V1 replacement airframe)
These machines all appeared in the summer of 1939. These aircraft

Finished overall in RLM 05 creme, D-AMSD was the 4th prototype and, like the V2 and V3, was powered by Junkers Jumo 211 engines. Though intended as a production prototype, the V4 had an insufficient bomb load capacity and poor performance.

incorporated further design changes to improve handling, but reverted to the original DB 601A engines. These aircraft were used to test handling while carrying various internal and external loads, and served as prototypes of the Do 217A-0 and C-0 pre-production machines (see below). By now it had become clear that with the engine types used so far, the Do 217 was underpowered and was not going to meet its specified performance requirements.

Do 217 V7 and V8
To overcome the shortage of power of the

earlier prototypes, these two prototypes were fitted with the new BMW 139 air-cooled radial engine rated at 1,550 hp (1156 kW). These aircraft first flew late in 1939, but although the new engine gave a useful increase in power it was prone to overheating and was unreliable.

Do 217 V9
This aircraft flew for the first time early in 1940. It was powered by the BMW 801 air-cooled radial engine rated at 1,580 hp (1179 kW), which offered slightly greater power than the BMW 139 but was also

Below: The Do 217 V11 was fitted with an additional pair of air brakes were mounted between the fuselage and each engine nacelle. These jammed during a test dive, causing the aircraft to crash, killing the crew.

Above and left: The cruciform dive brake fitted in the extreme tail of the Do 217 is shown here in the open and closed positions. During test flights, the pull-out from the steep dive attack was liable to lead to overstressing of the aircraft's rear fuselage. Though production aircraft were equipped with the brake, these were wired shut and never used.

significantly more reliable. Another important innovation incorporated into this aircraft was the redesigned fuselage, with deeper cross-section to incorporate a larger bomb bay. The V9 served as the basis of the Do 217E series of aircraft, the first variant to enter large-scale production.

Do 217 V11
Similar to the V9, in an effort to get an effective dive-bombing capability this prototype aircraft was fitted with an extra pair of air brakes comprising four parallel bars mounted between the fuselage and each engine nacelle. When not in use, the air brakes were locked in the normal, low-drag position. On entering the dive, the air brakes rotated through 90° to the high-drag position. To compensate for the resultant nose-down trim change, the elevator trim

tabs automatically applied a correcting force. This aircraft was lost when, following a test dive, the pilot returned the air brakes to the low-drag position but the trim tabs remained jammed in the 'up' position. The nose of the bomber rose sharply and it stalled and crashed. The experiment was not repeated.

Do 217 V13 and V14
Two Do 217Ms employed to flight test turbocharged DB 603 engines; plans to build a production variant so-equipped did not proceed.

Two of the Do 217 prototypes, including V7 (pictured), were powered by a pair of fan-cooled BMW 139 radials; these drove four-bladed propellers.

Do 217A

During 1940, in an effort to accelerate the development of this aircraft following the outbreak of war, the German Air Ministry instructed Dornier to build eight pre-production **Do 217A-0** reconnaissance aircraft based on the V5 and powered by DB 601A engines. These aircraft were similar to the early prototypes, apart from the under-fuselage bulge extending almost to the trailing edge of the wing to house a pair of vertically mounted cameras. The defensive armament comprised 7.9-mm MG 15 machine-guns in the nose, dorsal and ventral positions. In the summer of 1940 the Do 217A-0s were delivered to the secret reconnaissance unit subordinated directly to the Luftwaffe High Command (Aufklärungsgruppe der Oberkommando der Luftwaffe).

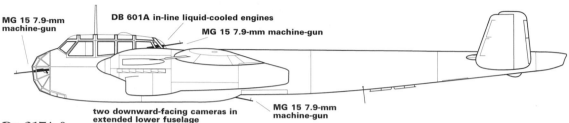

MG 15 7.9-mm machine-gun

DB 601A in-line liquid-cooled engines

MG 15 7.9-mm machine-gun

two downward-facing cameras in extended lower fuselage

MG 15 7.9-mm machine-gun

Do 217A-0

Below: Eight Do 217A-0s entered service with the Aufklärungsgruppe der Oberkommando der Luftwaffe late in 1940, flying clandestine photographic reconnaissance missions over the Soviet Union and Soviet-occupied Poland in preparation for the June 1941 invasion of the Soviet Union.

Above: This close-up of a Do 217A-0 shows the bulged fuselage housing the two vertically-mounted reconnaissance cameras.

Do 217C

Built in parallel with the Do 217A-0s, the Do 217 V1 and four **Do 217C-0** aircraft were pre-production bombers powered by DB 601A engines. These aircraft carried a defensive armament of five 7.9-mm machine-guns and one forward-firing 15-mm MG 151 cannon. There was provision to carry a bomb load of 6,600 lb. (3000 kg). The C-0s did not enter front-line service, and instead were employed in testing engines and other systems for the later Do 217 bomber variants.

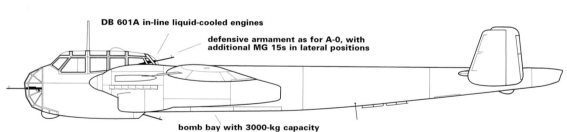

DB 601A in-line liquid-cooled engines

defensive armament as for A-0, with additional MG 15s in lateral positions

bomb bay with 3000-kg capacity

Do 217C-0

Left: This line-up of Dornier types, possibly at Lowenthal during 1939/40, includes an early Do 17 (nearest the camera) and 'CN+HL' – the Do 217 V1, which preceded another four Do 217C-0 pre-production bomber aircraft.

Do 217E

The first variant to enter large-scale production was the Do 217E, based on the V9 prototype. After a batch of pre-production Do 217E-0s was completed in the autumn of 1940, production shifted to the E-1 version, then the sub-types E-2 through E-5.

Do 217E-0
Pre-production bomber and reconnaissance variant based on the V9 prototype, powered by two 1,580-hp BMW 801MA engines. These aircraft carried a defensive armament of five 7.9-mm MG 15 machine-guns and one forward-firing 15-mm MG 151 cannon. The maximum bomb load was 8,800 pounds (4000 kg), of which 3,300 lb (1500 kg) was carried externally.

Do 217E-1
First full-scale production version of the Do 217, similar to E-0. This variant was not fitted with a dive brake. Deliveries started late in 1940. Number built: 94

Do 217E-2
Intended as dive-bomber, this variant actually followed the E-3 in production in 1941. Powered by BMW 801ML engines. The defensive armament comprised one 15-mm MG 151 cannon and a 7.9-mm MG 15 firing forwards, one 13-mm MG 131 machine-gun in the dorsal turret and another in the ventral position, and one MG 15 firing from each side of the cabin. Number built: 185

Do 217E-3
Following the operational experience with the E-1, this variant featured additional

armour protection for the pilot and dorsal gunner. Defensive armament as for the E-2, except that the forward firing cannon was upgraded to a 20-mm MG FF

Do 217E-4
Replaced the E-2 on the production line at the end of 1941. It was generally similar, but the troublesome dive brake installation had been deleted. Balloon cable-cutters were fitted along the leading edge of each wing. Number built (E-3 and E-4): 258

Do 217E-5
Similar to the E-4, but with provision to carry a single Henschel Hs 293A glider bomb, or an external fuel tank, under each outer wing panel. During operations, the normal configuration of this aircraft was a missile under the starboard wing counterbalanced by a drop tank under the port wing. Number built: 70

Rustätze field conversion kits
Greatly improving the versatility of the Do 217, at least 28 of these 'retrofit' kits were developed, mostly for the Do 217E though, in many cases these were also applicable to variants that followed. The most significant conversons are listed here; noted in brackets are the variants to which the conversion kit could be applied.

R1 (Do 217E-2, E-3)
Special carrier for SC1800 1800-kg (3,970-lb) bomb with annular fin

R2 (Do 217E-2, E-3)
External racks to fit under the outer wing panels, each to take one SC250 250-kg (550-lb) bomb

R4 (Do 217E-1, E-2, E-3, E-4, K-1)
Rack for a single LTF5 or F5B torpedo in the bomb bay. The installation included depth-setting and gyro-angling controls in the cockpit, and a system to provide heating for the bomb compartment

R5 (Do 217E-2, E-3)
Modification to fit a 30-mm Rheinmetall MK 101 cannon in the fuselage, firing forwards

R6 (Do 217E-1, E-2, E-4, K-1, M-1)
Modification to install reconnaissance cameras in the bomb bay

R7 (Do 217E-1, E-2, E-4, K-1)
Modification to install a four-man rubber dinghy in the rear fuselage aft of the wing

R8 (Do 217E-1)
Modification to install a 165-Imp gal (734-litre) auxiliary fuel tank in the forward end of the bomb bay

R9 (Do 217E-1)
Modification to install a 165-Imp gal (734-litre) auxiliary fuel tank in the aft end of the bomb bay

R10 (Do 217E-2, E-4, K-1)
Modification to install a rack to carry one Hs 293 glider bomb under each outer wing panel

R13 (Do 217E-2, E-4, K-1)
Alternative installation to fit an auxiliary

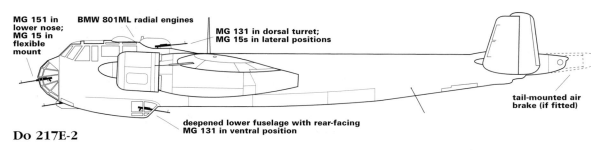

MG 151 in lower nose; MG 15 in flexible mount
BMW 801ML radial engines
MG 131 in dorsal turret; MG 15s in lateral positions
tail-mounted air brake (if fitted)
deepened lower fuselage with rear-facing MG 131 in ventral position

Do 217E-2

Above: The victory tally (including two factories, three barrage balloons and two aircraft) on the tail of this II./KG 40 Do 217E is typical of an aircraft that had taken part in nuisance daylight raids on targets in Great Britain. This view shows to advantage the fixed slot, built into the leading edge of the fin, to reduce the risk of the fin stalling if yaw was applied inadvertently while flying at low speed.

tank in the forward end of the bomb bay

R14 (Do 217E-2, E-4, K-1.
Alternative installation to fit an auxiliary tank in the aft end of the bomb bay

R15 (Do 217E-4, K-2)
Modification to install an Hs 293 glider bomb under each wing between the fuselage and the engine

R17 (Do 217E-4, K-2)
Modification to install a 255-Imp gal (1135-litre) auxiliary fuel tank in the forward end of the bomb bay

R19 (Do 217E-2, E-4, K-1, K-2, M-1, M-11)
This defensive system comprised a replacement tail cone for the Do 217, containing four 7.9-mm MG 81 machine-guns in a fixed installation pointing

Above: Do 217E-0 WNr 0800012 was the 12th pre-production aircraft. A number of Do 217E-0s were fitted with cameras and issued to a photo-reconnaissance unit based in Romania and engaged in reconnaissance of the Soviet border.

Below: A dedicated dive-bomber variant, the BMW 801ML-powered Do 217E-2 had an improved defensive armament fit, including an electrically-operated dorsal turret mounting a 13-mm MG 131 machine-gun with 500 rounds.

rearwards, each with 200 rounds of ammunition. The guns were fired electrically, by a button on the pilot's control column. Aiming was by means of a rearward-pointing periscope, which protruded from the cockpit roof, and extended downwards just forward and to the right of the pilot's face. The image was presented inverted: if the target was above the line of flight, it appeared below the centre of the sighting reticule; if to port, it appeared to the right of the reticule. Thus to aim the weapons the pilot had to steer the aircraft in the direction of the target. The system was used in action, but apart from its scare value it achieved little

R21 (Do 217E-3, E-4, K-1)
Modification to install a device to jettison auxiliary fuel tanks mounted externally

R22 (Do 217N-2)
Schräge Musik upward-firing cannon installation in the rear fuselage comprising four MG 151/20 cannon inclined at 70° to the aircraft's thrust line.

R25 (Do 217E-2, E-4, K-1, K-2, M-1, M-11 and P)
Modification to install braking parachute, to provide rapid deceleration to balk attacks by enemy fighters

Below: The R25 field modification kit provided for the installation of a tail-mounted fabric braking parachute. Not to be confused with the earlier unsuccessful dive-brake installation, the parachute could be opened and closed in flight and was intended to proved rapid deceleration to disrupt fighter attacks.

Above: This view of a Do 217E-2 with its bomb doors open, shows the unusual 'double fold' method employed. Had only a single fold been used, the doors would have served as additional keel surfaces, reducing manoeuvrability during a bomb run.

Above: The Do 217E-4 differed from earlier variants in being powered by BMW 801C engines in place of the similarly-rated BMW 801MA and ML. The E-4 was also fitted with a Kuto-Nase balloon cable-cutting device, applied to the leading edge of the wing.

Above: Carrying the markings of 9./KG 2, this Do 217E-4 is fitted with the tail-mounted gun installation comprising two sets of paired MG 81Z 7.9-mm machines. So equipped, this aircraft became a Do 217E-4/R19; note the pilot's periscopic sight above the cockpit.

Below: In many ways typical of German bomber cockpits of the period, that of the Do 217 provided its pilot with an excellent all-round view. Few German bombers had dual controls, though the aircraft could be flown by the observer in an emergency using the 'swing over' control yoke.

Henschel Hs 293 glider bomb

The Henschel Hs 293 glider-bomb was intended mainly for use against lightly-armoured warships and freighters. It resembled a miniature monoplane with a wingspan of just over 10 ft (3.05 m) and a 1,100-lb (500-kg) high-explosive warhead in the nose. After launch, the 1,300-lb (5.8-kN) thrust Walter HWK 109-507 liquid fuel rocket accelerated the missile to a speed of 370 mph (595 km/h) in 12 seconds. Then, with the fuel exhausted, the missile coasted in a shallow dive towards its target. The range of the weapon depended upon its altitude at release; typically, this was about 5 miles (8 km) if the launching aircraft was flying at about 4,500 feet (1400 m). A flare mounted on the tail of the missile enabled the navigator/bomb aimer to follow its path; he operated a small 'joy stick' controller which keyed the command transmitter with the required 'up-down-left-right' correction signals, which were radiated to the receiver in the glider-bomb. The crewman steered the missile's tracking flare until it appeared to be superimposed on the target, and held it there until the missile impacted.
Specification: weight at launch: 1,990 lb (902 kg); span: 10 ft 3½ in (3.14 m), length 11 ft 9 in (3.58 m).

Above: This close-up of an Hs 293 glider bomb shows the pylon fitted to the Do 217E-5 to allow carriage of the weapon. The duct near the front of the pylon carried hot air from the parent aircraft's de-icing system to the weapon's guidance and control systems, to prevent them freezing prior to launch.

Below: This Do 217E-5 of II./KG 100 is pictured with an Hs 293 glider bomb under its starboard wing. A drop tank mounted under the port wing served as a counterweight.

Ruhrstahl FX 1400 Fritz-X guided bomb

This weapon was intended for use against heavily-armoured naval targets such as battleships or cruisers. The Fritz-X looked like a normal bomb, with the addition of four stabilising 'wings' mounted mid-way along its body. The weapon weighed just under 3,500 lb (1588 kg), with a thick armour-piercing casing holding 660 lb (300 kg) of high explosive. Released from altitudes around 23,000 ft (7000 m), the unpowered weapon accelerated under gravity to reach an impact velocity close to the speed of sound. The Fritz-X was aimed like a normal bomb using the bombsight, and the bomb aimer transmitted the radio command signals to correct its trajectory only during the final part of the missile's fall.
Specification: weight: 3,462 lb (1570 kg); span (over 'wings'): 4 ft 4 in (1.35m); length 10 ft 8 in (3.26 m).

Left: This British Air Ministry illustration shows the arrangement of the Do 217E-3's defensive armament, including its ammunition supplies and sighting equipment. Forward-firing armament consisted of a fixed MG 151 15-mm cannon in the lower nose and an MG FF 20-mm cannon on a flexible mount. MG 131 13-mm machine-guns were installed in the dorsal turret and rear-facing ventral position, while MG 15 7.9-mm machine-guns occupied lateral positions to the rear of the cockpit.

Below: The addition of an electrically-operated EDL 131 turret in the dorsal position made the aircraft less vulnerable to attacks from above and behind. Its MG 131 13-mm gun was supplied with 500 rounds.

Do 217E defensive armament

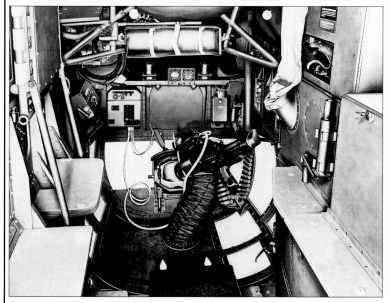

Below: This view shows that part of the aircraft's interior immediately below the dorsal turret. The gunner's seat has been removed.

Above and below: As may be appreciated in these views, the ventral gunner had a somewhat limited field of fire. His MG 131 was supplied with 1,000 rounds of ammunition, spent shells being ejected through the floor via the trunking shown. Just visible at the top of the above photograph is the dorsal gunner's seat mounting.

Right: The twin MG 81Z four-gun tail 'stinger' installation was of dubious value, its sighting system including a periscope which presented an inverted image to the pilot, who needed to hold his aircraft steady (when evasive action may have seemed more appropriate, not to mention attractive!) before firing the guns via a control column-mounted button. Used operationally, the system may have scared off attackers, if not actually causing significant damage to their aircraft.

Do 217H

Conversion of the 21st production Do 217E-1, the sole **Do 217H V1** was equipped with DB 601 engines fitted with experimental turbo-superchargers. The work was carried out by Daimler-Benz at Erprobungsstelle Echterdingen in September 1941 and high-altitude testing was conducted. The variant did not go into production.

Do 217J

In early 1942 a number of developments in RAF Bomber Command, including the introduction of the Avro Lancaster heavy bomber, the first use of the Gee navigation aid on a large scale and the first 'thousand-bomber' raid on Germany, put considerable pressure on the Nachtjagdflieger. With supplies of Junkers Ju 88C-6b night-fighters in short supply and the less than ideal Bf 110 also unavailable in the numbers required, an alternative basis for a night-fighter aircraft was sought. Having already had some success with conversions of Do 17Z and Do 215 aircraft, it was suggested that the Do 217E-2 might lend itself to conversion, despite its size and weight, in the interim. Two variants based on the Do 217E-2 followed.

Do 217J-1
Do 217E-2 converted for the night-intruder role, with a crew of three. This had the glazed nose replaced with a solid fairing housing a battery of four 20-mm MG FF cannon and four 7.9-mm MG 17 machine-guns. The dorsal and ventral gun positions were retained, each with a 13-mm MG 131 machine-gun. The aircraft could also carry eight 110-lb (50-kg) SC 50X bombs in its aft bomb bay; the forward bay carried a 255-Imp gal (1159-litre) auxiliary fuel tank.

Do 217J-2
Similar to J-1, but optimised for night air defence operations. The aft bomb bay was deleted and the aircraft was fitted with FuG 202 Lichtenstein BC airborne interception radar. Number built (J-1 and J-2): 130

Do 217J-2
FuG 202 radar set — **MG 131 in dorsal and ventral positions** — **Do 217E airframe (BMW 801 engines)**

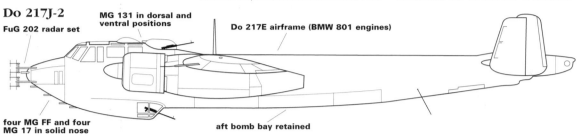

four MG FF and four MG 17 in solid nose — **aft bomb bay retained**

Below: The Do 217J was far from ideal for the night-fighter role, being much heavier than both the Bf 110 and Ju 88.

Above: The Matratzen ('mattress') array was used in association with the Do 217J-2's FuG 202 interception radar.

Do 217K

With the Luftwaffe's conventional bombing campaign increasingly restricted to the hours of darkness, Dornier redesigned the Do 217 so that it was better suited to the role. A continuously glazed, streamlined nose replaced the earlier stepped design, though the crew of four remained in the forward fuselage and the remainder of the aircraft's layout was as for the Do 217E. BMW 801D engines replaced the 801Cs of the late production Do 217E.

Do 217K V1
This aircraft was converted from an E-series aircraft and fitted with a redesigned forward fuselage with a glazed and rounded nose. For a short time it was fitted with a single fin and rudder.

Do 217K V2
As above, but retained the twin fins and rudders of the E-series aircraft. Served as the prototype for the Do 217K.

Do 217K V3
Similar to the V2. After its initial flight trials, this aircraft was modified to serve as a launching platform for the DFS 228 rocket-powered high-altitude reconnaissance aircraft; the latter was carried on a rigid mounting above the Dornier.

Do 217K-1
Night bomber fitted with the redesigned and glazed forward fuselage tested on the K-series prototypes. After the first production example flew on 31 March 1942, the variant replaced the E-series in production from September. Powered by two 1,700-hp (1268-kW) BMW 801D engines, the K-1 had a crew of four, its defensive armament comprising two 13-mm MG 131 and four (or six) 7.9-mm MG 81 machine-guns; the aircraft's maximum bomb load was 8,800 lb (4000 kg). Number built: approximately 300.

Do 217K-2
Similar to the K-1, the K-2 was intended to carry two FX 1400 Fritz-X guided bombs under the wings inboard of the engines, on ETC 2000/XII pylons. This variant had a mainplane of longer span – 81 ft 4⅓ in (24.8 m) compared with 62 ft 4 in (19 m) – and greater area to improve high-altitude performance and compensate for the weight of the Fritz-X weapons. Other equipment included FuG 203a *Kehl I* transmitter for the guided bomb's FuG 230a *Strassburg* receiver. Number built: approximately 40.

Do 217K-3
Similar to K-2, but fitted with either FuG 203c or 203d *Kehl IV* transmitter equipment to allow launch of either FX 1400 or Hs 293A. Number built: 40

Do 217K-1
MG 81 (later MG 81Z) in lateral positions — **Do 217E airframe aft of cockpit**

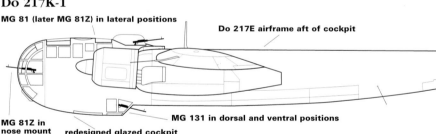

MG 81Z in nose mount — **redesigned glazed cockpit** — **MG 131 in dorsal and ventral positions** — **tail-mounted air brake (if fitted)**

Above: Though the significant redesign of the Do 217's forward fuselage allowed for a much roomier cockpit, there was little performance gain made with the more aerodynamic design. This example is finished in an experimental anti-shipping paint scheme.

As well as the transceiver antennas associated with the Fritz-X guided bombs it was intended to carry, the Do 217K-2 was equipped with long-span wings. This example is also fitted with the periscopic sighting equipment for tail-mounted, rearward-facing MG 81Z machine-guns.

Left: The cockpit layout of the Do 217K was markedly different to that of the earlier aircraft, with a clear emphasis on an unobstructed forward view. Prominent in this view of a K-1 cockpit is the MG 81Z twin 7.9-mm machine-gun mount to the right. The pilot's flight instruments are grouped on two small panels mounted on the left side of the cockpit. The sight mounted on the cockpit ceiling above the pilot's controls is the Stuvi 5B dive-bombing sight. The pilot's seat has been removed for clarity.

Above: A barrage balloon cable-cutter fits around the leading edge of the cockpit glazing.

Do 217L

Do 217L V1 and V2
These two prototypes appeared in spring 1943 and underwent testing at Dornier's Löwenthal plant. Based on the Do 217K, little is known of these machines other than that they had a rearranged cockpit and revised defensive armament. The Do 217L did not enter production.

Do 217M

Built alongside the Do 217K and entering service at much the same time, the Do 217M was powered by a pair of Daimler-Benz DB 603A engines as 'insurance' against production delays caused by shortages of BMW 801Ds.

Do 217M-1
Similar to the K-1, but powered by a pair of 1,750-hp (1305-kW) DB 603A in-line engines, the M-1 went into production in late 1942, the first example making its first flight on 16 July 1942. Number built: approximately 440

Do 217M-5
Similar to M-1, the M-5 had provision to carry a single Hs 293 glider bomb, semi-recessed beneath the fuselage. Ground clearance was poor, however, and it did not go into production.

Do 217M-11
Derived from the M-1, the M-11 incorporated the long-span wing of the K-2 and had provision to carry either one FX 1400 Fritz-X or one Hs 293 semi-recessed beneath the fuselage, as on the M-5. Not produced in quantity; only about 37 are believed to have been completed.

DB 603A engines with four-bladed VDM propellers

airframe and armament as for Do 217K

Do 217M-1

Do 217M production was drastically curtailed by the urgent need for night-fighters in the last months of Do 217 production. Consequently many Do 217Ms were completed as Do 217Ns and few standard Do 217M-1s, like this example finished in an overall black night bomber scheme, reached bomber units.

Do 217N

A development of the Do 217M-1 with the solid 'gun nose' of the Do 217J-2, the Do 217N flew for the first time on 31 July 1942. The N-1 variant entered service in April 1943.

Do 217N-1
Modification of the Do 217M-1 for the night-fighter role. Similar in appearance to the J-2, except for the Daimler-Benz DB 603A engines. Armament comprised four 20-mm MG 151/20 cannon and four 7.9-mm MG 17 machine-guns in the nose, and two 13-mm MG 131 machine-guns for rear defence. Some N-1s were fitted with the FuG 202 Lichtenstein airborne interception radar and many had the dorsal and ventral gun positions replaced with streamlined fairings. This work was carried out at Luftwaffe repair depots, modified aircraft being known by the designation

Do 217N-1/U1. Some of these aircraft were fitted with two or four 20-mm MG 151/20 cannon firing upward at 70° to

The first production Do 217N-2 awaits installation of its FuG 202 radar antennas. Despite attempts to reduce its weight, the Do 217N remained unpopular with crews, chiefly because of its lack of manoeuvrability. Its chief advantage was that it was exceptionally well-armed.

the horizontal, in a so-called *schräge Musik* installation to engage night bombers from below. These machines were designated

Do 217N-1/U3. Do 217N-1 production totalled approximately 240.

Do 217N-2
Refined version of N-1, with same forward-firing armament but with the rearward-facing weapons deleted and replaced by streamlined fairings. Fitted with the FuG 202 or 212 airborne interception radar, replaced in some cases (it is believed) by the later FuG 220 SN-2 radar. Some of these aircraft fitted with two or four 20-mm MG 151/20 cannon in *schräge Musik* installation, designated **Do 217N-2/R22**. Number built: approximately 95.

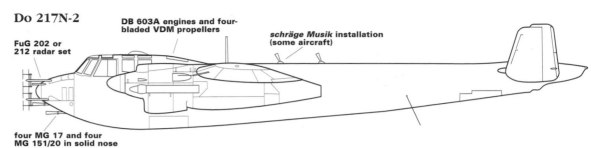

Do 217N-2

FuG 202 or 212 radar set

DB 603A engines and four-bladed VDM propellers

schräge Musik installation (some aircraft)

four MG 17 and four MG 151/20 in solid nose

Do 217P

Dornier adapted the basic Do 217 airframe for high-altitude bombing and reconnaissance as the Do 217P, building three prototypes and three pre-production aircraft before the Luftwaffe elected not to order the type into full-scale production

Do 217P V1
Three-seat high-altitude reconnaissance-bomber based on E-2, but featuring the new nose-section. Powered by two 1,750-hp (1305-kW) DB 603B engines, supercharged by a 1,400-hp (1044-kW) DB 605T engine buried in the fuselage. Began flight tests in June 1942. Attained altitude of 43,965 ft (13400 m) during test flying and later reached 52,597 ft (16032 m) service ceiling.

Do 217P V2 and V3
Similar to the V1, but with the increased span wing fitted to the Do 217K-2.

Do 217P-0
Three-seat high-altitude reconnaissance-bomber based on the Do 217P V2 prototype. Three pre-production P-0 aircraft built, carrying a defensive armament of two forward-firing and four aft-firing 7.9-mm MG 81 machine-guns. The aircraft also had provision to carry one 1,100-lb (500-kg) bomb under each outer wing section. Plans to produce this variant were abandoned in 1943.

In this view of the Do 217 V1 the underwing radiators for the DB 605T booster engine are visible, along with its bulged housing under the rear fuselage. The main DB 603B engines have deeper, squarer intakes and drive large broad-chord propellers. The Do 217P V2 and V3 differed in having longer wings.

pressurised cockpit cabin

MG 81Z in nose mount

supercharged DB 603B engines

DB 605T booster engine in fuselage

Do 217P-0

MG 81Z in dorsal and ventral positions

single Rb 20/30 and two Rb 75/30 cameras in lower fuselage

Do 317/Do 217R

In July 1939 the Luftwaffe issued its 'Bomber B' specification to selected manufacturers, calling for an advanced high performance twin-engined medium bomber. The aircraft was to have a pressurised cabin for high-altitude operations, with the defensive armament mounted in remotely controlled barbettes. A maximum speed of over 370 mph (595 km/h) was required, with a range of over 2,200 miles (3540 km) and the ability to carry a bomb load of up to 4,400 lb (1996 kg). During 1940, the Focke Wulf Fw 191 and the Junkers Ju 288 designs were chosen over the Do 317. Work continued on the latter at low priority, however, as insurance in case the winning designs ran into difficulties. Dornier offered two versions of the Do 317 to the Luftwaffe: the **Do 317A**, without the pressurised cabin and other refinements; and the **Do 317B** with the pressurised cabin and barbette armament. The Do 317B was to be powered by two Daimler Benz DB 610 engines, each rated at 2,870 hp (2140 kW) for take-off.

The **Do 317V1**, which first flew in 1943, came some way between the two

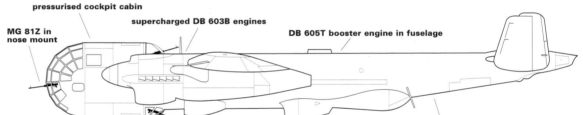

DB 603A engines

new all-metal two-spar wing

slotted ailerons interlinked with electric split flaps

redesigned twin tailplane

pressurised cockpit cabin

Do 317 V1

proposed variants. It featured a pressurised cabin, but it lacked any defensive armament. Power came from two Daimler Benz DB 603 engines similar to those fitted to the Do 217M. It closely resembled the latter type, apart from the revised tail unit with triangular fins and rudders. The Do 317 V1 did not perform impressively during trials, however, and at the end of 1943 the Do 317 programme was abandoned. By then work was well advanced on four Do 317A prototypes, V2 through V5, and these aircraft were completed. Redesignated as **Do 217R**s, they carried underwing racks for an Hs 293

glider bomb. These aircraft were issued to III./KG 100 in the summer of 1944, but they arrived shortly before the unit disbanded and it is doubtful whether any of them flew an operational sortie.

Below and below left: Intended as 'insurance' against the failure of Focke-Wulf and Junkers 'Bomber B' prototypes, the Do 317 V1 was the only example completed. Apart from its distinctive triangular-shaped tailfins, the Do 317 resembled the Do 217M.

Do 417

Yet another derivative of the Do 217, the Do 417 bomber was the subject of some study but failed to leave the drawing board. Radically different from its predecessors, the Do 417 was envisaged with a single fin and rudder and remotely-controlled armament. The **Do 417A** would have had BMW radial engines, while the **Do 417B** was to have been powered by Daimler-Benz DB 603s.

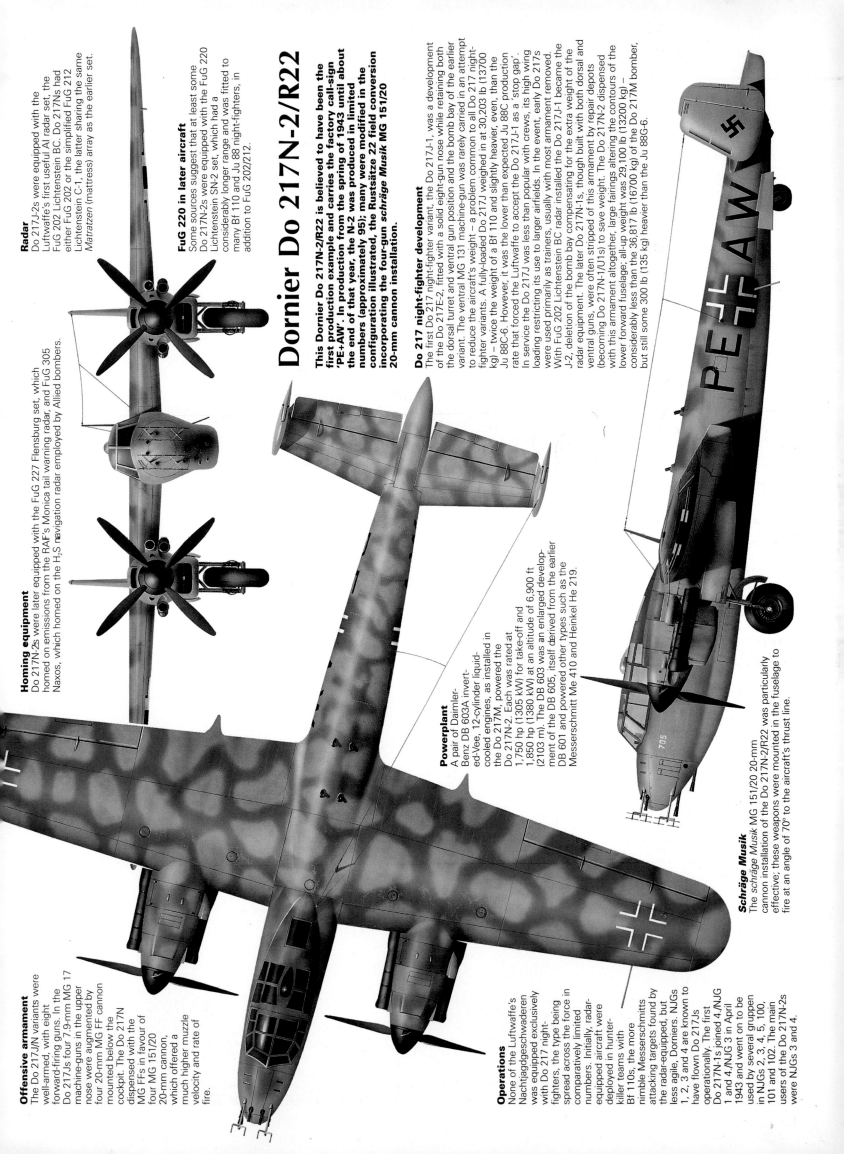

Offensive armament

The Do 217J/N variants were well-armed, with eight forward-firing guns. In the Do 217Js four 7.9-mm MG 17 machine-guns in the upper nose were augmented by four 20-mm MG FF cannon mounted below the cockpit. The Do 217N dispensed with the MG FFs in favour of four MG 151/20 20-mm cannon, which offered a much higher muzzle velocity and rate of fire.

Operations

None of the Luftwaffe's Nachtjagdgeschwaderen was equipped exclusively with Do 217 night-fighters, the type being spread across the force in comparatively limited numbers. Initially, radar-equipped aircraft were deployed in hunter-killer teams with Bf 110s, the more nimble Messerschmitts attacking targets found by the radar-equipped, but less agile, Dorniers. NJGs 1, 2, 3 and 4 are known to have flown Do 217Js operationally. The first Do 217N-1s joined 4./NJG 1 and 4./NJG 3 in April 1943 and went on to be used by several gruppen in NJGs 2, 3, 4, 5, 100, 101 and 102. The main users of the Do 217N-2s were NJGs 3 and 4.

Radar

Do 217J-2s were equipped with the Luftwaffe's first useful AI radar set, the FuG 202 Lichtenstein BC. Do 217Ns had either FuG 202 or the simplified FuG 212 Lichtenstein C-1, the latter sharing the same *Matratzen* (mattress) array as the earlier set.

FuG 220 in later aircraft

Some sources suggest that at least some Do 217N-2s were equipped with the FuG 220 Lichtenstein SN-2 set, which had a considerably longer range and was fitted to many Bf 110 and Ju 88 night-fighters, in addition to FuG 202/212.

Homing equipment

Do 217N-2s were later equipped with the FuG 227 Flensburg set, which homed on emissions from the RAF's Monica tail warning radar, and FuG 305 Naxos, which homed on the H₂S navigation radar employed by Allied bombers.

Dornier Do 217N-2/R22

This Dornier Do 217N-2/R22 is believed to have been the first production example and carries the factory call-sign 'PE+AW'. In production from the spring of 1943 until the end of that year, the N-2 was produced in limited numbers (approximately 95); many were modified in the configuration illustrated, the *Rustsätze 22* field conversion incorporating the four-gun *schräge Musik* MG 151/20 20-mm cannon installation.

Do 217 night-fighter development

The first Do 217 night-fighter variant, the Do 217J-1, was a development of the Do 217E-2, fitted with a solid eight-gun nose while retaining both the dorsal turret and ventral gun position and the bomb bay of the earlier variant. The ventral MG 131 machine-gun was rarely carried in an attempt to reduce the aircraft's weight – a problem common to all Do 217 night-fighter variants. A fully-loaded Do 217J weighed in at 30,203 lb (13700 kg) – twice the weight of a Bf 110 and slightly heavier, even, than the Ju 88C-6. However, it was the lower than expected Ju 88C production rate that forced the Luftwaffe to accept the Do 217J-1 as a 'stop gap'. In service the Do 217J was less popular with crews, its high wing loading restricting its use to larger airfields. In the event, early Do 217s were used primarily as trainers, usually with most armament removed. With FuG 202 Lichtenstein BC radar installed the Do 217J-1 became the J-2, deletion of the bomb bay compensating for the extra weight of the radar equipment. The later Do 217N-1s, though built with both dorsal and ventral guns, were often stripped of this armament by repair depots (becoming Do 217N-1/U1s) to save weight. The Do 217N-2 dispensed with this armament altogether, large fairings altering the contours of the lower forward fuselage; all-up weight was 29,100 lb (13200 kg) – considerably less than the 36,817 lb (16700 kg) of the Do 217M bomber, but still some 300 lb (135 kg) heavier than the Ju 88G-6.

Powerplant

A pair of Daimler-Benz DB 603A inverted-Vee, 12-cylinder liquid-cooled engines, as installed in the Do 217M, powered the Do 217N-2. Each was rated at 1,750 hp (1305 kW) for take-off and 1,850 hp (1380 kW) at an altitude of 6,900 ft (2103 m). The DB 603 was an enlarged development of the earlier DB 601 and powered other types such as the Messerschmitt Me 410 and Heinkel He 219.

Schräge Musik

The *schräge Musik* MG 151/20 20-mm cannon installation of the Do 217N-2/R22 was particularly effective; these weapons were mounted in the fuselage to fire at an angle of 70° to the aircraft's thrust line.

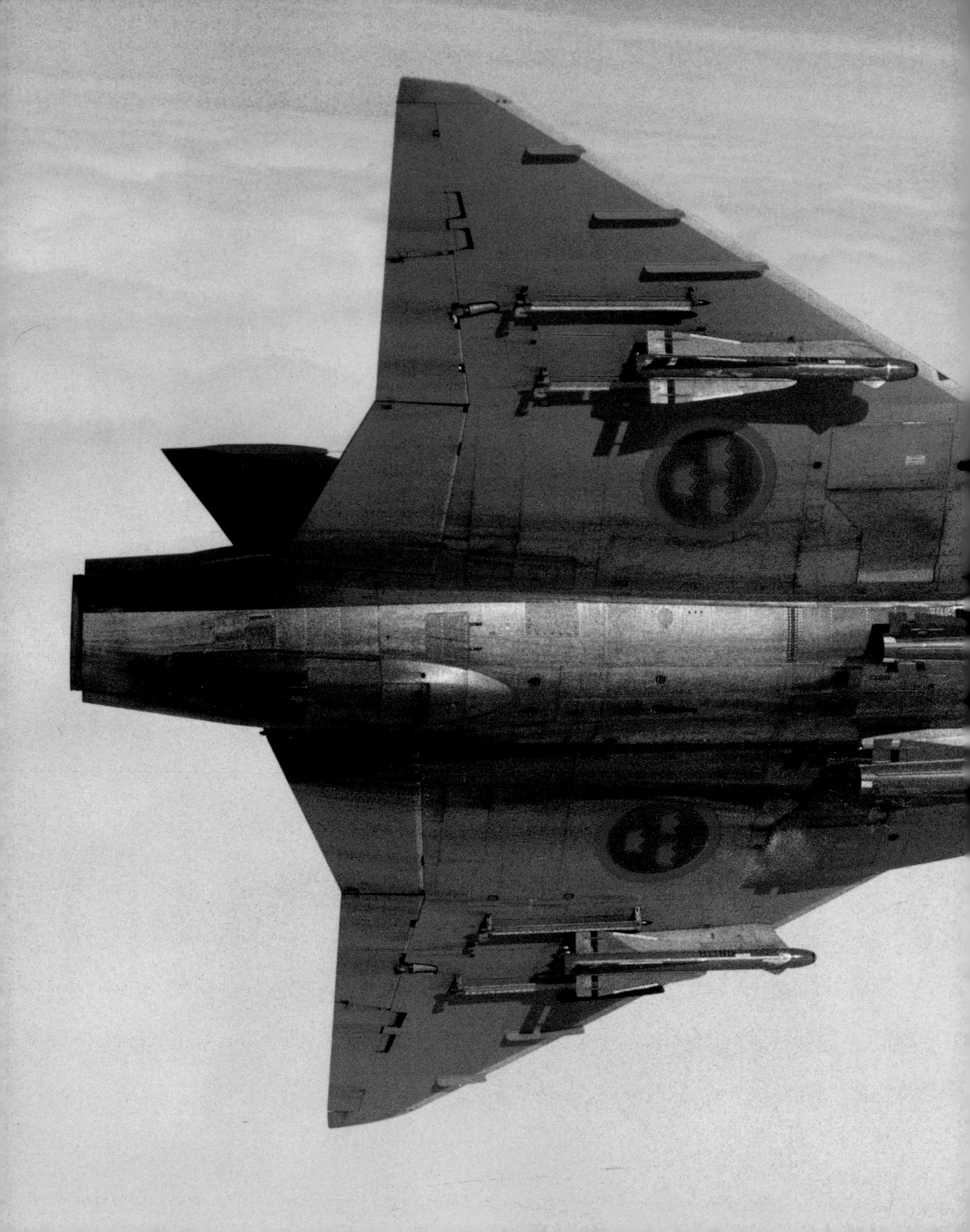

Saab 35 Draken

Sweden's double-delta dragon

During the decade that produced such important jet interceptors as France's Mirage III, the British Lightning, celebrated American 'Century Series' fighters and the Soviet MiG-21 'Fishbed', one of the most distinctive and enduring designs to emerge was the Saab Model 35. Entering service in 1960, the Draken (Dragon), as it was more popularly known, enjoyed almost 40 years service with Svenska Flygvapnet (the Swedish air force), undergoing progressive improvements to meet changing threats and gaining a deserved reputation as a 'pilot's aircraft'.

Saab's last production variant for Flygvapnet, the J 35F was the definitive Draken. Delivered between 1967 and 1972, the variant was the first to offer a truly credible air defence capability. Early examples were delivered to F13 Wing – the unit generally charged with introducing new Saab-built types into Flygvapnet service, thanks to its proximity to the Saab factory. This example, an F13 J 35F², is armed with a full complement of Rb 27/28 Falcon air-to-air missiles.

During the Saab 35 Draken's time in service, the military threat it has faced has changed constantly and, over the years, the '35 system' has been successfully adapted to match these different demands. The development of an improved engine, better missiles and sophisticated electronics have been important steps. As recently as 1980, a revolutionary modification programme was planned to re-role the Saab 35 for ground attack and would have included retractable canards mounted on the intakes, extended wingtips and advanced avionics. However, the decision was taken to buy the JAS 39 (Gripen) and the project was dropped. In the interim, 66 late F² series Drakens were modified with new wing pylons, allowing carriage of four drop tanks and two Sidewinder missiles. This modified variant was allocated a 'J' suffix in recognition of the fact that it was to be flown exclusively by the Swedish air force's F10 Wing (Flygvapnet wings being allocated callsigns according to their number and the corresponding letter of the alphabet, i.e. 10 = 'J' for JOHAN).

The Swedish air force kept the Draken on active duty until the end of the 1990s and, in 2002, the type is still flying with the Austrian air force. Thus the Draken has been in active service for over 40 years, a record for a Swedish fighter that will prove hard to beat. In 1999 the accumulated flight time by Swedish, Finnish, Danish, Austrian and American civilian-operated Drakens exceeded 1 million hours.

Draken genesis

Initial discussions regarding a fighter to replace the J 29 Tunnan in Flygvapen (Swedish air force) service began in the early autumn of 1949. The new aircraft was required to intercept small formations of nuclear-armed bombers at altitudes of around 30,000 ft (9144 m), travelling at speeds estimated to be in the transonic range (Mach 0.9). Thus, the requirement was for a day fighter with good take-off performance and supersonic climb capability.

Saab would build the new fighter as a follow-on from the Project 1100 Lansen, under the designation Project 1200; Erik Bratt – whose books *Silvervingar* (Silver Wings) and *Människor och flygande maskiner* (Men and Flying Machines) described the project's conception in 1949 – was awarded the job of Project Chief Engineer.

Project 1200 was to be a replacement for the Saab 29 and, as it was required to intercept bombers flying at approximately 1000 km/h (621 mph:540 kt) at high level, it would need to be capable of reaching a top speed of approximately 1600 km/h (994 mph:864 kt) at its operational altitude.

The initial specification requirement from the KFF (Royal Swedish Air Board) to Saab had anticipated use of the Swedish-built Dovern engine from Svenska Turbinfabriks AB (Swedish Turbine Factory Ltd) of Ljungström, which was also planned for use in the Saab 32 Lansen.

Project studies

In a memorandum from the KFF dated 21 September 1949, the new aircraft's specification began to take shape. Based on this memo, Saab completed a study on 7 December, presenting its findings to an air force conference the next day. (It is worth noting that the Saab 29 only started life in 1945; this conference was taking place just one year after its maiden flight on 1 September 1948.) The requirement now was for a supersonic missile-armed fighter that would intercept fast high-altitude bombers. At this stage, there was no experience of supersonic flying to draw upon and knowledge of supersonic flight was very limited. During this first presentation by Saab, three proposals were offered:

- A heavier development of the Saab 32 Lansen, with an uprated Dovern II engine to provide the necessary performance. Though this was not pursued, Saab 32 development continued as Project 1150. The first prototype made its maiden flight on 3 November 1952
- Project 1210 powered by a rocket engine and having a very light airframe. This concept was investigated further to obtain data for comparison with the Project 1220
- Project 1220, which became the leading contender, was expected to be equipped with the most powerful and efficient jet engine and afterburner combination available at the time of production, and have an estimated take-off weight of 8500 kg (18,740 lb)

The last named of these projects, dated December 1949, consisted of a single-seat, single-engined fighter to intercept fast, hostile bombers at high altitude. The aircraft was designed to be capable of supersonic flight and was armed with two wingtip-mounted Type (Rb) 321 air-to-air missiles – revolutionary at the time. (Robotbyrå – the Swedish Missile Bureau – began development of the Rb 321 in 1946, test firing beginning on 29 December 1948. Further tests of the missile for Project 1220 were carried out at Linköping using a Saab J 29B, but the project was abandoned in 1958.) The engine was assumed to be the size and weight of the STAL Dovern engine with a static thrust of 44.13 kN (9,917 lb), increasing by 37 per cent with afterburner. Fuel capacity was estimated at 2000 kg (4,409 lb).

On 23 January 1950 a memo from Frid Wänström at Saab floated the idea of building a test aircraft. The issue of

Draken evolution – towards the double delta

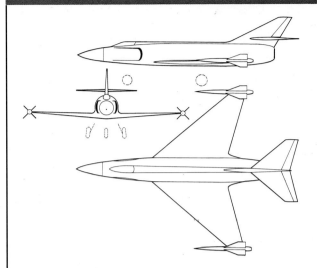

Type 1220 specification
Length: 13.36 m (43.83 ft)
Span: 9.10 m (29.86 ft)
Height: 4.20 m (13.78 ft)
Wing area: 28 m2 (302 sq ft)
Max. speed at 12000 m without
afterburner: (M1.02)
Max. speed at 12000 m with
afterburner: (M1.35)
Empty weight:
5848 kg (12,892 lb)
Max. take-off weight:
8390 kg (18,496 lb)
Engine thrust (dry):
44.13 kN (9,923 lb)

Rb 321 specification
Length: 4.00 m (13.12 ft)
Span: 0.90 m (2.95 ft)
Take-off weight: 150 kg (330 lb)
Speed: > Mach 1.0

Above: In terms of its basic layout the Type 1220 was clearly a derivative of the Saab 32 Lansen. Wingtip-mounted AAMs were a novelty in 1949.

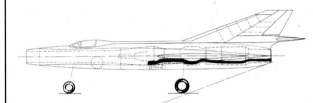

Type 1250 specification
Length: 1.70 m (41.67 ft)
Span: 9.80 m (32.15 ft)
Wing area: 51.38 m² (553 sq ft)
Take-off weight:
8660 kg (19,092 lb)
Landing weight:
7000 kg (15,432 lb)

Above: After General Jacobsson at the KFF declared that a tailless Type 1250 was simply not right, a sketch was prepared depicting a the design with the unnecessary tailplane added.

Right: In early 1950 the Type 1250 had progressed as far as this double-delta layout, with a central nose air intake for the engine. This design was later scaled down in size as the Type 1251, precursor of the Saab 210.

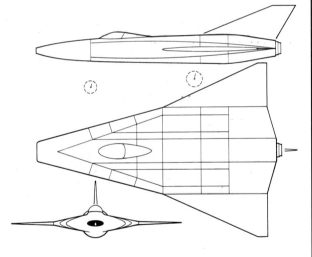

Left and below left: These later sketches of the Type 1250 show different intake configurations. Both carry the Rb 321 air-to-air missile on underwing pylons. The version with side intakes bore a strong resemblance to the final Type 35 Draken. Though no suitable radar for the aircraft existed at this time, it was noted that the adoption of a central air intake would have an impact on radar antenna size. This forward-thinking resulted in the selection of the final twin intake configuration.

In order to test the aerodynamics of the Saab 210, ¹/₇-scale swing-line models of the aircraft, powered by a small pulse jet engine, were test flown from a centre pedestal equipped with a recording camera.

This photograph of the Saab 210, taken immediately prior to its first flight from Linköping on a sunny 21 January 1952, shows the aircraft's original engine air intake configuration. This later gave way to a 'chisel' nose with more internal volume and greater intake area, but this posed insurmountable aerodynamic problems and was abandoned.

Lilldraken

The Saab 210 was powered by an Armstrong Siddeley Adder 1 rated at 3.92 kN (881 lb) of thrust – just sufficient to get the aircraft off the ground but barely adequate in the air.

Specification

Length:	8.80 m
Height:	2.78 m
Span:	6.35 m
Wing area (210A):	24.2 m²
Wing area (210B):	23.0 m²
Take-off weight:	1175 kg
Fuel capacity:	345 litres
Maximum speed:	545 km/h
Maximum range:	280 km

Pictured in May 1952, the 210 has tufting applied to its starboard wing to facilitate the recording of air flow patterns over the wing.

supersonic flight raised a number of questions to which the Saab team had few answers; while Saab engineers were experienced in most other areas, supersonic flight was new to all. Thus, Wänström was uneasy about investigating supersonic characteristics solely in a wind tunnel, and argued that live testing with a concept aircraft was the only way to collect reliable data. He wrote that it was a logical step to build an aircraft specifically designed to investigate the flying characteristics of a supersonic aircraft at various speeds. The aircraft he proposed was christened *Sound Barrier* and he envisaged that it would fly during the summer of 1951, achieve Mach 0.9 by September and (it

was hoped) break the sound barrier during the spring or summer of 1952. This schedule allowed time to finalise the aircraft's all-important wing configuration in July and that of the tail by November or December 1951.

In the meantime, the specification for Project 1250 was refined further. The aircraft would be a day fighter intended to intercept bombers flying at speeds just below Mach 1.0, attacking at altitudes between 10000 and 15000 m (32,808 and 49,213 ft) but possibly as high as 18000 m (59,055 m). The aircraft was to be armed with two Type 321 missiles or two rocket pods. A radar installation was not given much consideration at this time, as no suitable unit was available, though the availability of a suitable unit in the future influenced intake design. If the engine air intakes were split either side of the fuselage, there would then be space in the nose of the aircraft for a conventional antenna; a single centreline intake would force the use of a smaller lens antenna mounted within an intake centrebody.

The GLAN engine, producing a static thrust of 49.03 kN (11,018 lb), which increased by 40 per cent with afterburner, was proposed as the powerplant at this stage.

Finalising the wing configuration

Erik Bratt described the special advantages of the delta wing in supersonic flight in a memo dated 12 February 1951, adding a warning that next-generation bombers would operate at high subsonic speed.

He concluded that the only defence against such a threat was to fly supersonically and that the design of the aircraft's intakes would be among the biggest problems he would have to deal with as the aircraft's designer. He debated whether to place the intake in the nose (as in the Soviet MiG-21), free of airframe interference, with the ability to adjust the intake area according to the speed of the aircraft.

His analysis also highlighted the risk of stall during landing and, thus, the need for high landing speeds and sufficiently long runways. Optimistically, Bratt deduced that "fully developed stall will not occur below 30-40° AoA [angle of attack] and during approach Draken would operate at around 15° AoA giving a safe margin for any delta wing concept". (Unfortunately, the high AoA stall risk was not fully appreciated and, in the event, a number of Drakens were lost in so-called 'superstalls' after the type entered service. Surprisingly, however, few of these incidents occurred during approach and landing.)

Bratt concluded his memo with the words "a great deal of work and effort is required to develop this delta wing aircraft, but it should be done to allow us to develop the first viable conqueror of supersonic flight".

As Project 1250 gathered momentum, it became apparent that new and specific problems associated with supersonic flight would make this project a particularly challenging

The Saab 210 in its final guise as the 210B, with split oval intakes, looked much more like the 7/10-scale version of the final Saab 35 design. Note that the aircraft's undercarriage was, out of necessity, closer to full size and was only semi-recessed when retracted.

This view of a test pilot preparing to clamber into the 210B gives an excellent impression of the size of the aircraft. The 210 made made about 1,000 test flights, providing data vital to the successful execution of full-sized Draken design work.

one. The two main hurdles presented by supersonic flight were the need to reduce overall drag, particularly wave drag, (about which little was known), so as to allow supersonic flight using the thrust produced by the engines then available and, secondly, the provision of surfaces (and their hydraulic servos) strong enough to permit deflection proportional to stick movement by the pilot during supersonic flight.

The double delta solution

As far back as 1944, the German aerodynamicist Dr Alexander Lippisch had carried out wind tunnel tests indicating that a delta wing design was a feasible concept.

This was combined, by Saab engineers, with a slender, arrow-shaped wing with an extremely thin profile. After successfully flying paper darts down the corridors of the Saab design office, engineers built wooden models and tested them in Saab's wind tunnel. The result of their efforts was the decision in May 1950 to construct and fly the double delta Saab 210 testbed, powered by an Armstrong Siddeley Adder jet engine producing 3.92 kN (881 lb) static thrust.

The advantages of the double delta over swept wings or a pure delta included a closer correlation between the position of the aircraft's centre-of-gravity and centre-of pressure (thus improving stability), low supersonic drag, favourable low speed drag and a strong, stiff structure with more room for fuel and armament.

A small group of nine Saab personnel was behind the development and construction of the 210, which was accomplished in just nine months. Less ambitious than the *Sound Barrier*, the 210 was strictly a subsonic aircraft; supersonic testing would have to wait until the first full-size Draken was completed. That said, *Lilldraken* played a vital role in determining the handling characteristics of the double delta.

The radical layout of the Saab 210 was greeted with alarm by some at the KFF. Staff member General Jacobsson declared, "What the hell is this, an aircraft must have a damned tail, for heaven's sake." Everybody knew that the

design could not have a tail, but to satisfy the General and the KFF, a special sketch of the Type 1250 was produced with a small tail added.

To aid the team before the 210's first flight, a swing-line Draken model powered by a pulse jet engine was

Draken prototype 35-1 'Röd Urban' (Red U) made its first flight on 25 October 1955 (left), in the hands of test pilot Bengt Olow (accepting congratulations from designer Erik Bratt, above).

129

The second Draken test aircraft, 35-2 Blå Urban (Blue U), flew in March 1956 and was equipped with an afterburner from the outset. Like the 35-1 Röd Urban, it was unarmed and employed purely as a flight test airframe.

produced to gather more data. Swing-line models were flown extensively – even General Jacobsson's tail was tested; it made for a completely unstable aircraft. The models were ½-scale and controlled from a central pedestal that was also fitted with a high-speed recording camera. The noise of the pulse jet engine – a similar concept to the engines that powered German V-1 weapons in World War II – attracted the attention of both employees and the local population, making any attempts at keeping the project secret somewhat difficult. All were intrigued by these strange flying machines.

First flights of the 210

Finally, on 17 August 1951 the Saab 210 was lifted from its jig, ground testing beginning in early November. The Saab 210 left the ground for the first time on 10 December 1951 in the hands of Bengt Olow, but this short hop lasted only a few seconds. The official maiden flight, lasting 30 minutes, took place on 21 January 1952 in front of a specially invited audience. There followed an intensive period of flight testing, during which other pilots got the opportunity to test the qualities of the Saab 210. In addition to Saab's own pilots, Olle Klinker, Karl Erik Fernberg and Ulf Sundberg of Försökscentralen (Test Centre) also flew the aircraft.

Within weeks of the first flight of 35-2 both prototypes suffered accidents, setting the flight test programme back several weeks. On 29 March 1956 Bengt Olow retracted 35-2's undercarriage by mistake (upper right), while on 19 April 1956 K.E. Fernberg made a forced belly landing at Malmslätt (right) when the aircraft's 'undercarriage down and locked' cockpit indicator failed.

The aircraft had very low thrust, and as the summer temperature increased, the available thrust decreased, leading to long take-off runs and handling difficulties when airborne. Even so, on 6 July 1953 Olle Klinker gave a spectacular performance above central Stockholm to mark the city's 700th anniversary. As testing proceeded, several

Draken prototypes

In all, there were 13 Draken development aircraft, including nine new-build prototypes, as follows:

c/n	revised c/n	ff date	role/disposition (where known)
35-1	35101	25 Oct 55	first prototype; to technical school F14, 1964; later sold to private museum and restored
35-2	–	23 Mar 56	unarmed test aircraft; tested arrester hook for 35XD; nicknamed 'Old Granny'; destroyed in fire at FC, 1965
35-3	–	13 Sep 56	armed test aircraft; donated to Stockholm Technical High School (KTH), 1966
35-4	–	04 Jul 58	J 35A prototype; w/o 1968
35-5	–	15 Feb 58	J 35A prototype; avionics testing at FC; later modified to J 35B' standard; to KTH, 1966; preserved Flygvapnet Museum
35-6	35106	19 Oct 61	J 35D development; later flew with revised 35XD wing; fuselage reworked to J 35F standard; 35XD avionics and weapon testing; w/o 1973
35-7	35107	01 Oct 62	J 35F development aircraft; later to Defence Research Centre (FOA); w/o 1972
35-8	35108	16 Nov 62	J 35F radio and weapons development; w/o 1974
35-9	35109	18 Jun 63	J 35F development; w/o 1974
35-10	35110	(22 Dec 61)	converted from 35A 35082; J 35F development; test fired Rb 05 missile; w/o 1973
35-11	35111	(15 Jan 62)	converted from 35D 35275; J 35D avionics testbed; ECM testbed for Viggen programme; flown with dummy Rb 27/28 missiles; w/o 1975
35-12	35112	(09 Feb 62)	converted from 35A 35081; J 35D and S 35E trials aircraft
35-13	35113	(27 Dec 60)	converted from 35A 35013; J 35D prototype and avionics trials aircraft; Viggen ECM development, 1967; w/o 1969

All were operated by Saab and Försökscentralen (FC) at Malmslätt, with the exception of 35-1, which was operated by Saab throughout its flying career.

Draken 35-3 (nearest the camera) accompanies 35-5, the latter equipped with a revised cockpit canopy of the type intended for the production Draken (as shown on 35-4, first pre-production 35A prototype, inset). Note also the revised tailcone associated with the 35A's RM 6B engine and EBK 65 afterburner. 35-4's first flight took place on 4 July 1958, after that of 35-5, which had taken place during February.

modifications were made, in particular to the nose and the air intakes. The final design was designated the 210B (the original configurations becoming 210A), the aircraft completing over 1,000 flights before being retired to the Flygvapen Museum at Linköping.

Saab 35 takes to the air

The maiden flight of the Saab 35 Draken, which lasted 33 minutes, was made by Bengt Olow flying prototype 35-1 *Röd Urban* (Red Uniform) on 25 October 1955. The aircraft was powered by a non-afterburning Rolls-Royce Avon Mk 21 (RM 5A), manufactured by SFA. The flight was uneventful – a tribute to the extensive work carried out on the Saab 210. A period of intensive flight testing followed. In the autumn of 1955 the first of six Avon Mk 43 engines arrived from the UK and was immediately installed in 35-1; this engine was more powerful, though it still lacked an afterburner. On 26 January 1956, 35-1 made history by reaching Mach 1.0 and breaking the sound barrier in level flight. This was positive evidence that the delta wing was the right concept with which to break the sound barrier.

It is worth mentioning that the first Swedish aircraft to 'boom' had been the prototype Lansen (32-1), on 25 October 1953, although this was achieved in a dive. Later, on 7 July 1954, Frykholm exceeded Mach 1.0 in a J 29E with a modified wing. Mach 2.0 was achieved on 14 January 1960 in a Draken (35011) powered by an RM 6C.

The prototype 35-1 suffered a setback on 19 April 1956 when, on approach to the Saab airfield, the cockpit undercarriage indicators showed 'unlocked'. The test programme controllers advised the pilot, K-E Fernberg, to divert to Malmslätt airfield (F3 Wing) and carry out a belly landing on grass parallel to the runway. This he did, though unfortunately he landed on the wrong side of the runway; the aircraft was badly damaged as it slid over a concrete taxiway. Fernberg suffered a minor spinal injury.

The subsequent technical investigation showed that the indication system had been faulty and the undercarriage had functioned normally. After several months of repair, 35-1 returned to the test programme.

Five months after the maiden flight of 35-1, the second prototype, 35-2 *Blå Urban* (Blue Uniform), took to the air, on 23 March 1956. This aircraft was powered by a Rolls-Royce Avon Mk 46 engine equipped with a 33.8-in

(85.8-cm) afterburner. Unfortunately, only one week later, during landing, Bengt Olow inadvertently selected the undercarriage lever when reaching for the drag chute handle, and the undercarriage retracted. The aircraft was severely damaged during the subsequent belly landing but

Below: This rare view, believed to be dated around 1958, shows the first five Draken prototypes. Three of the aircraft carry 'FC' codes indicating service with the Försökscentralen (air force test centre).

As well as the nine prototype and trials aircraft, three J 35As and a J 35D were modified to serve as testbeds for equipment destined for production Drakens. Coded '42', 35-10 (formerly J 35A 35082) served as the J 35F prototype and was employed during testing of the Saab Rb 05 air-to-surface missile.

The sixth of nine Draken prototypes built from scratch, 35-6 first flew on 19 January 1961. Originally produced as a 35D testbed, with modified air intakes and the RM 6C engine, the aircraft is pictured equipped with the Falcon AAMs intended for the 35F. It was later modified further, with the strengthened wing designed for the Danish 35XD.

after a few months of repair was back in the air. A design fault in the positioning of the undercarriage lever was subsequently identified and the levers redesigned.

As mentioned above, the 35-1 suffered a belly landing shortly after 35-2's mishap and the Saab 35 programme was thus severely disrupted for some time.

Around this time, 35-2 was re-engined and would eventually fly with a Rolls-Royce-built Avon Mk 48 powerplant installed. Like 35-1, 35-2 was unarmed and remained a test vehicle, employed for arrester hook testing at Norrköping and subsequently 'superstall' trials, between 1961 and 1963. Ejection seat tests were also carried out with 'Old Granny', as the aircraft became nicknamed, though the aircraft came to a premature end in a fire at FC; it was officially written-off on 7 December 1965.

Test aircraft 35-3, *Gul Urban* (Yellow Uniform), was the first armed Draken, powered by an Avon Mk 46 engine with a 33.8-in (85.8-cm) afterburner and fitted only with guns at this stage. It flew for the first time on 13 September 1956 and was later used for technical and tactical develop-

ment work at FC. The aircraft finished useful test flying on 8 February 1963 and was then used as a stress and fatigue testbed with KTH (Technical High School) in Stockholm.

Test aircraft 35-4 was the first prototype of the 35A series to have a modified canopy and tailcone and an RM 6B engine – the first licence-built Avon Mk 48 engine. It flew for the first time on 4 July 1958.

Draken 35-5 was the first true test production-standard test aircraft for the 35A series. Following final assembly and inspection, it actually flew five months before 35-4, on 15 February 1958. It served with Saab until 11 August 1958 and subsequently with FC on test duties.

J 35A – first production variant

The first production-standard test aircraft, 35-5 was powered by a Rolls-Royce Avon Mk 48A engine. Later Draken aircraft were powered by the SFA-built variant of this engine, the RM 6 series. Aircraft 35-5 was subsequently re-engined with the RM 6B. A few months later, when 35-4 was airborne, both aircraft became A-model test aircraft.

Deliveries of the first J 35A to Flygvapnet began with aircraft 35002 in the spring of 1960 to F13 Wing Norrköping. The first production run of aircraft, up to serial number 35065, were powered by RM 6B engines with EBK 65 afterburners and were to be known as the 'short-tail' models.

At this stage, aircraft up to 35040 also lacked radar and a fire control system. Later, during 1962/63, 25 of the first J 35As were modified to become two-seat SK 35Cs. Aircraft 35010 became the first two-seat test machine 35800. Aircraft from 35041 were J 35As equipped with a PS-02/A radar and PN-793/A IFF, integrated into the French Thomson-CSF Cyrano S6 fire control system built under licence by

With the exception of the 12 aircraft assembled by Valmet in Finland (from kits supplied by Saab), all Drakens were completed at Saab's Linköping plant. These aircraft are part of the first batch of J 35As, pictured during final assembly in autumn 1959.

Ericsson. The FCS was complemented by an auto-pilot – the Lear L-14. Armament comprised two internally-mounted 30-mm ADEN cannon and up to four Rb 24 Sidewinder air-to-air missiles.

The ceiling of the J 35A – around 13000 m (42,650 ft) – was soon considered to be insufficient. In the autumn of 1957 comparative wind tunnel tests between a short-tail (35A1) and a modified long-tail (35A2) were made, showing that at 11000 m (36,089 ft) a clean long-tail aircraft should have a 25 per cent power advantage over the short-tail aircraft and that acceleration time to Mach 1.4 should decrease by 30 seconds. This was proved by flight testing with aircraft 35001, a long-tail model. These modifications to lengthen the tail and alter the afterburner improved high-altitude performance only marginally, but the medium-level improvement in performance was worthwhile, so the last of the A versions (35066-35090) were equipped with an EBK 66 afterburner and a lengthened tail that also incorporated a retractable tail wheel.

All of the J 35As were delivered to F13 Wing Norrköping during 1960/61 but were transferred to F16 Wing Uppsala between 1961 and 1965. In later years at F16 Wing, the J 35A was re-equipped with the S6B FCS, which was integrated with a nose-mounted ME *mörkerenhet* (infra-red sensor). The auto-pilot was upgraded to the Mk 51A (Lear 5107A) system, which was also used in the SK 35C.

Teething problems

Following deliveries of the J 35A to F13 in March 1960, conversion training for the first squadron began in earnest. This new aircraft was a very different machine to anything the Swedish air force had operated before, and some teething problems were bound to arise.

Spin testing had yet to be carried out and was scheduled for 1961, but before this had even begun, a number of the aforementioned 'superstall' crashes had occurred. It had become apparent during flight testing that the aircraft was very sensitive in pitch, particularly at low level and high speed. The first squadron pilots found that severe pilot-induced oscillations (PIOs) could be generated – some felt that their heartbeat was enough to trigger them! The PIOs could be uncontrollable and some were measured at a staggering +16*g*/-9*g* – forces that would have destroyed many aircraft. That the Draken withstood these loads is a testament to the inherent strength of the airframe but, not surprisingly, some pilots lost confidence in the machine and requested transfer off the type. The recommended procedure in the event of a PIO was to disconnect the auto-pilot (so as to disable its pitch-damping function), take hands off the stick and, finally, when the pitch oscillation ceased, abort the mission and land.

Obviously, these aerodynamic characteristics were unacceptable. A modified control system including stronger pitch- and yaw-damping modes for the auto-flight system, and a gearing mechanism to reduce control column sensitivity around the neutral position, had already been tested in 1959. These modifications were gradually introduced to the J 35A, B and SK 35C aircraft and solved some of the handling problems, though more were to occur later.

The J 35A continued in service with F16 Wing Uppsala until 1976, when it was replaced by the J 35F. The last 35As were retired in 1979, although a handful were rebuilt as trials aircraft.

The fully operational J 35B

In parallel with the construction and delivery of the J 35A, work continued apace to plan development of the Draken to cater for future requirements. This resulted in a

Above: Four J 35As of F13 Norrköping-Bråvalla taxi to the runway at the beginning of a training sortie during the spring of 1960.

The first production built J 35A (35001, left) first flew on 7th January 1959. This aircraft was never delivered to Flygvapnet but was instead used by Saab as the 'long tail' test aircraft. Note also the wing fences on the outer wing – an early anti-stall, vortex-generation modification.

Above: The first Draken equipped with radar was J 35A 35041, in which was installed a Swedish-built PS-02/A set, a development of the French Thomson-CSF Cyrano. Radar training was carried out at F13 using a Tp 83 Pembroke equipped with the radar. 'Long tail' J 35A 35077 is an example of a late-production aircraft equipped with radar and a gun sight. Note also the vortex generators on the underside of the wing. 35077's pilot is making full use of the Draken's tail wheel, in landing 'nose high' to assist aerodynamic breaking, without risking damage to its tail cone.

Left: F13 passed its J 35As to F16 in 1962/63, when it re-equipped with J 35Ds. Remarkably, the 35As remained in service into the mid-1970s. During their time with F16 the aircraft were re-equipped with the S6B fire control system, which incorporated an undernose IR scanner, as seen on this camouflaged machine.

This J 35B of F18 Tullinge is fully armed for the air interception role with two fuselage-mounted 75-mm FFAR rocket pods and wing-mounted Rb 24B Sidewinder AAMs. Equipped with the S7 fire-control system, the 35B was able to be fully integrated into Sweden's STRIL 60 semi-automatic air defence control system and, as such, was the first fully operational Draken. (STRIL was an acronym standing for Stridesledning och Luchbevakning, or Intercept Control and Early Warning).

Below: F18 was the only Wing to be fully equipped with the J 35B; a handful also served with 3./F16 in the early 1960s. Upon entering Flygvapnet service, early Drakens carried two- or three-digit codes based on their construction number; '226' was J 35B 35226.

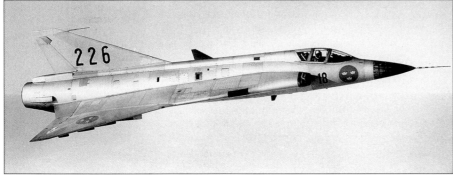

Below and below right: A cornerstone of Sweden's air defence strategy was the ability of each wing to operate from dispersed sites, often using public roads as runways. This F18 J 35B is landing on a typical road strip during autumn exercises in 1968. The drag 'chute could reduce the Draken's roll out distance to as little as 500 m (1,640 ft).

specification being drafted as early as January 1956 for the J 35B version. The principle improvements were to be increased thrust from a developed engine and afterburner, greater fuel capacity and improved avionics. The first B version test aircraft was the reworked J 35A 35011 and its maiden flight occurred on 29 November 1959.

Initial J 35B planning had anticipated use of the improved RM 6C engine and the new, Swedish-manufactured PS-03 radar. However, airframe development was complete before the engine was ready, and Ericsson was still working on the radar, so the first aircraft were delivered to F16 Wing in February 1962 without radar and

equipped with the RM 6B engine. Until the radar, a PS-03/A, and S7 fire-control system were retrofitted during 1964-1965, the aircraft was designated J 35B' (B prime). Flight tests of the Swedish-produced S7 FCS began during 1959 in an A 32 Lansen (32253) modified with a Draken nosecone.

The first production-built B version was 35201; it was never delivered to the air force, being retained by Saab and FC for further development testing of the S7 FCS and avionics. A total of 72 J 35B' aircraft (35202-35273) was delivered to the air force between February 1962 and March 1963. Of these, 69 were later converted to full J 35B specification.

Initial J 35B' deliveries were to F16 Wing. From June 1962, F18 Wing received its aircraft to replace the Hawker J 34 Hunter. Upgrades to full-specification J 35B were carried out later. Pilot training had been progressing with GFU (basic flight training unit) at F5 Wing Ljungbyhed, TIS (conversion training unit) at F16 Wing Uppsala and GFSU (basic air combat role training unit) at F18 Wing Tullinge. Advanced air combat role training then followed at the FFSU (F18 Wing).

From the autumn of 1966, F10 Wing Ängelholm received a number of J 35Bs (tail codes 42-69), the first 12 of which were immediately mothballed as operational reserves at F14 Wing Halmstad. These reserves were eventually withdrawn from inventory. Aircraft 54-69 continued in opera-

tional service until 1976, when F10 Wing re-equipped with the J 35F. F18 Wing Tullinge ceased operations with the J 35B in 1974 when it was disbanded.

J 35B armament was broadly similar to that of the 35A version. It retained the two internal 30-mm M55 (ADEN) cannon, and two or four Rb 24 Sidewinder missiles mounted on the outboard pylons. The 35A, B and later C versions carried a centreline drop tank, whereas subsequent models could carry two or four wing-mounted tanks. The 35A and B versions could mount 12 13.5-cm air-to-ground rockets, and to this capability the 35B added two belly-mounted pods each containing 19 7.5-cm folding-fin aerial rockets (FFARs). The original Mk 4E reflector gun sight fitted to the J 35B' was usable as a gyro-sight and served for both air-to-ground and air-to-air sighting of weaponry. With the upgrade to full J 35B standard, the PS-03/A radar and S7A FCS greatly improved weapons-aiming performance. In later years, the J 35B was also fitted with an improved ejection seat system featuring better low-level, high-speed capability.

Although the J 35B was withdrawn from active duty with Flygvapnet, it continued in service elsewhere; the Finnish air force acquired six aircraft in a lease/purchase arrangement in 1972, designating them Saab 35BS.

Draken as a display aircraft

At the beginning of March 1965, wing commanders Colonel Hedberg and Flying Major Torselius of F16 were directed by Major General Peyron (commander of the 3rd Group) to prepare for participation in the International Air Salon at Paris-Le Bourget in June 1965.

Planning called for the participation of two teams of J 35s, one from F16 Wing Uppsala with J 35As and the other from F18 Tullinge with J 35Bs. Captain Rolf Gustavsson was the leader of the F16 team, which had been the established 3rd Group display team since October 1964. The other 3rd Group display team, known as 'Acro Deltas', came from F18 Wing and was led by Captain Claes Jernow. At the F1 Wing Västerås air display in May 1964, their aircraft sported tail fins finished in the Swedish national colours of blue and yellow.

The 'Acro Deltas' and the F16 display team

Perhaps the most famous of the handful of Flygvapnet display teams equipped with the Draken was the 'Acro Deltas', formed by F18 during 1963 as replacement for the popular 'Acro Hunters'. Strengthened by the addition of aircraft and pilots from F16, the eight-ship 'Acro Deltas' made their international debut at the 1965 Paris Air Salon. Its aircraft were as follows:

c/n	code	type	wing	notes
35066	30	J 35A	F16	smoke generator fitted
35069	33	J 35A	F16	smoke generator fitted
35089	51	J 35A	F16	smoke generator fitted
35090	52	J 35A	F16	smoke generator fitted
35068	32	J 35A	F16	reserve
35070	34	J 35A	F16	reserve, painted
35214	24	J 35B	F18	smoke generator fitted
35256	56	J 35B	F18	smoke generator fitted
35258	58	J 35B	F18	smoke generator fitted
35263	63	J 35B	F18	smoke generator fitted
35257	57	J 35B	F18	reserve
35267	67	J 35B	F18	reserve

Although the 'Acro Deltas' disbanded at the end of 1965, F16 continued to field a display team for the next 13 years. Seven aircraft were used, though they were only able to be distinguished from a standard squadron aircraft by a small yellow Rocka (stingray) marking on the tip of their tailfins.

A mixed formation of J 35A and B aircraft displays the 'Acro Deltas' tail marking adopted prior to the team's appearance at the 1965 Paris Air Salon, led by Captain Claes Jernow.

J 35A '52' (35090) was one of the aircraft assigned to the F16 Wing display team in the 1970s. Note the discreet yellow stingray marking on the fin of the aircraft.

Four SK 35Cs from Ängelholm formate in an echelon to port. Note the pair of ventral fins under the wings of each aircraft fitted to correct a yaw instability problem. Also evident on each aircraft is the stereoscopic periscope above the canopy, fitted to improve the instructor's otherwise poor forward visibility.

All 26 SK 35Cs were converted from 'short-tail' J 35A airframes, the first example (35800, pictured) having been built as 35010 and flown for the first time, as a two-seater, on 30 December 1959. Space was made for the second cockpit by reducing the size of the forward fuselage fuel tank; extra wing tanks were fitted in place of gun armament to compensate.

In order to educate pilots in the stall and spin characteristics peculiar to the Draken (the so-called 'superstall'), an anti-spin parachute was installed in some SK 35Cs. Standard drag 'chutes were strengthened and provided with an extended cable to an overall length of about 50 m (164 ft). The system could be activated from either seat using a trigger on the stick. The ordinary drag 'chute handle was used to detach the 'chute once the aircraft was once more under control.

Competition was fierce, as F13 Wing Norrköping had been flying a four-ship formation display team led by Captain Per Fogde since 1960.

In an order from the C-in-C dated 6 April 1965, Lieutenant General Thunberg confirmed that the Swedish air force would participate in the 1965 Paris Air Salon with no fewer than 11 Draken aircraft (eight display and three reserve aircraft). When Hedberg and Torselius were directed to prepare for the air show, the 'Acro Deltas' of F18 had not worked up for an air display since the previous September and the F16 team had only just started its training. Initially, each team practised independently and then, for two weeks in April 1965, at F15 Wing Söderhamn training began to integrate the two teams into one unit. More independent team training led to a second joint training camp between 10 and 14 May at F10 Wing, where the teams got another opportunity to co-ordinate manoeuvres.

Conditions at Le Bourget were closely scrutinised and the possibility of having to take-off with a tail wind noted. Accordingly, the teams practised four-ship formation take-offs with up to 30-km/h (19-mph) tailwinds. The total runway length was only 2000 m (6,562 ft), so the first formation lined up 100-200 m (328-656 ft) from the threshold.

The final rehearsal before the Paris show took place at an air display hosted by F15 Wing on 22 May 1965. For the first time, the aircraft were equipped with a smoke-producing system consisting of a drop tank filled with diesel oil and a pipe, with a diffuser, directed into the exhaust.

After the Paris show, things returned to normal, i.e., the display team was at F16 with its J 35As under the leadership of Captain Rolf Gustavsson, and 'Boris' Bjuremalm, who transferred from F18 to F16, took over training. The team performed a number of different displays, such as Denmark in 1971 and Moscow in 1972. The pilots were instructors from the TIS squadron (conversion training unit) at F16. All the display aircraft wore camouflage and had a small yellow stingray emblem painted on the top of the tail fin.

After 1978 two four-ship formations displayed on different occasions: the 'Ghost Rider' team (1./F10) with team leader Major Mats Lindskoog (J 35F/J 35J) and a Danish team led by Major Stephen Solomon (RF-35).

Since 1985 one of the largest military air shows in the world has taken place at RAF Fairford, England. The Royal International Air Tattoo (RIAT) regularly attracts nearly 400 participating aircraft from around the world. In 1994 Sweden participated with an unprecedentedly large group, comprising three J 35Js (F10), two Tp 84 Hercules (F 7), one J 32E (F16M) and two HKP 9s (AF 2). These increased numbers can be attributed to heightened Swedish air force involvement in the international scene, brought on by rapid changes in the political climate in Europe.

Captain Ingemar Axelsson, a 36-year old with 1,400 hours in Drakens, succeeded in winning the Super Kings Trophy, awarded to the best solo jet display, when he beat all participating pilots from 54 countries, including F-16s from Belgium, F-4s from Germany, Mirage 2000s from France and MiG-29s from the Czech Republic.

SK 35C trainer variant

When designing the J 35, Swedish air force authorities also planned for a two-seat trainer version. One of the reasons for this, was the number of crashes that had occurred when the J 29 was introduced, for which no two-seater had been built. Moreover, experience from abroad

F10's J 35As were replaced by J 35Ds during 1962/63. These two early production examples (35301 and 35304) are pictured during March 1964. Apart from its new RM 6C engine, which offered almost 20 per cent more thrust in afterburner, important features of the J 35D were its PS-03/A radar (below left) and associated S7A fire control system and a much improved ejection seat (the Saab 73SE-F), which was functional at zero altitude and at speeds above 100 km/h (62 mph). These aircraft also carry 12 13.5-cm air-to-ground rockets, mounted in three pairs beneath each wing.

also suggested that the development of a conversion trainer variant would be worthwhile.

Test pilots who had flown the Draken at Saab and FC considered that its landing characteristics, compared to the J 29, were such that there was no need for a Draken trainer version. Later, ordinary air force pilots beginning their operational training on the type took a totally different view. Faced with 'superstall' crashes and the risks of entering uncontrollable situations, the SK 35C became invaluable in basic training and in training pilots in 'superstall' recovery techniques. Furthermore, it is very unlikely that foreign sales would have been possible without a dual-seat trainer version.

Preliminary specifications for the SK 35C were presented in October 1954. Initially, the trainer version was to be a variant of the original J 35, with the intention that SK 35Cs could later be reconverted into fighters. This was never realised, primarily for economic reasons; instead, the 35C was converted from the original 35A with the short tail. Experience has shown that the SK 35C programme was money well spent. By the mid-1990s, the TIS conversion unit had not lost a single aircraft flown by a student coming directly from GSFU. Interestingly, during the early stages of the Saab 37 Viggen programme, it was decided that the SK 35C would be used to train Viggen pilots; in the event a two-seat Sk 37 was later built for this purpose instead.

A total of 26 short-tail J 35As was converted to SK 35Cs and received construction numbers from 35800 to 35825. 35800 (formerly 35010) was the test machine and flew for the first time on 30 December 1959, remaining with Saab as a test airframe. When Saab chose the SK 35C to be a test-bed aircraft for JAS 39 Gripen equipment, the type proved equal to the task and had no problems with loadings of up to 10g during manoeuvres.

The conversion from short-tail J 35A to SK 35C was made at CVV Västerås. The work began in August 1961 and lasted until June 1963, by which time all 24 SK 35Cs had been delivered. Aircraft 35814 crashed into Lake Roxen on 7 January 1963 before it was delivered. The crash was caused by engine malfunction and a problem with the cabin air system. Pilot Lieutenant Lars-Erik Jetzén, a test pilot at FC, and engineer Hans Pettersson, also from FC, both ejected safely. It was the first completely successful escape since the Draken begun operational duty in 1960.

All the aircraft were modified at CVV Västerås in May 1966 as follows:

- installation of new PN 793/A navigation radar and a new Fr 14 radio
- Modification of Fr 13 radio
- Installation of a variable gear in the flight control system
- Modification of the emergency power unit

- Lowering of the cabin floor and installation of new ejection seats
- Mounting of a stall fin under the outer delta wing to improve yaw stability

SK 35Cs were delivered to F16 from May 1962 to June 1963. In 1986, 10 of the remaining SK 35Cs were transferred to F10 Ängelholm together with TIS 35. Number 35810 returned to service in 1989 with TIS 35 at F10 Ängelholm. Numbers 35816 and 35825 were reduced to spare parts for the remaining 'silver birds', as they were known at F10. In

Drakens 35382 (Urban 42) and 35389 (Urban 43, pictured) were modified at Flottilj 21 Luleå-Kallax with the bulged J 35F canopy. They remained unique in the J 35D fleet, however, because of the high cost of the modification.

RM 6B: Draken 35A, 35B, 35C

When the first Saab 35 Draken test aircraft made its maiden flight on 25 October 1955, it was powered by a non-afterburning RM 5A produced by Svenska Flygmotor AB (SFA). This was a licence-built Rolls-Royce Avon Mk 21 producing 33.93 kN (7,630 lb) of thrust as originally employed to power the Saab 32A and 32C Lansen. It had been the intention that an example of the Avon 200, developing 47.96 kN (19,783 lb) thrust, power the first Draken, but Rolls-Royce was unable to deliver an example in time for the Draken prototype's first flight.

The first Avon 200s delivered to Sweden were six non-afterburning Mk 43s, one of which was installed in Draken 35-1 in 1955. Development continued, via the Mk 46 and Mk 47A. The latter was known locally as the RM 6AE (or 6AS if built by SFA) and was fitted with a locally-designed EBK 61 afterburner for use in the J 32B Lansen. This, in turn, led to the Mk 48, three of which were bought for use in Draken test aircraft. These three engines were later fitted with an EBK 65 afterburner to Mk 48A standard; it was this engine that was licence-built by SFA (and, later, Volvo Flygmotor AB) as the RM 6BS (S stood for Swedish-built, E for English-built) for the J 35A, J 35B and SK 35C Drakens.

On 15 February 1958 the fifth Draken test aircraft (35-5) flew for the first time with a Rolls-Royce-manufactured RM 6BE (Avon Mk 48A) rated at 64.09 kN (14,410 lb) thrust with afterburner. From 1958, Flygmotor built a total of 186 RM 6BSs; production ceased in 1961.

After the original batch of three Avon Mk 48s, Rolls-Royce delivered another 16 Mk 48A (RM 6BE) engines. The RM 6BE and 6BS were fully interchangeable.

The RM 6B turbojet had a 15-stage axial compressor driving a twin-stage turbine. Between the compressor and turbine were eight combustors mounted in a joint combustion chamber. The afterburner was connected directly after the turbine. The EBK 65 afterburner was used on J 35As and SK 35Cs serialled 35002-35065. The EBK 66 afterburner was employed in the long-tail J 35As (35001, 35066-35090) and the J 35B, and offered higher thrust and minor improvements in climb and high-altitude performance.

RM 6C: Draken 35D, 35E, 35F, 35J

On 29 November 1959 test aircraft 35011 took off with an example of the Avon Mk 60, a version of the newly-developed Avon 300 engine, which developed 78.45 kN (17,640 lb) of thrust with afterburner. Initially designated RM 7, this engine was soon renamed RM 6C. The Draken's engine bay had been designed with an extra 10 cm (3.94 in) of lengthwise space built in, in case the RM 6B installation needed to be altered to allow centre of gravity adjustments. This proved to be unnecessary, but raised the possibility of installing the longer Avon 300 engine (RM 6C) in the Draken.

Fitted with an EBK 67 afterburner, the Avon Mk 60 was built under licence by Volvo Flygmotor between 1961 and 1971; some 445 were produced and powered the J 35D (serials 35274-35393), S 35E (35901-35960) and J 35F (35401-35630), as well as the J 35Ds modified to J 35ÖE standard for Austria, and the later J 35J variant.

The RM 6C has an axial compressor with 16 stages, a combustion system consisting of eight combustors, a two-stage turbine and an outlet comprising the afterburner with a two-position nozzle.

To prevent the afterburner gas stream from reaching a critical speed (Mach 1.0) in the afterburner, the nozzle area doubles automatically open to increase the outlet area. The strain on the engine and higher fuel

The Volvo Flygmotor RM 6C (above) with EBK 67 afterburner fitted was described as "the longest jet engine in the world". The RM 6C differed from the earlier RM 6BS (below) in having an extra frontal compressor stage ('0-stage') which added about 10 cm (3.94 in) to its length. The EBK 67 afterburner was also considerably longer than the EBK 65 fitted to the RM 6B.

consumption limit the use of afterburner; in the Draken the flight manual prescribes that maximum use of afterburner should be no more than 10 minutes per sortie. Take-off and climb to 15000 m (49,213 ft) by a J 35F with a take-off weight of 10550 kg (23,258 lb) with afterburner uses 45 per cent of the aircraft's internal fuel load.

RM 6BS and RM 6C compared

type	dry thrust	thrust with a/b	rpm	frontal area	all-up weight
RM 6BS	46.58 kN	62.17 kN	8,000 rpm	0.95 m²	1720 kg
	(10,474 lb)	(13,980 lb)		(10.23 sq ft)	(3,792 lb)*
RM 6C	55.41 kN	76.05 kN	8,100 rpm	0.95 m²	1770 kg
	(12,459 lb)	(17,100 lb)		(10.23 sq ft)	(3,902 lb)

* type 66 afterburner

Engine problems

Over the years, RR Avons installed in both the Lansen and Draken have thrown up a number problems. The combination of a small frontal area, the need for reasonable fuel efficiency and a good rate of thrust have led to these powerplants becoming prone to compressor surging. Factors that could initiate surging included excessively violent throttle movements, flying the aircraft near its stall limits, and violent manoeuvring that may lead to air disturbances at the engine air intake. This can be exacerbated by high revs and low temperatures and high altitude. A number of aircraft and their pilots have been lost in accidents caused by engine malfunction, though component modification programmes have gone some way to reducing the number engine-related incidents experienced by Draken crews.

Oddly the RM 6Bs installed in the SK 35Cs which were specially equipped for 'superstall' training suffered no engine surges or flame-outs during spin training. The Danish 'superstall' training aircraft, by contrast, have experienced two flame-outs, perhaps due to the differences in the powerplant; the TF-35 was a derivative of the J 35F which was powered by the RM 6C, with its different inlets and an extra compressor stage.

J 35J Draken: RM 6C/EBK 67 installation

1   Oil tank
2   Engine inlet and compressor
3   Air intake to engine de-icer
4   Alternator
5   Alternator drive unit
6   Warm air valve for engine de-icer
7   Firewall
8   Engine mounts
9   Cooling device for alternator drive unit
10 Cooling air duct
11 Thermostatic element
12 Fire extinguisher
13 Throttle lever
14 Linkage to engine governor
15 Fuel tank for engine starter
16 Engine starter pump unit
17 Engine starter
18 Engine starter exhaust
19 Engine mounts
20 Governor for engine and afterburner
21 Accessories gearbox
22 Hydraulic pumps
23 Cooling air exhaust
24 Bleed air duct from combustion chamber
25 Bleed air duct from afterburner connection
26 Intermediate duct between engine and afterburner
27 Afterburner fuel pump
28 Afterburner mounts
29 Afterburner nozzle actuator
30 Two-position nozzle

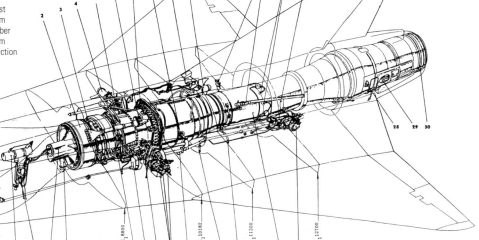

In time-honoured fashion Saab produced this publicity photograph of a J 35B to demonstrate what it called 'the outstanding versatility of the Saab 35 Draken weapon system'. The stores arrayed before the aircraft are (from left): a 500-litre (132-US gal) drop tank, a rocket pod (with a frangible nose cone) containing 19 75-mm FFARs, an Rb 324 (later designated Rb 24B) Sidewinder AAM, a 500-kg bomb, a 250-kg bomb, six 135-mm air-to-surface rockets, three 80-kg bombs and a 30-mm M55 ADEN cannon.

Below: Apart from the Flygvapnet's SK 35C two-seaters and S 35E reconnaissance aircraft, all Drakens carry at least one 30-mm ADEN cannon, each supplied with 90 rounds. Mounted in the wing, the cannon was easily accessible to ground crews and presented no gun gas ingestion problems.

With the J 35F came the option of Hughes AIM-4 Falcon AAMs, the Flygvapnet opting for locally-built (by Saab and Bofors) AIM-26B semi-active radar-homing (known as Rb 27) and IR-guided AIM-4C (Rb 28) missiles. Up to four Falcons could be carried at once; in this view (upper right), Rb 27s are carried on pylons either side of the centreline with Rb 28s on wing pylons. Rb 24J (AIM-9J) Sidewinders were later procured and were carried alongside earlier Rb 24Bs (foreground).

Right: Draken prototype 35-8 fires an Rb 27 Falcon AAM as part of J 35F systems testing during 1965/66, when about 80 Falcons were launched.

Below: Drakens were also utilised to test weapons for other aircraft. Here the SK 35C prototype (35800) carries an Rb 05 air-to-surface missile, under development by Saab for the AJ37 Viggen.

Exter
J 35A,
Draker
limitec
single
(132-U
tank o
hardpc
could
once b
and la
thoug
to limi
arman
Danisl
able to
1275-l
(337-U
tanks.

Camc
Standa
Finnisl
blue-g
unpair
olive c
two-se
Sk 35(
painte
serials
white
the Fir
Ilmavc
low-vi:
to rece

Unde
Amon
retract
centre
require
(gener
The la
100 m

S 35E prototype 35901 (converted from J 35D 35278) flew for the first time on 27 June 1963 and was used exclusively as a test airframe. Twenty-nine new S 35Es and 30 converted from J 35Ds followed, equipping both squadrons of reconnaissance wing F11 and one of the two squadrons in composite wing F21.

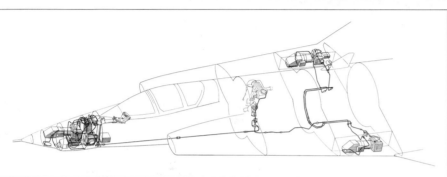

Above: The layout of the S 35E's camera installation made for excellent accessibility. This drawing of the camera fit includes the pilot's sighting equipment and the heating system provided for the camera housings and ports.

Right: The camera equipment fitted in the nose and wings of an S 35E (prior to 'apparat 15' upgrade) was as follows: ❶, ❺ (both wings) SKA 24-600 high-altitude vertical camera; ❷ SKA 16B forward oblique camera; ❸ SKA 24-44 low-altitude vertical camera; ❹ two SKA 24-100 side-looking oblique cameras. Port ❻ was used by the pilot's sighting equipment, linked to a scope in the cockpit which replaced the radar scope found in Draken fighter variants.

Visible in this view of S 35E 35937, modified to 'apparat 15' standard in the 1970s, are a small ECM pod under the outer wing and a flare dispenser faired into the afterburner air intakes. From the outset the S 35E was also able to carry a chaff dispenser, in place of a drop tank on the starboard inner wing pylon.

January 1995, nine SK 35Cs were still operational at F10, each with around 200 hours left to fly before their final retirement. During 1993-1994, the remaining SK 35Cs flew 550 hours at F10. The type served with F10 until 1998, continuing to perform an important role providing operational conversion training for Swedish and Austrian pilots and 'superstall' training for all Draken pilots (Swedish, Finnish and Austrian).

Five SK 35Cs were sold to Finland and four have been lost in crashes. The first was lost as mentioned above during a pre-delivery test flight in 1963. The second was so badly damaged in a starter explosion that it was retired later (1977), having been deemed beyond economic repair. The other two aircraft were lost in crashes, in 1982 (pilot disorientation at low level) and in 1984 (engine problems).

J 35D – bigger engine, more fuel

In August 1957 the specification for the 35B2 variant was confirmed though, before long, the 35B2 was redesignated 35D. The improved RM 6C engine was now ready for introduction, together with the new EBK 67 afterburner. Compared to the earlier A and B versions, the D could carry an additional 600 litres (132 Imp gal:172 US gal) of fuel internally and a second 500-litre (110-Imp gal: 132 US gal) drop tank. Flight testing with the new RM 6C engine was carried out with test aircraft 35011, which flew for the first time on 29 November 1959. External differences between the A, B and D version consisted mainly of longer,

S 35E camera equipment

Camera fit, as built
    one SKA 16B forward oblique camera (nose)
    three SKA 24-600 high-altitude vertical cameras (nose and wings)
    one SKA 24-44 low-altitude vertical camera (nose)
    two SKA 24-100 side-looking oblique cameras (nose)

Camera fit, after 1970s upgrade
    one SKA 16B low-altitude camera (nose)
    one SKA 24B-57 low- and high-altitude camera (nose)
    one SKA 24B 120V (left) low-altitude camera (nose)
    one SKA 24B 120H (right) low-altitude camera (nose)
    one SKA 24-600 high-altitude camera (nose)
    one SKA 24-600V high-altitude camera (left wing)
    one SKA 24-600H high-altitude camera (right wing)
    (one 'Blue Baron' pod with two Vinten SKA 34 low-altitude cameras and flashlight illumination equipment for night operations, from 1973)

more slender engine intakes and a redesigned fin-top with a pitot tube – also features of the following E, F and J variants. The 35D offered significant operational improvements in climb performance and acceleration, but its absolute ceiling remained unchanged. It was for this reason that a new type of air-to-air missile with which to intercept an enemy at up to 15000 m (49,213 ft) altitude was sought. However, this did not become reality until the J 35F was developed, and the 35D had to rely on its short-range, IR-homing Rb 24 Sidewinder missiles.

The first true D version prototype flew on 27 December 1960. Serial production started in 1962 and delivery to F13 began the following year. The official debut of the J 35D was at a press briefing on 30 September 1963. The first production aircraft (35274) was used for ongoing trials at Saab and FC, and was not delivered to the air force until 1967; by then it had been modified as an S 35E (number 35960). Subsequent aircraft in the first D series batch of 30 flew for only a brief period of time with F13 before they were mothballed; they lacked radar, since the PS-03 set was still under test at this point. These aircraft were transferred to Saab in 1966 and were later converted to S 35E standard at CVM and Saab. With few exceptions, the original straight cockpit canopy was retained on this version.

The second serial batch comprised 35305-35328, excluding 35310. These were given the production designation J 35D1 and were delivered to F13 in the spring of 1964. When F13 later received J 35Fs, their J 35D1s were transferred to F10, from as early as the summer of 1964.

The third and last batch of this variant, comprising numbers 35329-35393, was known by the designation J 35D2. These aircraft had a complete set of electronic equipment, including the S7A FCS, PS-03A radar and 05 auto-pilot with Mach-holding mode. These aircraft were delivered to F13, F10, F3, F4 and F21. Most examples of the J 35D were replaced by J 35Fs, though F4 and F21 operated their J 35Ds until 1984, when they received Viggens.

D1 standard aircraft delivered to the wings were later modified to D2 standard at CVV Hässlö. Aircraft 35305-35353 were modified at CVV, with the exception for 35306, 35309, 35310, 35316, 35319 and 35331, which were altered by their home wings. The modification incorporated not only D2 equipment but also a new ejection seat, which required that the cockpit floor was lowered by 5 cm (1.97 in). All modifications were to have been finished by 1 January 1968, but the work lasted to the end of that year. The sub-designations D1 and D2 were then dropped, all aircraft becoming known simply as J 35Ds.

Of the rejected early series J 35Ds, Saab bought back 24 aircraft from the air force for conversion to 35OE standard and export to Austria. They were all modified with the 35F-type canopy.

Reconnaissance Draken – S 35E

During the spring of 1960, work began on a reconnaissance version of the Saab 35. The first test aircraft (number 35901) made its maiden flight on 27 June 1963, piloted by Ceylon Utterborn. This aircraft, never delivered to the air force, was a converted J 35D (number 35278) and was used as a test machine at both Saab and FC until it was finally

Projected Draken developments

JA 35 – the fighter/attack Draken

In the late 1960s, Saab proposed the JA 35 (JA for 'fighter/attack') with a larger radar and IR scanner unit. Preliminary designs were completed and wind tunnel tests conducted in April 1969. However, the results of these tests with a lengthened nose, enlarged radome and IR scanner indicated a significant increase in drag and reduction in yaw stability. Saab put the JA 35 forward as an alternative to the fighter version of the Viggen (later designated JA 37). The decision to pursue the JA 37 was taken some years later and the JA 35 was abandoned.

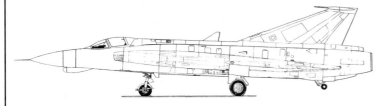

J79-powered AS 35X for export

As part of the on-going effort to export the Draken this radically reworked version was proposed. The AS 35X (AS for 'attack/reconnaissance') was to be powered by a General Electric J79/J1S turbojet giving 44.13 kN (9,917 lb) of dry thrust dry and 73.55 kN (16,528 lb) with afterburner. The J79 was shorter and had an outboard afterburner nozzle, allowing the AS 35X to have a short tail and dispense with its tail wheel. Wing span was increased by 0.6 m (1.96 ft) and the outer wing sported a 'dog tooth' leading edge. A new, more conical radome reduced drag.

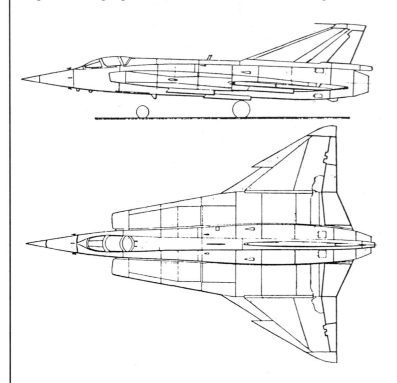

AJ 35 and 35 MOD

By the end of the 1970s considerable efforts were being made to find a way to upgrade Flygvapnet's remaining S 35Es and J 35Fs. The intention was to make the Draken compatible with Sweden's new air base system (BAS 90) and to provide an air-to-ground attack capability, pending the delivery of a successor to the AJ 37.

Features of the most advanced AJ 35 proposal (designated 'level B' or 'level 4') included a newly-redesigned outer wing (with a 'dog tooth' leading edge) and a pair of retractable canards. To manage a considerable increase in external load, the undercarriage was strengthened to a similar standard as that applied to Denmark's 35XD. Maximum take-off weight was estimated to be approximately 15000 kg (33,070 lb); by way of comparison the Danish Draken had an MTOW of 17000 kg (37,480 lb). The 'level 4' AJ 35 was also expected to carry a pair of Rb 15 anti-ship missiles, though this stipulation led to a problem with ground clearance, particularly when landing with the Rb 15s at an AUW of around 11000 kg (24,250 lb). Simulations showed that the clearance between the missile and the ground was only 44 mm (1.7 in).

The AJ 35's wingspan, with its new outer portion, was increased by around 1 m (3.28 ft), increasing the aircraft's turn capability. Combined with the retractable canard this feature also reduced approach and landing speeds at more desirable angles of attack. To increase yaw stability of the aircraft when armed with Rb 15s, the fin was raised slightly.

The less ambitious 'level 1B' modification programme included new underwing hardpoints, improved electronics, a larger drag 'chute, and new wheels and brakes. This modification formed the basis of the later J 35J conversions.

When the decision was made to develop the JAS 39 Gripen, the more complex

modifications to the Draken were abandoned. As a result, both the AJ 35 and the 35 MOD were halted at the design stage. Work on lesser modifications continued at FMV and Saab.

A number of different canard configurations were considered for AJ 35/35 MOD. These preliminary sketches show (below left, top to bottom) three proposed non-retractable canard configurations, all mounted on the engine intakes: swept canards, a shorter 'squarer' canard with a moving control surface and an all-moving canard, and two retractable canard configurations (below right), one mounted in a lengthened nose and the other mounted in fairings fitted to engine air intakes.

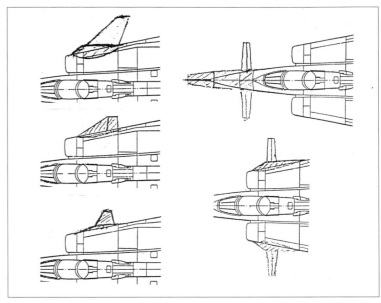

AJ 35 Draken project specification (as at 1 June 1979)

Length:	14.28 m (46.85 ft)
Height:	4.09 m (13.42 ft)
Wing span:	10.56 m (34.65 ft)
Wing area:	53.84 m² (580 sq ft)
Powerplant:	VFA RM 6C turbojet
Thrust (dry):	55.41 kN (12,452 lb)
Thrust (with afterburner):	76.05 kN (17,090 lb)
Internal fuel capacity:	2745 kg (6,051 lb)
External fuel:	2 x 500 litres (110 Imp gal)
Normal take-off weight:	11800 kg (26,014 lb)
Maximum take-off weight:	14800 kg (32,628 lb)
Take-off run:	750 m (2,460 ft)
Landing run (with drag 'chute):	600 m (1,968 ft)
Landing run (without drag 'chute):	800 m (2,625 ft)
Combat radius (lo-lo-lo):	303 km (188 miles)

Armament alternatives: Rb 75 (Maverick ASM), Rb 24J/Rb 74 (Sidewinder AAM), podded air-to-ground rockets, Rb 28 (Falcon IR AAM), Rb 15 (AShM), ECM pods and one internal 30-mm cannon

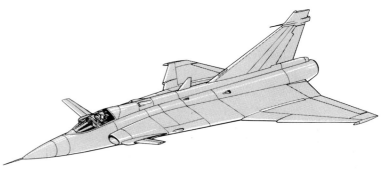

These drawings show the configuration envisaged for 35 MOD with the preferred canard configuration; the lower view shows an aircraft with 'level 4' modifications, carrying two Rb 15 AShMs, two Rb 74 Sidewinder AAMs and a jamming pod and chaff dispenser.

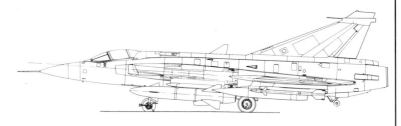

retired on 24 June 1974. The first new-build S 35E (number 35902) made its first test flight on 13 May 1965 and was delivered to F11 that summer.

Installation of the cameras in the Draken's slim nose posed a challenge. The French company OMERA/Segid developed the OMERA 31 camera to Swedish specifications, making it possible to install five cameras in the nose of a Draken and two downward-looking cameras in the wings. A vertical Jungner sight was used for aiming. For easy access to the cameras the nosecone could be pushed forward on rails. In addition, SRA installed a tape recorder that registered all signals and the pilot's voice, thereby aiding mission evaluation.

In 1965/66, 30 S 35Es (numbers 35902-35931) were delivered to F11 and F21. Those aircraft from number 35912 were fitted with the same type of canopy as the J 35F and all were equipped with the more powerful RM 6C engine (with EBK 67 afterburner). The reconnaissance Draken was also equipped with the same 'zero-zero' ejection seat system as used in the J 35F, J 35J and AJ 37 Viggen.

Operational conversion training on Drakens with F11 was well under way in April 1966, and to facilitate matters F11 borrowed two SK 35Cs from TIS 35 (F16). The S 29Cs of the 2nd and 3rd Squadrons were gradually replaced by S 35Es, although the 1st Squadron continued to operate S 32C Lansens. From autumn 1966 through 1967, an addi-

Of the three squadrons in F10, 3. Jaktflygdivision operated the J 35J for the shortest period, between 1989 and 1993. These 3./F10 aircraft, on a routine sortie, each carry a single Rb 74 round and 500-litre drop tanks.

Below right: As Saab made concerted efforts during 1960 to find an export customer for the Draken, the company designed oversized ferry tanks. A set of these tanks was experimentally fitted to J 35A 35017, though it is thought that no Draken ever flew with them. Each is believed to have had a capacity of around 2000 litres (440 Imp gal/ 528 US gal).

Below: Draken 35011 is pictured at Malmen on 1 September 1962, having been relegated to ground instructional status after being demonstrated to the Swiss two years earlier. Built as a 35A, the aircraft had been converted to 35D standard and further modified as the 35H demonstrator.

tional 29 S 35Es were delivered to F11 and F21. These aircraft (numbers 35932-35960) were modified J 35Ds and kept the original canopy. The first test aircraft in this second batch of S 35Es (35932) flew for the first time on 21 January 1964.

Reconnaissance Drakens could fly at supersonic speeds at low altitudes with up to four drop tanks holding 2000 litres (440 Imp gal/528 US gal) of fuel. These high-speed, low-level tactics led to a higher rate of failures in this version, mainly involving problems with the RM 6C engine, which suffered structural cracks and vane fractures. The S 35E served with F11 until the Wing was disbanded in 1979, and with F21 until SF/SH 37 Viggens began to be delivered to F21, F13 and F17.

Eighteen S 35Es were lost in accidents; in 1980 five of the original S 35Es were sold to Denmark, where they served as ground instruction and spares aircraft.

S 35Es were heavily modified during their final years of service. They were equipped with an extensive ECM system known as 'apparat 15', consisting of a radar-jamming unit, a radar warning receiver and chaff/flare dispensers. This equipment was similar to that fitted as part of the NATO WDNS (Weapon Delivery and Navigation System) modification applied to Danish Drakens. The radar-jamming unit comprised two small pods mounted under the outer delta wings. Flare dispensers were installed in the back of the extended afterburner cooling inlets.

The definitive Draken fighter – J 35F

The Saab J 35F was developed directly from the RM 6C-powered 35D version, but differed in the important respect that it was to be armed with a new air-to-air missile. In March 1959, the air force decided to buy Hughes Falcon

AAMs to equip future Draken production aircraft. Thus, J 35F armament consisted of one 30-mm gun in the starboard wing (the port gun having given way to an air data computer); the Rb 27 radar-homing missile (Hughes HM 55) for all-weather, medium- to high-altitude intercepts and the IR-homing Rb 28 (Hughes HM 58) for visual day and night medium- to low-altitude intercepts from behind. Rb 24s (Sidewinders) could still be carried but were not integrated with the radar/FCS. The J 35F also retained the air-to-ground capability of earlier Drakens.

The type's new PS-01/A intercept radar, with its integrated S7B FCS, was developed by Ericsson. Other equipment included the PN-594/A navigation radar and the PN-793/A IFF (Identification Friend-or-Foe).

In the J 35F, for the first time, the Draken pilot was a 'system operator' and faced a much increased workload. Most important information regarding the heading, distance and altitude of a target was transmitted from STRIL 60 ground stations via a datalink and presented to the pilot on his instruments and radarscope. This system was first introduced with the J 35D and was, at that time, advanced by international standards. As all information was digital and 'aimed' at an aircraft with considerable accuracy (a precursor of modern datalinks), it was very difficult to jam. There was no need for radio conversation between the fighter controller and the pilot, and the fighter controller could even send short messages via the datalink.

Once a target was found, the pilot aimed and locked the radar on the target, using a control column on the left of the cockpit. He could then choose the best weapon and direct the aircraft's FCS computer to make the required calculations and give the pilot correct firing instructions.

J 35F system tests were finished by late 1965/early 1966, by which time some 80 Falcon missiles had been fired. Saab and Bofors manufactured the Rb 27 and Rb 28 under licence from the American manufacturer, the Hughes Aircraft Company.

Initially, the new Draken fighter variant was designated J 35B3 and the first test aircraft 35-10, a modified J 35A (35082), flew for the first time on 22 December 1961. Most of the other test aircraft (35-6, 35-7, 35-8, 35-9, 35-11, 35-12 and 35-13) were used for later tests of armament and electronics.

Saab began series production of the J 35F at the same time that the final J 35Ds were being assembled. The first

First export attempt – Saab 35H

In the late 1950s a close battle was fought between Dassault and Saab for the contract to supply Switzerland with aircraft to replace its de Havilland Vampires. While work commenced on designing a Draken to suit Swiss requirements, the fifth prototype (35-5) visited Switzerland in 1958, giving a spectacular display over Basle on 7 September.

To fill the requirement for a new fighter the Saab 35H was proposed by the company later that year and a J 35A (35011) converted as a demonstrator. Based on the 35D, the 35H was to be equipped with a British Ferranti AI.23 Airpass radar, as installed in the contemporary English Electric Lightning. 35011 first flew in its modified state on 29 November 1959 and was demonstrated together with its main competitor, the Mirage IIIC, in November 1960. Unfortunately this early attempt to sell the Draken abroad came to nought, Switzerland opting for the locally-built Mirage IIIS.

Saab J35 J Draken
3./F10 Ängelholm 1993

'Johan' 35586 was built as a J 35F² and entered Flygvapnet service on 30 October 1969. Its entire career was spent with F10, initially coded '06' and, later, '68' before entering the J 35J conversion programme. Its swordfish wing markings are derived from the 3. Division emblem and were applied to '62' in its role as F10 Wing's display aircraft.

Designation

F10's three constituent squadrons were Johan Röd (1./F10, red), Johan Blå (2./F10, blue) and Johan Gul (3./F10, yellow), corresponding to the 10th letter of the Swedish phonetic alphabet. Because these aircraft would be operated solely by F10, air force high command intervened to ensure that the upgraded J 35F²s be designated J 35J.

Draken in the 1990s

Modified to J 35J standard, the Draken was able to function more effectively in the new tactical environment of the 1990s. Increased armament capacity (more rounds for the gun and up to six missile pylon stations), improved target tracking capability for head-on intercepts at low level, improved ability to counter ECM and track manoeuvring targets, and increased flexibility and speed in the handling of all weapons, both in 'head-up' and 'head-down' modes, were the key areas of improvement. The aircraft's existing ability to withstand high g loads, compared to more modern aircraft, remained an important advantage. During a mission lasting approximately 50 minutes, a fully armed and fuelled J 35J could achieve very good results for half the operating costs of a Viggen. The cost of a Draken flight hour during FY 1993/94 was calculated at SEK12,500 (£1,250) compared to nearly SEK25,000 (£2,500) for the JA 37 Viggen.

Armament

The addition of two new hardpoints under the engine intakes (wired to carry Rb 74 Sidewinder, but not Rb 27/28 Falcon) was an important improvement. It allowed the aircraft to carry two drop tanks and four air-to-air missiles, or four tanks and a pair of Rb 74 Sidewinders. The J 35F, from which the J 35J was converted, was restricted to either four missiles and no external fuel or two missiles and two drop tanks. Other armament changes included improvements to the firing mechanism of the single ADEN cannon and an increase in the number of rounds carried.

Ejection seat

Beginning with late-production J 35Ds, the Draken was equipped with a Saab RS-35 'zero-zero' ejection seat. This replaced the earlier 73SE-F design which, while it had a 'zero altitude' capability, required that the aircraft had a forward speed of at least 100 km/h (62 mph) in order to function correctly.

Right: Saab 35BS DK-202, one of the original batch of 35Bs leased in 1972, was photographed during its last flight in September 1993. By then it was finished in the later version of the Ilmavoimat's Draken livery, with small national insignia and small black serials applied to the forward fuselage, above the roundel.

Identifiable by its undernose IR sensor, the 35S was the only Draken variant assembled outside Sweden. A derivative of the J 35F², the 35S was able to carry both IR- and radar-guided versions of the Falcon AAM.

Parallel to the manufacture of the first 100 J 35Fs was an intensive development of electronic and IR techniques. The result was that the F version, from number 35501, was equipped with an IR sensor (S71N) under its nose slaved to the radar antenna. The scanner was designed by Hughes and licence-built by Ericsson.

Using this scanner, heat radiation from the target was converted into electrical signals by the FCS computer, which then presented steering information to the pilot on his radar scope. This was a great improvement, making it possible to approach an enemy from behind in clear weather without being given away by radar emissions. Only when the aircraft was within firing range did the radar have to be switched on to obtain accurate target data. If radar-homing missiles were to be used, it was necessary to give the missile a reflected homing signal, but if IR-homing missiles were selected a totally 'silent' intercept was possible. The aircraft's radar, after modifications to incorporate the IR sensor, was redesignated PS-11/A.

The new F variant was designated with a superscript '2' suffix (J 35F²) and production examples were delivered between October 1967 to June 1972, when Draken production finally ceased. The original J 35F was retroactively designated J 35F¹. Saab delivered a total of 130 J 35F² aircraft (35501-35630), originally to F13, F10, F3 and F1.

The modifications introduced in the F² enhanced the value of the J 35F, especially where there was a major electronic jamming threat and, particularly, at night. The IR

three production J 35Fs (35401-35403) were used for ongoing testing, number 35404 being the first genuine J 35F. It made its first pre-delivery flight on 9 February 1965 and received approval three months later when it was delivered to F13.

F13 Wing at Norrköping received around 40 J 35Fs up to November 1966, as successors to its J 35Ds. The next wing to get the F version was F1 Västerås, which received its new aircraft between October 1966 and September 1967. These aircraft were intended to replace the obsolete J 32B Lansen. Another 10 J 35Fs were delivered to F13 in 1967. A total of 100 J 35Fs (35401-35500) had been manufactured and delivered to Flygvapnet by the mid-1980s. Most of the first J 35As and Bs delivered had been retired; others were sold to Finland.

To augment the Finnish Draken fleet after the delivery of a dozen 35S aircraft, the Ilmavoimat acquired two batches of ex-Flygvapnet J 35F¹'s totalling 24 machines. These were known locally by the 35FS designation. DK-237 is the nearest aircraft to the camera in this May 1992 view.

Finland's 35CS two-seaters were simply ex-Flygvapnet SK 35Cs with minor changes made to communications and navigation equipment. Though a post-World War II peace treaty restricts the Ilmavoimat to 60 combat aircraft, the 35CS two-seat and 35BS single-seat aircraft were classified as trainers and were therefore not included in the total.

sensor had the additional benefit in that it offered a limited 'look-down' capability; the Draken's radar was not designed with any such ability.

The last Drakens – J 35J 'Johan'

In the late 1980s, 67 J 35F²s were modified to what would be the ultimate fighter Draken standard – J 35J. The aircraft chosen for modification were 'low-time' examples from those mothballed. The remaining 63 J 35F²s were retired with expended fatigue life or with accident damage. Typically, a Draken would accumulate between 2,300 and 2,600 flying hours before retirement.

In 1985 the Swedish government decided to modify 55 J 35F²s to J 35J standard (including a prototype). In 1987,

Finnish Draken variants and serial numbers

J 35A
One aircraft (35026) was received in August 1976 as a gift for the Rovaniemi vocational school. Later, this aircraft was also used as a static demonstration airframe in northern Finland and was marked DK-200. The tailcone was probably used to repair DK-264.

35S (35XS)
Twelve examples of this version were manufactured by AB Valmet in 1974/75 and delivered to HävLLv 11. Main components were delivered from Saab at Linköping and the aircraft were assembled in Finland. This variant broadly corresponded to the Swedish J 35F² except for minor changes in the electronic system. Its missile armament consisted of Falcons manufactured by Saab at Linköping. These aircraft were normally equipped with an IR sensor and were continuously upgraded until, by the time of their retirement, they had a capability close to that of the Swedish J 35J.

35BS (J 35B)
Six 35BS aircraft were leased from Sweden in 1972. Following an engine fire that caused one 35BS to be scrapped, a replacement was leased. In 1976 the Finnish Air Force purchased all these aircraft. With the exception of modified electronics, the BS corresponded to the Swedish J 35B. They had a low level of equipment, lacking such items as radar, and were used for basic training until retired in 1991-92.

35CS (SK 35C)
Five Swedish SK 35Cs were purchased for Finnish conversion training and underwent minor changes in the communication and navigation systems.

35FS (J 35F¹)
In 1976 six used J 35F¹s were purchased, followed by 18 delivered in 1984/85. Some were used for basic flight training and technical education.

To increase range and endurance, Finnish Drakens were able to employ 1275-litre (280-Imp gal/337-US gal) external fuel tanks (as used by their Danish counterparts). Gradually, the airframe life of the Finnish Drakens was extended from 1,500 hours to 2,700 hours and all were modified with second-generation rocket ejection seats that included an emergency survival kit.

Finland's Drakens were serialled in an unusual manner, as the accompanying table illustrates. Even numbers were allocated to those aircraft employed as trainers, including the leased Flygvapnet J 35Bs, a J 35A acquired as an instructional airframe and, later, a batch of two-seat 35CSs delivered in 1976. Odd numbers were applied to the 'combat' aircraft, i.e. the 35S aircraft assembled by Valmet and to all subsequent deliveries of ex-Flygvapnet single-seat Drakens.

variant	serial	c/n	delivery (variant as built)
J 35A	DK-200	35026	1976
35S	DK-201	351301	1974
35BS	DK-202	35265	1972 (J 35B)
35S	DK-203	351302	1974
35BS	DK-204	35261	1972 (J 35B)
35S	DK-205	351303	1974
35BS	DK-206	35266	1972 (J 35B) later DK-942
35BS	DK-206	35245	1984 (J 35B, replacement)
35S	DK-207	351304	1974
35BS	DK-208	35214	1972 (J 35B)
35S	DK-209	351305	1974
35BS	DK-210	35243	1972 (J 35B)
35S	DK-211	351306	1975
35BS	DK-212	35257	1972 (J 35B)
35S	DK-213	351307	1975
35S	DK-215	351308	1975
35S	DK-217	351309	1975
35S	DK-219	351310	1976
35S	DK-221	351311	1975
35S	DK-223	351312	1975
35FS	DK-225	35417	1984 (J 35F)
35FS	DK-227	35425	1986 (J 35F)
35FS	DK-229	35481	1986 (J 35F)
35FS	DK-231	35416	1986 (J 35F)
35FS	DK-233	35443	1985 (J 35F)
35FS	DK-235	35444	1986 (J 35F)
35FS	DK-237	35446	1985 (J 35F)
35FS	DK-239	35447	1985 (J 35F)
35FS	DK-241	35448	1986 (J 35F)
35FS	DK-243	35450	1985 (J 35F)
35FS	DK-245	35451	1985 (J 35F)
35FS	DK-247	35441	1986 (J 35F)
35FS	DK-249	35455	1986 (J 35F)
35FS	DK-251	35458	1985 (J 35F)
35FS	DK-253	35462	1985 (J 35F)
35FS	DK-255	35483	1986 (J 35F)
35FS	DK-257	35487	1984 (J 35F)
35FS	DK-259	35499	1985 (J 35F)
35FS	DK-261	35460	1976 (J 35F)
35CS	DK-262	35823	1976 (SK 35C)
35FS	DK-263	35412	1976 (J 35F)
35CS	DK-264	35820	1976 (SK 35C)
35FS	DK-265	35489	1976 (J 35F)
35CS	DK-266	35803	1976 (SK 35C)
35FS	DK-267	35449	1976 (J 35F)
35CS	DK-268	35807	1984 (SK 35C)
35FS	DK-269	35445	1976 (J 35F)
35CS	DK-270	35812	1984 (SK 35C)
35FS	DK-271	35493	1976 (J 35F)
35BS	DK-942	35266	1972 (J 35B) ex DK-206
J 35B	-	35252	1976 static demonstrator

The Danish soldier emblem on the tail of the much modified Draken prototype 35-6 indicates service as a trials airframe during development of the 35XD variant for Denmark. Test flying of the aircraft, already utilised during development of the J 35D and 35F, was carried out at FC during 1967.

Lower right: Construction of the second 35XD (F-35) is well under way in this view of the Saab production line at Linköping in 1969/70.

Danish 35XD Draken variants

F-35

The F-35 was a fighter-bomber with equipment similar to the Swedish J 35F but a substantially increased load capacity. The avionics were adapted to meet Danish requirements and were therefore not as extensive as in the J 35F; for instance, the F-35 lacked a radar installation. Internal fuel capacity was increased to 4034 litres (887 Imp gal/1,065 US gal) (30 per cent more than in the J 35F) and two external fuel tanks, each with a capacity of 1275 litres (280 Imp gal/337 US gal), could also be used.

Armament was carried on three hardpoints under each wing and three under the belly. Unlike the J 35F, the F-35 was armed with two 30-mm ADEN guns. For training purposes, the port gun could be replaced by a 12.7-mm gun installation. Various types of external armament could be carried, including AIM-9B and AIM-9N-3 Sidewinder AAMs, bombs, AGM-12B Bullpup air-to-surface missiles and rocket pods. The F-35's ejection system was the same as that fitted to Swedish J 35Fs.

With a maximum take-off weight of nearly 17 tons, this was without doubt the heaviest Draken version built. It had a take-off speed, when fully loaded, of nearly 370 km/h (230 mph). In order to manage the increased take-off weight, the Saab 35XD was equipped with a strengthened undercarriage. A new stall warning system was developed as a result of a direct request from the Danes and was later installed in all Swedish Drakens. Additionally, all Saab 35XDs had a modified auto-pilot system.

RF-35

This reconnaissance version had five Vinten 360 and 544 cameras mounted in the nose, like the Swedish S 35E. Unlike the unarmed Swedish aircraft, however, RF-35s were fully armed and did not carry wing cameras. Like the F-35, the RF-35 had 18 internal fuel tanks with a total capacity of 4034 litres (887 Imp gal/1,065 US gal) plus two 1275-litre (280-Imp gal/337 US gal) external fuel tanks. The RF-35 was capable of carrying the Swedish 'Red Baron' IRLS (infra-red line scan) pod.

TF-35

The twin-seat TF-35 was developed from the J 35F Draken and was equipped with the same tactical equipment as the F-35, with the exception of carrying only one gun, in the port wing. It was fitted for A-38DK ECM pods. The TF-35 had only 16 internal fuel tanks, giving a capacity of 3227 litres (710 Imp gal/852 US gal), but retained the two 1275-litre (280-Imp gal) drop tanks. The Danes modified some drop tanks into so-called 'TIC tanks', with a compartment for the pilot's personal luggage.

Right: From the outset the Danish 35XD Drakens were to be multi-role aircraft, though with a much greater emphasis placed on their ground-attack capability. A strengthened wing, with greater internal fuel capacity, was able to carry much increased weapon loads over longer distances, especially when twin 1275-litre drop tanks (pictured) were carried. This F-35 carries four 250-kg (551-lb) bombs on four of its six underwing pylons. With the retirement of Denmark's Drakens in 1993, the 1275-litre tanks were purchased by Saab and refurbished for use on Flygvapnet Gripens.

after the shape of Sweden's future defence needs had been formalised, a follow-on order for the modification of 12 aircraft was placed. The J 35J was intended to fill a gap in Sweden's air defence capability pending the first deliveries of the JAS 39 Gripen.

The commander of the air force decided that the version letter should be a 'J' (instead of the more logical 'G') as 'J' is the 10th letter of the alphabet and F10 Ängelholm was to be the only wing to operate the Draken. By then F10 was responsible for all Draken maintenance, for the TIS and for ongoing support provision to foreign operators.

The decision to modify the J 35F[2] was preceded by several years of investigation and analysis by FMV (Försvarets Materielverk, formerly the KFF). The main purpose of the 35J modification was to bring the type's electronic equipment up-to-date, thereby increasing its operational effectiveness; extensive modifications were envisaged.

Among the enhancements introduced were an improved radar/FCS and IFF system, a modified 71N IR scanner with thermo-electric cooling, a new air data computer system (LD 8), a new clock, backup instrumentation (altimeter, air speed indicator), a new g meter, an improved cannon installation and two extra hardpoints fitted below the aircraft's engine intakes.

F10 first prepared the aircraft selected by the FMV for conversion; these were then delivered to FFV Aerotech at Linköping which dismantled the airframe. The forward section was transported by road to Saab-Scania at Linköping for the installation of electronic and armament equipment, while the aft section remained at FFV for refurbishment. When work on the forward section had been completed by Saab, it was returned to FFV for re-assembly and testing.

Aircraft 35598, designated J-0, was the first prototype and was tested at Försökscentralen, coded 'FC-18'. Later, 35619 and 35576 were employed in further testing at FC. Tactical trials and evaluation were undertaken by F10 during 1987/88, and on 17 December 1987 FMV was issued with a type certificate for the J 35J. Excluding J-0, some 66 J 35F²s became J 35Js. The first two aircraft were handed over to the air force at a ceremony at Saab on 3 March 1987, and the last to F10 on 21 August 1991.

In 1986, the TIS 35 was transferred from F16 Uppsala to F10 Ängelholm. Four to six student pilots each year were trained on the remaining SK 35Cs. The old SK 35C simulator at F16 was also transferred to F10 and the J 35F simulator was modified to J standard during 1987/88. The first training of J 35J pilots began in 1988.

It was estimated that the J 35J would be in service until 1997, beyond which time Draken operations could not be guaranteed. During spring 1992, despite calls from some

Denmark's first F-35, A-001, was retired in 1992, when Flyvevåbnet concentrated all Draken operations with Esk 729. In 2002 it is displayed at the recently opened aviation museum in Helsingør, Denmark. Note that the Danish aircraft lack the dorsal fin fitted to all other Drakens.

Danish Draken variants

Denmark's 35XD Drakens were the only new-build examples exported from the Saab factory; the second batch of five TF-35s, delivered in 1977, were the last Drakens built, 22 years after the type's first flight. The S 35Es and J 35Fs delivered during the 1980s were acquired as a source of spare parts.

variant	serial	c/n	delivery date
F-35	A-001/A-020	351001-351020	1970-1971
RF-35	AR-101/AR-120	351101-351120	1971-1972
TF-35	AT-151/AT-161	351151-351161	1971-1977
S 35E	–	35905	1980
S 35E	–	35922	1980
S 35E	–	35925	1980
S 35E	–	35929	1980
S 35E	–	35931	1980
J 35F	–	35420	1987
J 35F	–	35552	1987

Progressively upgraded, the Danish Drakens were equipped with fin-top and wing-tip RWR antennas, chaff/flare dispensers and formation lights during the 1970s, prior to the comprehensive WDNS upgrade undertaken during the 1980s. In this view of F-35 A-018 its reshaped nose, housing the Ferranti LRMTS, is evident.

Above: Denmark's TF-35s were the only two-seaters built from scratch and, as they were derived from the F-35 and thus equipped with RM 6C engines, were the most powerful of the 'twin-stick' Drakens. Apart from having only one ADEN cannon, the TF-35 had the same weapons carrying and delivery capability as the single-seat aircraft and were upgraded to the same standard.

Above: This well-armed Viking-like character symbolised the new offensive and defensive capabilities of the Danish WDNS-equipped Drakens. The programme logo also acknowledged the four contractors supplying avionics for the upgrade.

politicians for the disbanding of F10, the government pointed out that disbanding F10 was out of the question so long as Sweden was party to binding contracts with those countries that had bought Drakens – Denmark, Finland and Austria. These contracts ran until 1995 and each country had the option of buying continued services after that date.

The result of this political wrangling was a decision to disband Viggen units F6 Karlsborg and F13 Norrköping, which meant that F10 Ängelholm had to find room for different versions of Viggens and the personnel from these wings. In the end, the 'Ghost Riders' of 1./F10 re-equipped with AJS 37 Viggens, which led to the mothballing of a large number of J 35J aircraft, mainly due to a lack of hangar space. These aircraft went into storage in the under-

The WDNS upgrade incorporated a new ECM suite, the centrepiece of which was the AN/ALQ-162 jamming pod.

ground hangar of F9 Säve (Göteborg). Most of these aircraft were scrapped; a handful were donated to museums.

In 1994, great political changes – both in Sweden and the rest of Europe – led to an alteration in the air force's future development plans. From 1999, the JAS 39 Gripen would replace both Drakens and AJ 37s. Furthermore, for economic and political reasons, it was decided to disband F10 at the end of 2002, even though Ängelholm had been completely upgraded to operate the JAS 39 Gripen, at considerable expense.

So it was that, on 12 December 1998, the Draken was officially retired by Flygvapnet, the final Draken sortie by F10 taking place on 18 January 1999. A single Draken (J 35J 35556) was retained in airworthy condition as part of the Flygvapnet historic flight at Ängelholm.

Finland's Drakens

As early as the beginning of the 1960s, the Finnish air force (Ilmavoimat) entered negotiations with Sweden for the purchase of modern aviation equipment, but it was a decade before these plans were realised. On 1 October 1970 a parliamentary defence committee was appointed to evaluate the political situation regarding defence and to create guidelines for future development, completing its work on 23 June 1971. Two other committees were appointed on 27 March 1975 and 20 December 1979, the latter reporting on 5 March 1981.

Despite calls from communist elements in the Finnish parliament for close co-operation with the Soviet Union, on the pretext that Finland should not become dependent on Western technology, the Finnish cabinet took the decision, on 8 April 1970, to buy Saab 35S aircraft (S for Suomi/Finland; referred to as 35XS in some sources) for local assembly. (The first parliamentary defence committee later proposed the purchase of fighters to fill two fighter wings.) On 25 June 1970 the Finnish parliament approved the purchase by a majority of 127 to 35 against. The deal included weapons and armament systems, spares and 12 Saab 35S, to be assembled at the AB Valmet factory in Kuorevesi. The total cost was FIM200 million (SEK100 million). Like the J 35F (from which it was derived), the 35S was armed with Falcon air-to-air missiles.

Swedish equipment continued to be purchased well into the 1980s. In addition to the 12 Saab 35Ss assembled in Finland between 1974 and 1975, six Saab 35Bs were leased on a long-term basis in 1972 and designated 35BS. These aircraft were purchased outright in 1976, at the same time as another batch of six used Saab J 35F's (known locally as 35FSs) and three Saab SK 35C (35CS) trainers were purchased. The cost of these 15 used aircraft was FIM63

Denmark's surviving Drakens were gathered together at Karup in May 1991 on the occasion of the Esk 725's 40th anniversary. Here the aircraft are seen, lined up in serial number order, with LISBON 725 (A-009) standing out, eighth from the camera. (A-003 had been lost in an accident on 20 September 1974).

million (around SEK80 million) and also included a simulator, service contract and one J 35A for the training school at Rovaniemi. Financing was included in the budgets for FYs 1976 to 1980.

On 29 March 1984 it was decided to purchase two additional 35CSs and 18 Saab 35FSs, delivered between 1984 and 1986. This purchase included one J 35B to replace DK-206, which had to be written-off with fire damage. These Drakens were based at HävLLv 21 at Tampere-Pirkkala Air Base in southeastern Finland. AB Valmet (later renamed Finavitec) was responsible for the maintenance, conversion and modification of Finland's Drakens and also serviced the Swedish-made Falcon missiles carried by the 35S and 35FS aircraft.

During the autumn of 1990, Requests For Proposals were made with a view to finding a successor to the Saab 35. The JAS 39 Gripen was considered, but on 6 May 1992 the Finnish government opted for the McDonnell Douglas F/A-18C/D Hornet.

HävLLv 21 became the first Hornet-equipped unit, receiving its first aircraft in 1995. The squadron's Drakens bowed out in September 1997 as its new F/A-18s became fully operational, the final Draken sortie (by two 35FSs and a 35CS) taking place on 30 November. The redundant aircraft were transferred to HävLLv 11 at Rovaniemi and remained on active duty until the second half of 2000, final retirement taking effect on 16 August. The Draken had completed 29 years of service with the Ilmavoimat.

Drakens in Denmark

In the spring of 1967 Saab began developing a fighter-bomber version of the Saab 35X (X for export). It was known that the Flyvevåbnet (Danish air force) was looking to purchase a new aircraft to handle all types of mission, including interception, ground-attack and bombing. The test aircraft 35-6, which had flown for the first time on 19 January 1961, undertook extensive 35X testing. This aircraft had already been employed as a test vehicle for 35D and later the 35F variants of the Swedish Draken and, to this end, had been fitted with a strengthened wing with more hardpoints. In 1967 a series of test flights was made with substantially increased external loads. Two drop tanks each of 1275-litre (280-Imp gal) capacity were carried, extending range to 3000 km (1,864 miles).

Above: During 1986 RF-35 AR-117 was adorned with a Danish flag as a gesture of support for the Danish national soccer team, participating in the World Cup in Mexico. The aircraft (callsign DYNAMITE 6-1) made two flights around Denmark to celebrate wins in the opening round of the tournament. Upon its return to Karup the water-soluble paint was to be removed, but this proved impossible and the Draken completed some 20 hours of operational flying before falling due for an overhaul and repaint!

Left: 'Superstall' training for Danish Draken crews was carried out using TF-35 AT-154. This two-seater was equipped with a new tailcone (from one of the ex-Flygvapnet S 35Es acquired in 1980) on which was mounted an anti-spin parachute in place of the standard drag 'chute. The Danish anti-spin installation was tested for the first time on 7 March 1984. Previously 'superstall' training had been carried out using Swedish aircraft; with a modified TF-35 it was now possible to provide more realistic training as the Danish aircraft were considerably heavier than their Swedish cousins.

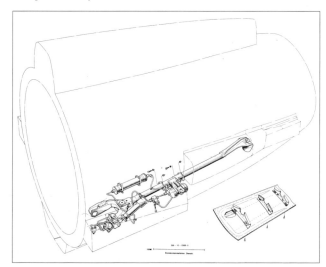

A feature peculiar to the Danish Drakens (though it was offered by Saab as an option on all aircraft) was an arrester hook, housed in the extreme rear of the aircraft and allowing an arrested landing on a suitably equipped airfield, should the need have arisen.

The first two of 24 Saab 350Es refurbished for the Austrian air force were flown directly to Vienna from Sweden. Such a flight pushed the aircraft to the outer limits of their range and there was little margin for error if weather conditions deteriorated. Though the flight was completed without mishap, the risks were recognised and all subsequent delivery flights were made over two legs, with a refuelling stop in West Germany. Pictured off Sweden's southern coast is the last of the Austrian aircraft – c/n 351424, formerly the last production J 35D c/n 35393, built in 1965 and a veteran of service with Flygvapnet wings F13, F3 and F21.

Far right: An Austrian pilot steps into his Draken at the beginning of a sortie. Note that his flying suit is of Swedish origin and even includes a Flygvapnet shoulder patch. As the poorly-equipped Austrian air force had not operated a 'modern' supersonic fighter type before the Draken, the contract for the supply of 24 aircraft included all manner of associated equipment, including flying suits.

Below: Armourers replenish the 30-mm ammunition in an Austrian Draken. Until civil war broke out in neighbouring Yugoslavia and restrictions imposed by the Staatsvertrag were lifted the Austria's Drakens were armed solely with ADEN cannon.

On 29 March 1968 a contract was signed in Copenhagen by the Swedish and Danish Ministers of Defence covering the delivery of 23 Saab 35XDs, including three two-seat conversion trainers. This first batch would cost DKK250 million ($US35 million).

Competing against the Draken for the Danish order were the French Dassault Mirage 5 and the American Northrop F-5. Final bids, made on 25 March 1968, for the supply of two squadrons of aircraft (around 50 machines) were as follows:

- Draken   DKK450 million   ($US64.3 million)
- Mirage 5   DKK497 million   ($US71.0 million)
- F-5   DKK514 million   ($US73.4 million)

The F-5 was soon rejected on the grounds of cost and because it was judged as unsuitable. Finally, a combination of price, compensation agreements and projected running costs tipped the balance in favour of the Draken. When deliveries started in 1970, the 35XD (known as the F-35 in Danish service) replaced one of three squadrons of ageing F-100 Super Sabres. When the first order was signed in 1968, another contract for an optional 20 RF-35s (a photo-reconnaissance variant) was signed, with a view to replacing the Flyvevåbnet's RF-84F Thunderflashes. Within three months, the option was taken up as a firm order, to which was added an order for three additional TF-35s.

On 1 September 1970 the first three F-35s (A-002 to -004) were delivered to Esk 725 at Karup; deliveries were completed in May 1971. The three TF-35s arrived in June (AT-152 and -153) and October 1971 (AT-151).

F-35 A-001 (which flew for the first time on 29 January 1970) was used as a test aircraft and went through tactical trials at FC in Sweden before it was flown to Denmark on 1 April 1971. A small number of Danish pilots underwent conversion training with Flygvapnet in order to serve as instructors when training began in Denmark.

Between May 1971 and May 1972, 20 RF-35 were delivered to Esk 729 at Karup. Another three TF-35s were delivered to each of Esk 725 and 729 between December 1971 to April 1972. Five additional TF-35s, ordered later, arrived between April 1976 to April 1977.

In 1980 five retired Swedish S 35Es were purchased as a source of spare parts. Two J 35Fs followed (one in 1987); the complete nose section of one of these airframes was used to build a simulator.

Austria's 35OEs

Austria's 24 refurbished ex-Flygvapnet J 35Ds (redesignated 35OE) were delivered between 1987 and 1989. All were given new construction numbers in the range 351401-351424. Their former identities were as follows: 35313, 35314, 35315, 35317, 35323, 35324, 35328, 35335, 35336, 35338, 35340, 35341, 35342, 35347, 35351, 35360, 35368, 35370, 35373, 35378, 35382, 35384, 35386 and 35393. In addition, five ex-Flygvapnet J 35Fs were purchased in the late 1990s as a source of spares.

One of two J 35Fs sold in the United States for use in motion pictures, J 35F² N543J (ex-35543) was maintained in airworthy condition and appeared in the movie Wings of the Apache. *Owned by James Merizan and operated by Flight Test Dynamics Inc., the aircraft was later resprayed in the house colours of the Miller Brewing Company with 'Miller Genuine Draft' titles and logos, as part of the brewing company's sponsorship of a team competing in the Indianapolis 500 motor race.*

As elsewhere, the Flyvevåbnet Drakens were continuously modified during their service. An AIMS altimeter replaced the original Swedish altimeter and a fatigue gauge replaced the Swedish strength and load registration system. In the first half of the 1970s, a new RWR system (AN/ALR-45) was introduced; in the second half, a chaff and flare dispenser system was fitted. The latter was updated during the later WDNS programme. The AN/ALR-69 RWR later replaced ALR-45, its installation comprising six new antennas on the fin, two on the nose (under the pitot tube) and one in each wingtip. The tailcone housed chaff and flare dispensers.

WDNS upgrade

In the mid-1980s all Danish Drakens went through the comprehensive WDNS (Weapon Delivery and Navigation System) modification that included a new Lear-Siegler navigation/attack computer for mission planning and execution, a Marconi Series 900 head-up display and a Singer-Kearfott inertial navigation system. The F-35 and TF-35 were equipped with a nose-mounted Ferranti Laser Rangefinder and Marked Target Seeker (LRMTS) and there-

'Draken No. 1' – the first prototype 35-1 – made its last flight on 4 February 1964 and finished up at a Flygvapnet technical school at F14 Halmstad as an instructional airframe. It was later sold to become the centrepiece of the Svedino Air & Car Mueum at Ugglarp (north of Falkenberg), founded by former Draken pilot Björn Svedfelt.

after sported reworked nose cones not unlike those of the RF-35 reconnaissance aircraft. A new AN/ALQ-162 podded jamming system completed the package.

F-35 AR-109 was to be the WDNS test aircraft and made its first test flight on 17 August 1981. All modification work (4,500 man-hours per aircraft) was carried out by the air force at Karup AB and was completed on 30 May 1986 when the last (A-014) of 43 Drakens rolled from the workshop.

According to the Flyvevåbnet, the F-35 fitted with the WDNS had a similar combat capability in the ground-attack

An air display at Ängelholm on 10 June 1990 saw a flypast by 28 F10 Drakens in a giant 'Draken' formation. By this time F10 was the last Flygvapen Draken wing, reprieved as a stopgap pending the arrival of JAS 39 Gripens and continuing to provide support for foreign Draken operators (a role that Sweden was obliged to fill until 1995).

In 2002 Sweden's air force historic flight continued to operate the last J 35J conversion (35556) for display purposes. The flight was based at Ängelholm, alongside the JAS 39 Gripens of 1. and 2./F10. (The red AJS 37 Viggen pictured was finished in an overall red scheme to mark the retirement of the type by 1./F10 in June 2001.)

role to the Lockheed Martin F-16 and was superior in the reconnaissance role.

When the Draken entered Danish service the aircraft's useful service life was projected to end in 1985, but as had been done elsewhere, the type's service life was extended through modification.

For economic reasons, Denmark was forced to concentrate Draken activities in only one squadron from early 1992 and therefore all remaining Danish Drakens flew with Esk 729 at Karup until early 1994, when their duties were taken over by Flyvevåbnet F-16s . A farewell ceremony was held in early December 1993 to mark the Draken's retirement and the disbandment of Esk 729. By this time there were seven RF-35s, two TF-35s and up to eight F-35s (including reserves) in use.

Between 1970 and 1994, a total of 143,944 flying hours was amassed in the Danish Drakens. Nine aircraft were lost and three pilots killed in fatal accidents. Up to 1992 average crash frequency per 10,000 flight hours was 0.56 and 0.21 for fatal crashes. This compares favourably with the

equivalent Swedish figures of 1.58 and 0.51, respectively, up to 31 December 1992.

A number of the remaining TF-35s and RF-35s were sold to the National Test Pilot School and Flight Test Dynamics Inc., both based in California. Four TF-35s were delivered to the NTPS for DKK18 million ($US3 million), the deal including a 12-year contract for the training of Danish pilots at this school.

35OE for Austria

Austria originally considered purchasing Drakens as early as 1967, the Swedish fighter joining a shortlist that included the Soviet MiG-21, the French Mirage III and the American Douglas A-4 Skyhawk. However, no purchase was made for financial reasons.

In order to temporarily make good the lack of supersonic fighters in the Österreichische Luftstreitkräfte (Austrian air force), the Ministry of Defence resumed discussions regarding the purchase of up to 24 Saab 35s on 14 July 1982. Evaluation and negotiations began in 1983 and continued

Commemorative colours

As the Drakens of F10 Ängelholm approached retirement during the 1990s, each of the wing's three squadrons resprayed aircraft in special colour schemes to mark the end of their association with the Draken.

J 35Js 35540 (right) and 35541 (far right) were painted in squadron colours to mark the retirement of the Draken by 2./F10 and 3./F10, respectively. 35540 flew the last operational sorties by a Flygvapnet Draken during December 1998. Both aircraft were subsequently preserved by the Flygvapnet museum at Linköping.

Right: JOHAN 66, F10's last J 35F (35468), was flown by display pilot Captain Vincent Ahlin at F10 Ängelholm's June 1990 open day. It was the aircraft's last display before retirement.

16 August 2000 saw the final retirement of Finland's last Drakens. To mark the occasion, the last Draken unit in the Ilmavoimat (HävLLv 11 of Lapland Air Command) resprayed this 35S in an overall yellow scheme incorporating a wisent (European bison) design – part of the squadron's badge.

for another two years (in the face of considerable public debate in Austria as to the necessity of such a purchase) until, on 26 March 1985, the Landsverteidigungsrat unanimously decided to buy the aircraft. On 21 May the final contract for 24 refurbished ex-Flygvapnet Drakens was signed in Vienna. The order included spare engines and other components, ground equipment, a flight simulator, pilot equipment, documentation and technician training. Initial pilot training was to be undertaken at F10 Ängelholm. The deal was valued at SEK1.1 billion ($US550 million).

The best 24 of 50 ex-Flygvapnet J 35Ds were selected for modification to 35OE standard at Saab, Linköping. The aircraft were bought back by Saab and underwent extensive radar and avionics modifications (20,000 man-hours per aircraft) before being delivered. Each aircraft was delivered with an expected fatigue life of 1,000 hours and received a new construction number in the range 351401-351424. The Saab 35OEs were delivered between 25 June 1987 and 18 May 1989, entering service with the two squadrons of Fliegerregiment 2 at Zeltweg and Graz-Thalerhof.

Initially, a lack of suitably trained aircrew meant that the type's utilisation rate was low and this, coupled with inadequate maintenance, led to several groundings. A crash involving an Austrian pilot flying a Flygvapnet Draken during November 1986, was blamed on human error and led to major revisions in the Austrian selection procedure for Draken pilot training, largely based on Swedish experience.

In the 1990s political instability in neighbouring countries and the eruption of the Yugoslavian crisis led to the Drakens being seen in a new light. When purchased the Draken's were restricted to the standard pair of ADEN cannon as their only armament; no air-to-air missiles were carried. The Austrian armed forces were forbidden to oper-

ate any type of guided missiles by the Staatsvertrag (State Treaty) signed by Austria and the four Allied powers. However, a number of border violations by JRV aircraft during the fight for Slovenian independence in 1991 saw the restriction lifted, Austria ordering a batch of AIM-9P Sidewinders, which were delivered from 1993. Sidewinder training was carried out at F10 Ängelholm.

Apart from the addition of Sidewinder missiles, the only upgrading work carried out on the Austrian aircraft has been the installation by Valmet of RWR and chaff/flare dispensers salvaged from retired Danish Drakens in the mid-1990s.

The 35OE was scheduled for replacement in 1996, contenders for the follow-on type having been narrowed down to the Lockheed Martin F-16, Boeing F/A-18, Mirage 2000-5, Saab JAS 39 and Mikoyan MiG-29SE. Naturally, a public debate on the necessity for such an aircraft ensued and plans have since been revised. Political will and funding permitting, a new 'light combat aircraft' type will enter service in 2005. In the meantime, planning calls for No. 2 Squadron at Graz to disband in 2003, its remaining aircraft transferring to Zeltweg.

Finale

As the career of the Draken finally drew to a close, a significant milestone was passed on 28 October 1999 by a Draken pilot of the Austrian air force. Captain Grossmaier flew the one millionth flight hour by a Draken, 44 years after the type's first flight. A ceremony was held at Zeltweg Air Base to mark the occasion, the flight taking place before representatives of the four air forces that operated the Draken. Guest of honour at the ceremony was none other than Dr Erik Bratt, the Draken's chief designer.

Bo Widfeldt and Stefan Wembrand

A pair of F10's J 35Js streak low over the sea along Sweden's west coast, their afterburners lit. As well as providing Flygvapnet with its first supersonic interceptor, the Draken programme was of utmost importance to Sweden's technological development, in particular in the fields of jet propulsion, electronics, ejection seat systems and flight safety.

SWEDEN
Svenska Flygvapnet

Saab 35 Draken production amounted to just over 600 examples. Of these all but 63 were delivered to Flygvapnet (apart from a handful of test aircraft that were retained by Saab). The remaining new-build aircraft comprised 12 aircraft assembled in Finland for the Ilmavoimat and 51 aircraft built in Sweden for Denmark. Another 61 ex-Flygvapnet aircraft were refurbished for export to existing foreign operators; airframes exported for component recovery are not listed in the accompanying table.

The Draken served the Flygvapnet for almost 40 years. For 20 years the type was the backbone of its fighter force, until supplanted by the Viggen and finally phased out with the introduction of the JAS 39 Gripen in the late 1990s. At peak strength in the late 1970s the Draken equipped 17 *divisioner* (squadrons), but cuts in Sweden's defence budget and the introduction of the JA 37 Viggen during 1980 brought a steady reduction in Draken numbers;

F12 was the first J 35 *flygflottilj* (wing) to disband, in 1978.

Nine Draken *divisioner* remained by the spring of 1984, when it was announced that six of these (in F4, F16 and F21) would convert to Viggens during 1984/85. Insufficient funding and the short remaining airframe life of some early production Viggens meant that F10 would remain a Draken unit to the end, equipped with J 35Js and Sk 35Cs inherited from F16.

Apart from front-line units, the Försökscentralen (Test Centre) or FC at Malmslätt has also been provided with a number of Drakens for research and trials purposes.

Draken production and delivery 1955-1989

variant	c/n	no.	delivered	notes
35-1/35-9	–	9	1955-1963	prototypes
35-10	35110	(1)	1961	prototype (converted J 35A 35082)
35-11	35111	(1)	1962	prototype (converted J 35D 35275)
35-12	35112	(1)	1962	prototype (converted J 35A 35081)
35-13	35113	(1)	1960	prototype (converted J 35A 35013)
J 35A	35001-35065	65	1959-1961	short tail
J 35A	35066-35090	25	1961	long tail
J 35B	35201-35273	73	1962-1963	
Sk 35C	35800	(1)	1959	Sk 35C prototype (converted J 35A)
Sk 35C	35801-35825	(25)	1962-1963	converted from J 35A
J 35D	35274-35393	120	1962-1965	30 later converted to S 35E
S 35E	35901	(1)	1963	prototype, converted J 35D
S 35E	35902-931	30	1965-1966	
S 35E	35932	(1)	1966	prototype, converted J 35D
S 35E	35933-35960	(28)	1966-1968	converted from J 35D
J 35F[1]	35401-35500	100	1964-1967	
J 35F[2]	35501-35630	130	1967-1972	
J 35J	35598	(1)	1987	prototype, modified J 35F[2]
J 35J		(66)	1987-1991	modified J 35F[2]
35XD	351001-351020	20	1970-1971	F-35, to Denmark
35XD	351101-351120	20	1971-1972	RF-35, to Denmark
35XD	351151-351161	11	1971-1972	TF-35, to Denmark
35XS	351301-351312	12	1974-1975	35S assembled by Valmet, Finland
35BS		(8)	1972-1984	35BS, to Finland, former J 35B
35FS		(24)	1976-1985	35FS, to Finland, former J 35F[1]
35CS		(5)	1976-1984	35CS, to Finland, former Sk 35C
35OE	351401-351424	(24)	1987-1989	export to Austria, refurbished J 35D

Total new-build Drakens: **615 (excluding modified or refurbished aircraft)**

Left: This J 35J, F10's display aircraft during the early 1990s, is finished in the camouflage scheme worn by Flygvapnet Drakens for much of their career. The swordfish marking on the upper surface of the wing is based on the badge of 3. Jaktflygdivision/Flottilj 10.

Below: To mark the 35th anniversary of the Draken's entry into service, this impressive line-up of 28 F10 aircraft was arranged at Ängelholm on 10 June 1990. By then large numbers of Drakens had been retired in the mid-1980s, leaving the J 35Js of F10 as the last Draken unit.

Draken units of Flygvapnet, 1959-1998

wing/sqn	J 35A	J 35B	SK 35C	J 35D	S 35E	J 35F	J 35J
Flottilj 1 Västerås-Hässlo							
1. Jaktflygdivision	-	-	-	-	-	1966-83	-
2. Jaktflygdivision	-	-	-	-	-	1966-83	-
3. Jaktflygdivision	-	-	-	-	-	1966-83	-
Flottilj 3 Linköping-Malmslätt							
1. Jaktflygdivision	-	-	-	1965-70	-	1970-73	-
2. Jaktflygdivision	-	-	-	1965-70	-	1970-73	-
3. Jaktflygdivision	-	-	-	1965-70	-	-	-
Flottilj 4 Östersund-Frösön							
1. Jaktflygdivision	-	-	-	1969-84	-	-	-
2. Jaktflygdivision	-	-	-	1974-84	-	-	-
Flottilj 10 Ängelholm							
1. Jaktflygdivision	-	-	-	? (1967)	-	1969-87	1987-96
2. Jaktflygdivision	-	-	-	-	-	1970-88	1988-98
3. Jaktflygdivision	-	-	-	1964-70	-	1970-89	1989-93
TIS 35	-	-	1986-98	-	-	-	-
Flottilj 11 Nyköping-Skavsta							
1. Spaningsdivision	-	-	-	-	1964-79	-	-
2. Spaningsdivision	-	-	-	-	1966-79	-	-
Flottilj 12 Kalmar							
1. Jaktflygdivision	-	-	-	-	-	1969-79	-
2. Jaktflygdivision	-	-	-	-	-	1968-79	-
Flottilj 13 Norrköping-Bråvalla							
1. Jaktflygdivision	1960-62	-	-	1962-65	-	1965-75	-
2. Jaktflygdivision	1960-62	-	-	1963-65	-	1965-78	-
3. Jaktflygdivision	1960-63	-	-	1963-66	-	1966-78	-
Flottilj 16 Uppsala							
1. Jaktflygdivision	1961-76	-	-	-	-	1976-85	-
2. Jaktflygdivision	1963-76	-	-	-	-	1976-86	-
3. Jaktflygdivision	1964-76	1962-64	-	-	-	1976-87	-
TIS 35	-	-	1962-86	-	-	-	-
Flottilj 17 Ronneby							
1. Jaktflygdivision	-	-	-	-	-	1972-81	-
2. Jaktflygdivision	-	-	-	-	-	1973-77	-
Flottilj 18 Tullinge							
1. Jaktflygdivision	-	1962-74	-	-	-	-	-
2. Jaktflygdivision	-	1962-74	-	-	-	-	-
3. Jaktflygdivision	-	1962-76	-	-	-	-	-
Flottilj 21 Luleå-Kallax							
1. Spaningsdivision	-	-	-	-	1966-79	-	-
2. Jaktflygdivision	-	-	-	-	1969-84	-	-

Saab 35 Draken

The emblems of Flygflottilj 10's three squadrons – (from left to right) 1. Jaktflygdivision, 2. Jaktflygdivision and 3. Jaktflygdivision – were made famous during the Draken's last years of service in the Flyvapnet. Despite having only recently swapped its Viggens for JAS 39 Gripens, F10 is due to disband in 2002 with the closure of Ängelholm.

The first J 35As off the Saab production line were delivered to F13 Norrköping in late 1959. Lacking radar, these aircraft were of little operational use and were generally confined to advanced training tasks.

Above: Newly delivered J 35B '68' (35250) awaits its next sortie in a dispersal at F10 Ängelholm. The J 35Bs were camouflaged shortly after entering service.

Below: During F10's transition to the J 35J during the late 1980s, the wing operated three Draken variants: Sk 35C, J 35F and J 35J.

Above: F3 Linköping enjoyed a comparatively short eight-year stint operating Drakens. Equipped with J 35Ds in 1965, it transitioned to the 35F (pictured) in 1970, before succumbing to budget cuts that saw three wings disbanded during 1973/74.

Below: Checkerboard markings such as these, usually in red or yellow, were often applied to Flygvapnet aircraft during tactical exercises.

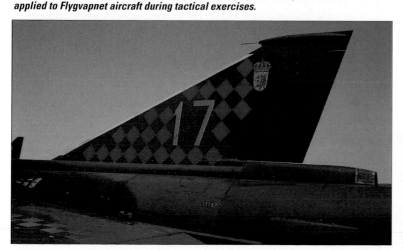

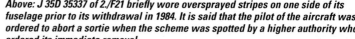

Above: J 35D 35337 of 2./F21 briefly wore oversprayed stripes on one side of its fuselage prior to its withdrawal in 1984. It is said that the pilot of the aircraft was ordered to abort a sortie when the scheme was spotted by a higher authority who ordered its immediate removal.

Below: This J 35J of 3./F10 is typical of the Flygvapnet's last Drakens in their final months of service in the late 1990s. By then the overall grey colour scheme had been adopted and large Dayglo codes applied to upper wing surfaces.

AUSTRIA
Österreichische Luftstreitkräfte

Austria's Saab 35OEs were delivered between June 1987 and May 1989 for service with Fliegerregiment 2, headquatered at Zeltweg. One of three Fliegerregiments (Aviation Regiments) in the Österreichische Luftstreitkräfte, Fliegerregiment 2 includes an Überwachungsgeschwader (Surveillance Wing) comprising two Staffelen (Squadrons) equipped with Drakens. 1. Staffel is based at Zeltweg, while 2. Staffel is stationed further east at Graz-Thalerhof (until it disbands in 2003).

Equipped with air-to-air missiles (AIM-9 Sidewinder) for the first time from 1993 and upgraded to a limited degree with second-hand RWR and chaff/flare dispensing equipment, the 35OEs are destined to remain in use until at least 2005, assuming a decision can be made regarding a replacement.

Pictured during an August 2001 display at the Constanta, Romania, 35OE '08' carries a display scheme clearly inspired by the national flag. Note the fin-top RWR antenna and chaff/flare dispenser fitted aft of the afterburner air intake; this equipment was salvaged from the Danish Drakens and installed by Valmet.

Above: A pair of Österreichische Luftstreitkräfte Drakens gets airborne from its base. Originally split between two bases, the Austrian Draken fleet will be concentrated at Zeltweg from 2003.

Below: The 35OEs were delivered in a three-tone grey finish and have changed little in outward appearance since entering service, apart from the adoption of overwing Dayglo codes, à la Flygvapnet aircraft.

DENMARK
Kongelige Danske Flyvevåbnet

Delivered during 1970/71, the KDF's 20 F-35 Drakens replaced F-100 Super Sabres in the ground-attack role with Eskadrille 725 at Karup AB. The follow-up order for 20 RF-35 reconnaissance platforms was placed to enable the retirement of the last RF-84F Thunderflashes in Danish service. These were delivered shortly after the F-35s and equipped Eskadrille 729, also at Karup. Three batches of TF-35 trainers, totalling 11 aircraft, were split between the two units, which formed part of the Flyvevåbnet's Flyvertaktisk Kommando (Tactical Air Command)

Defence budget constraints forced the amalgamation of the two Draken units in 1991; Esk 725 disbanded and a number of the surviving aircraft were grounded. In 1993 Esk 729 followed suit, the Draken's duties being taken over by Flyvevåbnet F-16s.

On 31 July 1991 Flyvevåbnet Major Frode M. Sveistrup became the first Draken pilot of any nation to achieve 3,000 flying hours 'on type'. The record-breaking sortie was flown in F-35 (A-018) of Esk 725. Sveistrup had a total of 3,003 flight hours on the Draken when it was finally retired from Danish service in 1993.

With dehumidifiers pumping air into their intakes as a corrosion prevention measure, F-35 A-004 and RF-35 AR-120 rest in a revetment at Karup AB. Karup was the base for the entire Danish Draken fleet and was closed when the type was retired.

Above: With the retirement of the Flyvevåbnet's last seven active RF-35s in 1993, the service's Tac/R role was taken over by Esk 726 F-16s equipped with camera pods. AR-117, the former DYNAMITE 6-1, was later sold to Flight Research Inc in the US.

Below: TF-35 AT-160 taxies back to its revetment at Karup AB after a training sortie. This aircraft was later donated to the Flygvapnet museum at Linköping, Sweden.

FINLAND
Suomen Ilmavoimat

Finland's Drakens were initially delivered to the two flights of Hävittäjälentolaivue (Fighter Squadron, or HävLLv) 11 of the Hämeen Lennosto (Häme Wing) based at Tikkakoski in central Finland. Later HävLLv 11 relocated to the extreme north to the base at Rovaniemi in Lapin Lennosto (Lapland Wing).

In 1984 a second Draken squadron was formed. HävLLv 21 of Satakunnan Lennosto (Satakunta Wing) was established at Tampere-Pirkkala in southwest Finland, also with two flights.

By the end of 1995 the last of the 35BS single-seat 'trainers' had been retired and only a single two-seat 35CS remained in use. During 1997 HävLLv 21 completed conversion to the F/A-18 and its remaining 35S and 35FS were passed to HävLLv 11, which

Large black serial numbers and full-size national insignia were applied during the early years of Finnish Draken operation. Camouflage, to the same pattern/colour as Swedish aircraft, was adopted later (with either black or white serials); low-visibility black serials and smaller national insignia were used from the late 1980s/early 1990s. DK-225, a 35S (right), is finished in the standard camouflage scheme with white serials; DK-262 (below right), an ex-Flygvapnet 35CS, carries the later small serials and insignia. The ex-Swedish 35CSs (Sk 35Cs) remained unpainted, as they had been in Flygvapnet service.

continued operations with the Draken until the second half of 2000. Final retirement took place on 16 August.

Drakens were also attached to the Koulutuslentolaivue (Flight Training Squadron) at Kuorevesi-Halli and the Ilmavoimien Koelentokeskus (Air Force Flight Test Centre).

CIVILIAN OPERATORS

Apart from a number examples preserved by museums, mainly in Europe, a number of retired Drakens have been purchased for use by civilian organisations with a view to maintaining the aircraft in airworthy condition. In early 2002 there was one airworthy civil-registered example in Europe and no fewer than 15 registered in the United States, though not all of these aircraft were airworthy. The latter included J 35D 35350, registered to Flight Materials Inc. of McClellan, California. This aircraft was previously registered in the UK as G-BOAR; its current status is unknown.

Flight Research Inc.
National Test Pilot School

These associated companies, based at Mojave, California, provide a range of aircraft, systems and weapon test and pilot training facilities, respectively. They share a common fleet of varied aircraft, including six single- and two-seat Saab 35XD Drakens acquired from the Danish

air force. These comprise RF-35 N110FR (formerly AR-110), TF-35 N166TP (AT-151), TF-35 N167TP (AT-153), TF-35 N168TP (AT-154), TF-35 N169TP (AT-157) and RF-35 N217FR (AR-117).

Above: The National Test Pilot School has four TF-35s at its disposal, including N168TP. As soon as these aircraft entered service with the NTPS in 1994, the Danish air force signed a 12-year training contract.

Above: Pictured at an Edwards AFB open house in October 1997, N217FR is one of two single-seat Drakens usually employed by Flight Research, Inc. for aircraft and weapon testing and certification tasks.

Flight Test Dynamics LLC

Flight Test Dynamics, with facilities at Inyokern, California, has five Drakens on the US civil register, though it is unclear how many of these aircraft are airworthy in 2002. All are ex-Danish air force aircraft and represent all three 35XD variants operated by the service. Their identities are as follows: RF-35 N106XD (formerly AR-106), RF-35 N116XD (AR-116), TF-35 N155XD (AT-155), TF-35 N156XD (AT-156) and

F-35 N20XD (A-020). In addition, RF-35 N111XD (AR-111) and RF-35 N119XD (AR-119), registered to Vortex Inc. of Newport News, Virginia, are believed to part of the Flight Test Dynamics stable, along with ex-Flygvapnet J 35F^2 N543J (35543), originally registered by Jim Merizan in 1994 and employed in at least one motion picture and as a 'flying billboard' by the Miller Brewing Company.

Above: During 1997 Flight Test Dynamics' hangar at Inyokern held at least five Drakens, including AT-155 (foreground). In 1999 a number of these airframes, including AT-155, were noted outdoors where their external condition appeared to have deteriorated.

Scandinavian Historic Flight

The Scandinavian Historic Flight operates a number of warbirds on the European air show circuit from bases in Norway, Sweden and Denmark, including examples of the P-51, A-26, Harvard, Hunter and Vampire. TF-35 Draken AT-158 was donated to the SHF by the Danish air force in 1994. Although the aircraft was in excellent

condition a considerable effort – both technical and bureaucratic – was required to bring the aircraft up to a standard to allow civil certification. This work included the installation of a civil avionics suite. Based at Karup Air Base in Denmark, the Draken is maintained by SHF members, a number of whom are ex-Danish air force personnel.

Above: After much bureaucratic wrangling the SHF has been able to operate this TF-35 (AT-158) on the European air show circuit – the first truly supersonic type operated by a warbird organisation in Europe.

'Experimental Bomber, Long Range'
Boeing XB-15 and Douglas XB-19

General Billy Mitchell died in 1936 before his theories regarding the use of long-range bombers were proved right, during World War II. Had he lived another year he might have witnessed the maiden flight of the first of two new bombers which represented the Air Corps' first attempts at building such an aircraft.

By the standards of the day the XB-19 was a behemoth, though by the time it flew, in 1941, its size had become a liability and its military significance had been lost. Too big and too slow, the aircraft lacked important features such as self-sealing fuel tanks and armour plating; the addition of these items would have simply slowed the aircraft still further.

By 1934, the young air power devotees who had been captains during the Billy Mitchell court martial of 1926 had risen to positions of power within the US Army Air Corps. Mitchell's idea that America should have extremely-long-range bombers to strike at a potential enemy's heartland was no longer considered far-fetched foolishness.

When Mitchell left the service, his ideas about air power were officially discounted at the highest levels, but they were alive among a growing group of Air Corps junior officers – including men such as Henry Harley 'Hap' Arnold – who had an eye on the future.

Large, four-engined flying-boats were routinely flying the Atlantic and dramatic long-distance flights were grabbing the headlines. Large, long-range bombers were also evolving. Between September and November 1929, a large globe-circling Soviet Tupolev TB-1 (ANT-4) bomber had crossed the northern tier of the United States on a round-the-world demonstration flight. This aircraft was of some interest, but it was only slightly larger than the Keystone bombers that the Air Corps was flying. Four years later, its sister ship was an entirely different aircraft.

On 3 July 1933, the enormous Soviet Antonov TB-4 (ANT-16) bomber made its much-publicised maiden flight. Its wing span of 177 ft (54 m) was more than double that of the Keystones. Though its range was not significantly greater than the TB-1 – and about the same as the Keystones – this fact was not known at the time. People who had read about a Soviet bomber flying to and across the United States now discovered that the Soviets had a

Though it flew after the first Model 299 (B-17), the Model 294 (XB-15) pre-dated the famous Flying Fortress, which mirrored the giant bomber in terms of its engine arrangement, fuselage cross-section and the positioning of crew stations and military equipment.

bomber that was twice as big – and both were modern metal monoplanes at a time when the US Army Air Corps was operating fabric-covered biplanes.

For those who understood the potential of long-range bombers in warfare, it was a technological wake-up call almost like that which *Sputnik* would be in 1957. Even the old horse soldiers on the US Army General Staff could see that the times were, indeed, changing.

It was time for America to have a big bomber. At Wright Field, near Dayton, Ohio, the US Army Air Corps Materiel Division began seriously studying the notion of a long-range bomber such as Billy Mitchell had advocated more than a decade before.

'Project A'

Finally, in the spring of 1934, the General Staff in Washington acquiesced to the concept, and the Materiel Division was authorised to proceed with a fully-fledged, albeit secret, feasibility study. The new project, the first of its kind, was officially and cryptically designated as 'Project A'.

On 14 April, General Conger Pratt, the chief of the Materiel Division, issued a formal request for proposals under 'Project A' that was humbly entitled *Long Range Aircraft Suitable For Military Purposes*.

Ten days later, Pratt opened his morning paper to another bombshell from the East. On 24 April, the Soviet Union took delivery of what was – by far – the largest aircraft in the world. With a wing span of 210 ft (64 m), it dwarfed the TB-4 and every aircraft then flying in the United States. Designed by Tupolev and designated as ANT-20, it was named *Maxim Gorkii* after the widely celebrated Russian author. The eight-engined behemoth touted a kitchen, an onboard movie theatre and a printing plant.

The ANT-20 had been built as a propaganda stunt, and the stunt worked. There was no hangar in the world where the conversation did not turn to a discussion of *Maxim Gorkii*.

At that same moment, the aircraft manufacturers in the United States who had received the secret 'Project A' memo from the Air Corps were studying the specifications. They were seeing that when General Pratt and his staff said "Long Range Aircraft Suitable For Military Purposes", they meant not simply a 'long-range' aircraft, but an aircraft with a maximum range of 5,000 miles (8050 km).

The urgency of the task was underscored by the roll-out of *Maxim Gorkii*, but no one in the United States realised that the 'Project A' range specification was more than six times the range capability of the big Soviet aircraft. This state secret was conveniently omitted from the publicity churned out by the Moscow propaganda mill.

'Project A' was a daunting task at a time when aircraft ranges were typically measured in hundreds, not thousands, of miles. However, both the Boeing Airplane Company of Seattle, Washington, and the Glenn L. Martin Company of Baltimore, Maryland, submitted proposals that were deemed worthy of further elaboration.

On 14 May, General Pratt convened a meeting at Wright Field where he sat down with Boeing president Clairmont 'Claire' Egtvedt and

With Major Stanley M. Ulmstead at the controls, the XB-19 made its first flight from Clover Field, Santa Monica, on 27 June 1941. The 55-minute flight took the aircraft to March Field via Douglas's new Long Beach plant, where a low-level pass was made.

C.A. Van Dusen of Martin to discuss 'Project A'. Pratt's aide, Captain Leonard 'Jake' Harmon, a devotee of Billy Mitchell who was destined to play a key role in developing strategic air power during World War II, was also present. Harmon underscored the need for a gross weight of 60,000 lb (27216 kg) and explained the 5,000-mile (8000-km) range requirement in the context of wanting an aircraft that could carry out defensive operations in Alaska, Hawaii and the Panama Canal Zone from bases in the continental United States.

Egtvedt and Van Dusen were invited to submit plans for this mammoth by the middle of June. Although neither company had yet built a four-engined aircraft, they both had such projects on the drawing board, and their engineers clearly understood the technical nuances of large aircraft. The Boeing Model 294 proposal was to be designed under the designation XB-15, while the Martin Model 145 would reach the blueprint stage having been designated as XB-16. Allison V-1710-3 inline engines, each delivering 1,000 hp (746 kW), were specified for both aircraft.

The Boeing Model 294 was a conventional

Martin Model 145 (XB-16)

Martin's XB-16 proposal was perhaps the most radical of the three long-range bomber designs in terms of its layout. The Model 145 would have been a twin-boom, four-engined aircraft, power coming from Allison V-1710s (though some sources suggest that Pratt & Whitney R-1830s were to be employed). The outer engines would have been equipped with pusher propellers, their nacelles extending ahead of the wing leading edge to provide a position for a remotely-controlled barbette mounting two guns. Other defensive gun positions were provided on each tail boom, in the nose and at the extreme rear of the central fuselage. A bomb load of 12,000 lb (5443 kg) was envisaged.

With other more lucrative projects, like the DST airliner, providing more than enough work for the company, and with inadequate funding from the Air Corps holding up the completion of an increasingly anachronistic aircraft, Douglas was all for cancelling the XB-19 programme and cutting its losses. Here the aircraft is seen under construction in the company's Santa Monica plant.

mid-wing, all-metal monoplane with a single tail, with its engines on the leading edges of the wings. The Martin Model 145 was a more unusual design, featuring a twin-boom fuselage and a twin-rudder tail. There would be four of the Allison engines, two on the leading edge with tractor propellers and two on the trailing edge with pusher props. The Martin design would also feature a tricycle landing gear, which was unusual for 1934.

The Martin Model 145 was the larger of the two aircraft, with a gross weight in excess of 100,000 lb (45360 kg), compared to about 65,000 lb (29500 kg) for the Boeing entry. The Model 145 was designed with a wing span of 173 ft (52.7 m), compared to 149 ft (45.4 m) for the Model 294.

Both were smaller than the TB-4 or *Maxim Gorkii*, but they were still among the largest aircraft that had ever been conceived. Certainly, they were among the largest yet designed by American manufacturers.

When the plans came in, the XB-16 was rejected, but Martin engineers used their experience on the project for other projects. The company was already proceeding with its Model 130 commercial flying-boat. During World War II, Martin built its huge four-engined PB2M-1 Mars flying-boat. With a 200-ft (61-m) wing span, the Mars was the largest operational aircraft flown in the United States until the end of World War II.

On 28 June 1934 – 11 days after *Maxim Gorkii* took to the air for the first time – the Materiel Division issued Boeing a contract to build a single XB-15. A team headed by project manager Jack Kylstra began work on an undertaking that would consume roughly 670,000 man-hours over the coming three years.

By coincidence, several of the men working on the Model 294 project literally had 'inside knowledge' of the big airframe work that had

The XB-15's structure was similar to that of earlier Boeing monoplanes such as the Monomail, though the wing was fabric-covered from the main spar aft. Among the novel features of the XB-15 was a 110-volt AC electrical supply provided by two alternators driven by auxiliary petrol engines.

been created by the Tupolev design bureau. Five years earlier, when the TB-1 had made its flight across the United States, it had made the flight from Siberia through Alaska as a seaplane. When it arrived in Seattle in mid-October 1929, it needed to have its floats removed and wheels attached for the cross-country leg of the trip. To whom would one turn for floatplane work in Seattle in 1929, other than the largest maker of floatplanes in the northwest? While the Boeing crews were working on the big aircraft, they had found and repaired some frayed control cables. In the course of doing this, they had become intimately acquainted with the handiwork of the Junkers and Tupolev designers who had created the TB-1.

During the summer of 1934, new tooling was constructed, and gradually the major structural elements of the Model 294 took shape. Work began at Boeing's Plant One, but the airframe

was subsequently barged up the Duwamish River to Plant Two, where a larger space was available. The same scenario was pursued a few years – and a technological generation – later, when Boeing developed the B-29 Superfortress.

'Project D'

Even as Boeing was moving ahead with work on the XB-15, the Air Corps Materiel Division was emboldened to think of even more grandiose notions of long-range aircraft. On 5 February 1935, the Air Corps announced its 'Project D'. Like 'Project A', 'Project D' envisaged a very-long-range warplane, but it was a great deal more conceptual than 'Project A' in specifying just how long that range should be.

'Project A' had asked for 5,000 miles (8000 km), but the authors of 'Project D' explained that they were issuing their request for proposals "in an effort to further the advancement of military aviation by investigating the maximum feasible range into the future."

It was, in short, an experimental programme aimed at letting manufacturers do whatever they could to push the performance envelope as far as possible.

As with 'Project A', two acceptable initial proposals were submitted. One was from the Douglas Aircraft Company of Santa Monica, California, which was about to emerge as a major player in commercial aviation but had yet to build a large four-engined aircraft.

The other company responding to the 'Project D' invitation was the Sikorsky Aviation Corporation of Stratford, Connecticut, which had been founded by Russian émigré Igor Sikorsky. As a pioneer aircraft maker in Russia, Sikorsky had built warplanes for Tsar

Nicholas II during World War I, one of which was the astounding bomber named after the legendary hero Ilya Mourometz. This four-engined bomber had made its first flight six months before World War I began. Though few in number and not available for large massed raids, the big Sikorskys made more than 400 attacks against Germany during the war, clearly garnering the notice of men such as Billy Mitchell.

After the Russian Revolution of 1917 toppled the Tsar, Sikorsky relocated to America, and had already begun building four-engined, long-range commercial flying-boats when the 'Project D' memo came went out.

Bomber, Long Range

By the summer of 1935, the Air Corps decided to merge 'Project D' with the earlier 'Project A' as the Bomber, Long Range (BLR) programme, a name that obviously described both projects. The aircraft being ordered under both projects would be designated with the 'BLR' prefix rather than the more generic 'B for bomber' prefix. The Boeing Model 294, which was already in the works under the designation XB-15, was redesignated as XBLR-1. The Douglas bomber became XBLR-2, and the Sikorsky project was designated as XLBR-3. The Martin project had already been terminated, so there was no XBLR-4.

On 5 June 1935, representatives from both Douglas and Sikorsky sat down with the Air Corps Materiel Division people at Wright Field to discuss Type Specification X-203 for the 'Project D' bomber, which required, among other things, the use of 1,600-hp (1194-kW) Allison XV-3420-1 engines in the prototype aircraft.

It was decided in the meeting that the two firms would submit their preliminary designs in seven weeks, by 31 July. The detailed work would begin by 31 January 1936, and a completed example of the XBLR-2 and of the XBLR-3 would be delivered by 31 March 1938.

Meanwhile, the formal contract to build the XBLR-1 was issued to Boeing on 29 June, three weeks after the Wright Field conference. All three aircraft were now in process.

The Air Corps Materiel Division inspected wooden mock-ups of the XBLR-2 and XBLR-3 in March 1936, and the decision was made to cancel the Sikorsky XBLR-3 and proceed to prototype only with the Boeing and Douglas aircraft. After an auspicious beginning as one of the world pioneers in large, four-engined bombers, Igor Sikorsky and his company left that field permanently.

Also in 1936, the Air Corps rethought the idea of the XBLR designations and dropped them in favour of merging these aircraft back into the 'B for bomber' lineage. The Boeing XLBR-1 became the XB-15 again, and the Douglas XLBR-2 was redesignated as XB-19. Another change came when the Air Corps changed the powerplant specification from the Allison XV-3420-1 water-cooled engines to 2,000-hp (1492-kW) Wright R-3350 radials for the XB-19 and Pratt & Whitney R-1830 Twin Wasps for the Boeing XB-15.

With a two-year head start, the XB-15 was moving ahead rather quickly by this time and the prototype was already taking shape in the hangar at Boeing Field. Progress on the XB-19 was slow, however. Air Corps funding was stretched thin during 1936 and 1937 and, at the

Inside the XB-15

The XB-15's spacious flight deck (above) accommodated five of the aircraft's crew of 10. Seated behind the two pilots, on the left and right sides of the aircraft, respectively, were the radio operator and navigator. The flight engineer had a flip-up seat behind the radio operator (right), and an instrument panel fixed to the side of the fuselage and arranged around a small doorway which gave access to the inside of the wing. This allowed the engineer to inspect and service the engines' accessories in flight, if necessary. To the right of the flight engineer's position was a sliding door providing access to the rest of the aircraft. Behind the cockpit were a toilet, bunks and a galley, the latter including a hot plate, soup heater, coffee percolator and ice box. Access to the nose from the cockpit was gained via a double-leaf door between the pilots' seats (above right). This feature necessitated the provision of a throttle quadrant for each pilot.

The XB-15's defensive stations comprised blister fairings either side of the fuselage (below) and on the lower fuselage, fore and aft. Turrets were fitted in the upper half of the glazed nose (below right) and in the upper fuselage, above the bomb bay. The latter was equipped with a pair of 0.5-in machine-guns; all other stations mounted a single 0.3-in machine-gun.

The XB-15's three bomb bays – the main, so-called, 'body' bomb bay (below) and two 'wing stub' bomb bays (below left) in the wing between the fuselage and the inner engines – typically held 8,000 lb (3629 kg) of bombs.

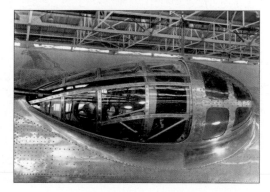

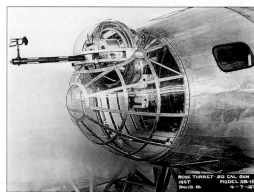

299 was smaller than the XB-15, but it was also powered by four engines. Submitted for the approval of the Air Corps, the Boeing Model 299 was ordered under the designation B-17; it became a legendary World War II warplane, and is now regarded as one of the half dozen greatest aircraft of its type ever flown.

The B-17, dubbed Flying Fortress at the time of its roll-out, was not originally conceived as a bomber capable of attacking an enemy at great distances, as were the 'Project A' bombers. Rather, it was seen by the Air Corps as a coastal defence aircraft capable of attacking enemy warships – something that was of great interest to the Air Corps in the late 1930s. Of course, history has shown that the Flying Fortress became the ultimate 'poster child' for the mission envisaged by 'Project A' and 'Project D'.

Being smaller than the XB-15, the Model 299 did not push the limits of engine technology. Therefore, it demonstrated exceptional performance, making it an easy sale, and it was already in series production by the time that the XB-15 made its first flight.

The XB-15 was longer and had a wing span roughly half again greater than the Model 299 aircraft. Its gross weight of 70,706 lb (32072 kg) was double that of the early Model 299s. The flight deck was the size of the living room in a small house, larger than anything most aircrews had ever seen. There were sound-proofed sleeping areas, a kitchen and a lavatory – something unheard-of in military aircraft. As with the

same time, Douglas had other projects underway, including production of its extremely popular DC-3 airliner and the B-18 Bolo twin-engined bomber. Not until 8 March 1938 did the Air Corps allocate adequate funds to complete the single XB-19 prototype.

By this time, there was concern at Douglas that precious resources were being devoted to a project that they saw as fast becoming an enormous white elephant. At the end of August 1938, the company noted that the schedule had slipped so far that the XB-19 was going to be obsolete before it was built, and argued that the obsolescence should be grounds for cancellation. The Material Command disagreed and Douglas was ordered to finish the big aircraft.

Arrival of the XB-15

The first of the 'Project A'/'Project D' bombers made its maiden flight from Seattle's Boeing Field on 15 October 1937 with Edmund T. 'Eddie' Allen at the controls. The XB-15, serial number 35-277, made history as the largest aircraft yet built and flown in America. The 1,000-hp (746-kW) air-cooled Pratt & Whitney R-1830 Twin Wasp engines – which had been used instead of the Allison V-1710s originally specified – were among the most powerful then available.

The science of aeronautics had come a long way in just over 30 years. The XB-15 makes a low-level pass over the Wright Brothers Memorial Pylon at Kitty Hawk, North Carolina, during August 1938.

However, by the autumn of 1937, the big aircraft was overshadowed by a smaller sibling. More than two years earlier, another important first flight had occurred at Boeing Field. On 28 July 1935, just a month after the Air Corps gave Boeing the go-ahead for its Model 294 under the XB-15 designation, the Seattle aircraft manufacturer had flown the prototype of its Model 299. Built at company expense, the Model

The XB-19 was among the first applications of Wright's new R-3350 two-row 18-cylinder radial engine, the first examples of which had been bench-tested in 1937. The Cyclone 18 produced twice the power of the Pratt & Whitney R-1830s that equipped the XB-15, though the new powerplant displayed a tendency to overheat and suffer reduction gear failures during the early years of its use. Once these 'snags' were rectified the R-3350 went on to power a number of well-known types, including the Boeing B-29.

contemporary Boeing Model 314 commercial flying-boat, the wings were large enough to accommodate passageways that engineers could transit in order to inspect the engines from inside the wing while the aircraft was in flight. The range with 2,500 lb (1134 kg) of bombs was calculated to be 3,400 miles (5470 km).

The size of the XB-15 was both impressive and problematic. The problems came in the design stage, because no one in the United States had yet built an aircraft so large. The engines, while powerful, were not powerful enough for the XB-15's enormous bulk. The big aircraft had a top speed of just under 200 mph (322 km/h), which was about three-quarters the speed of the first-generation B-17s.

Two service test aircraft were ordered with higher-performance Pratt & Whitney R-2180 engines under the service test designation Y1B-20, but these aircraft were cancelled. The Air Corps reassigned this designation number to the Boeing Model 316, which was also cancelled. The latter project did, however, eventually evolve into the Boeing Model 345, which was ordered under the designation B-29 – and which later was directly copied by the Tupolev design bureau to create its Tu-4.

XB-15 in service

In August 1938, 10 months after its debut, the XB-15 completed its flight test programme and was officially delivered to the US Army Air Corps' 2nd Bombardment Group, commanded by Colonel Robert C. Olds and based at Langley Field in Virginia. Coincidentally, the 2nd Bombardment Group had been one of the units under Billy Mitchell's command during the

A crew of 16 was envisaged for the operational version of the XB-19 (pictured here on its 27 June 1941 first flight), comprising a pilot, co-pilot, aircraft commander, navigator, flight engineer, radio operator, bombardier and no fewer than nine gunners. Six bunks and eight seats were provided for a relief crew.

great aerial assaults conducted in support of the Meuse-Argonne offensive in October 1918.

When the XB-15 arrived at Langley, the Group was thoroughly involved in the evaluation of the XB-15's sister ship, the Y1B-17. Indeed, all of the operational Y1B-17s in the Air Corps were assigned to the 2nd Bombardment Group. In February 1938, those aircraft had made their much publicised demonstration flight from Langley Field to Buenos Aires in Argentina.

In October 1942, the 2nd Bombardment Group, along with its Flying Fortresses and its XB-15, were transferred across the continent to the Army Air Field at Ephrata in the high desert

country of central Washington state, just 150 miles (240 km) due east of where they had all been built. The XB-15, which had been delivered in natural metal finish, was painted in the standard heavy bomber finish of the era, which was olive drab on the upper surfaces and neutral grey below. By this time, the XB-15 had taken on a sort of 'mascot' status, being referred to amiably as *Grandpappy*.

Despite the lack of performance from the R-1830 Wasps, the Air Corps had used the XB-15 to capture several payload records. Through July 1939, the big aircraft had lifted 71,167 lb (32281 kg) to an altitude of 8,200 ft (2500 m), and carried a 4,400-lb (1996-kg) load for 3,107 miles (5000 km).

When World War II began, there were thoughts of using *Grandpappy* as a long-range bomber, but its lack of speed to avoid enemy interceptors led to such notions being dismissed. In 1943, however, it was decided that the big aircraft's range and load-carrying ability would make it useful as a transport. It was redesignated as XC-105 and assigned to the 20th Troop Carrier Squadron and based in the Panama Canal Zone for use in the Caribbean.

In May 1945, even before the war had ended, *Grandpappy* was flown to the surplus aircraft collection point at Kelly Field, near San Antonio, Texas and cut up for scrap.

Debut of the XB-19

When it was completed, the Douglas XB-19 (serial number 38-471) superseded the Boeing XB-15 as the largest aircraft yet built in the United States, and no other American aircraft

Shortly after the Japanese attack on Pearl Harbor, the XB-19 was given a coat of olive drab/neutral grey paint and carried defensive armament on all future test flights. Armament comprised a nose turret with 37-mm and 0.3-in guns, 0.3-in guns in positions either side of the lower nose, an upper turret behind the cockpit mounting 37-mm and 0.3-in guns, 0.5-in guns in a belly and two waist positions and in an upper rear turret, 0.3-in guns in positions either side of the rear fuselage below the tailplane and a final '.50-calibre' in the tail gunner's position. Bomb loads of up to 18,000 lb (8165 kg) were also envisaged.

built before the end of World War II was larger. The aircraft – which its maker had tried to have cancelled in 1938 – rolled out at Santa Monica in May 1941. Nearly four years had passed since the debut of the XB-15, and now it was more than three years after the Air Corps had formally authorised Douglas to proceed with construction. During that time, Douglas had given a fairly low priority to the project. Apparently, the Air Corps had started to come around to the company's assessment, and in 1940 the 'Secret' classification had been removed from the XB-19 project.

Taxi tests began on 6 May, with a first flight originally pencilled in for 17 May. However, the taxi tests revealed a number of problems. First it was the brakes, then the engines and then the variable-pitch propellers. It seemed as if each time the Douglas engineers thought they had the problems fixed, something else would arise. The delays required to correct the brakes and engines, and then to rewire the propeller

controls, pushed back the first flight schedule more than five weeks.

On 27 June, exactly one week after the US Army Air Corps officially became the US Army Air Forces (USAAF), the XB-19 was finally ready. Carrying a crew of seven, with USAAF Major Stanley M. Ulmstead at the controls in the left seat of the spacious flight deck, the huge aircraft thundered down the runway and lifted off from Santa Monica's Clover Field. After a low-level pass over the new Douglas facility at Long Beach, Ulmstead set a course for March Army Air Field, about 65 miles (105 km) due east of Santa Monica. The first flight lasted 55 minutes and was described as having been 'uneventful'.

Roll-out and first flight

The 'event' was in the media. The aircraft was no longer classified, so Douglas was able to use the roll-out and first flight as publicity occasions, which were heavily covered in the popular press. World War II was raging in Europe, and although the US was not yet in the war, the phrase 'National Defence' was becoming popular. Even President Franklin D. Roosevelt wired company-founder Donald Douglas to congratulate him.

Although the XB-19 was a one-of-a-kind aircraft that was no longer state-of-the-art, it was portrayed as a symbol of American might because of its size. In that respect, it was, indeed, mighty. With a wing span of 212 ft (64.6 m) – half again longer than that of the

still-to-come Boeing B-29 – and a gross weight of 162,000 lb (73483 kg), the XB-19 was the largest aircraft in the world.

Like the XB-15, but unlike other bombers of the era, there were bunks where the crew could rest and a galley in which to prepare hot meals. The crew complement was officially 16, plus provision for a pair of flight mechanics and a six-man relief crew. This meant that the XB-19 could accommodate more people than a Douglas DC-3 airliner in its DST configuration with sleeper berths.

Flight testing took place during the summer and autumn of 1941, and the aircraft was provisionally accepted – pending further tests – in October. After the attack on Pearl Harbor in December, there was a fear that the West Coast of the United States would be attacked, so the USAAF decided to move the huge symbol of air power – which it had not yet formally accepted – inland to Wright Field, Ohio, for safe keeping.

The 'further testing' continued at Wright Field, and formal acceptance of the XB-19 came in June 1942, a year after its first flight. Formal acceptance meant a final payment: including progress payments going back to 1935, the USAAF reportedly paid a total of $US1.4 million ($US15.2 million in today's dollars) on the XB-19 programme up to the aircraft's acceptance. Douglas, however, is estimated to have spent more than two and a half times that sum on the project.

Having repainted the XB-19, like the XB-15, in olive drab and neutral grey, the USAAF might have considered using the big aircraft as a bomber. It had a maximum range of 7,710 miles (12408 km), which was much greater than any other bomber in the USAAF. During the early months of World War II, USAAF planners had anxiously been trying to devise ways of attacking Tokyo. In April 1942, they sent General Jimmy Doolittle to lead 16 North American B-25 medium bombers to do the job, launching from an aircraft-carrier. In contrast, the XB-19 had more than three times their range and could have made a round trip to Tokyo from the tip of the Aleutian Island chain or from northern

Specifications

	Boeing XB-15	Douglas XB-19	Douglas XB-19A
Powerplant:	4 x 1,000-hp (746-kW) P&W R-1830	4 x 2,000-hp (1492-kW) Wright R-3350-5	4 x 2,600-hp (1940-kW) Allison V-3420-11
Span:	149 ft (45.4 m)	212 ft (64.62 m)	212 ft (64.62 m)
Length:	87 ft 7 in (26.7 m)	132 ft 4 in (40.34 m)	132 ft 4 in (40.34 m)
Height:	18 ft 1 in (5.5 m)	42 ft (12.8 m)	42 ft (12.8 m)
Wing area:	2,780 sq ft (258.3 m²)	4,285 sq ft (398 m²)	4,285 sq ft (398 m2)
Empty weight:	37,709 lb (17105 kg)	86,000 lb (39010 kg)	92,400 lb (41913 kg)
Gross weight:	70,706 lb (32072 kg)	162,000 lb (72483 kg)	140,230 lb (63608 kg)
Cruising speed:	152 mph (245 km/h)	135 mph (217 km/h)	185 mph (298 km/h)
Maximum speed:	200 mph (322 km/h)	224 mph (360 km/h)	265 mph (426 km/h)
Service ceiling:	18,900 ft (5760 m)	23,000 ft (7010 m)	39,000 ft (11887 m)
Range:	5,130 miles (8255 km)	7,710 miles (12408 km)	4,200 miles (6759 km)

Above: Photographed during 1943, before conversion as the XC-105 transport, the XB-15 carries a camouflage scheme applied for stateside exercises in 1940.

Left: Sixth Air Force personnel admire their handiwork after applying an appropriate emblem, incorporating the Grandpappy nickname coined while the aircraft was still a bomber, to the XC-105.

China. It would be two years before the capital of Imperial Japan was hit again after Doolittle's raid, and then by aircraft that did not exist in April 1942.

Had it been used as a bomber, the XB-19 would have carried impressive armament on its long-range strike. The bomb bay could have accommodated eight 2,000-lb (907-kg) bombs, and 10 wing racks capable of carrying such ordnance were theoretically possible. It was designed to be fitted out with 10 gunner positions (manned by nine gunners and the bombardier) for operational missions, including two with 37-mm cannons.

However, the Materiel Division decided to use it strictly as a test aircraft. In 1943, to eliminate a nagging cooling problem with the Wright radials, the aircraft was refitted with four Allison V-3420-11 inline engines, each delivering 2,600 hp (1940 kW). Redesignated as

XB-19A, the aircraft continued in its test programme. Like an orphan being shuffled from foster home to foster home, the XB-19A was based temporarily at many of the archipelago of airfields and bases in the Dayton, Ohio, area that were satellite fields for Wright Field.

World War II – the war that was a looming threat on the horizon when the aircraft was designed – ended without the XB-19A venturing overseas to do battle. Unlike the XB-15, it survived the war, but a year after the war ended, the USAAF officially declared it 'surplus'. On 17 August 1946, it touched down at Davis-Monthan Field near Tucson, Arizona, having made its last flight. Coincidentally, just nine days

earlier, the Convair B-36 – the only American bomber ever built with a wing span greater than the XB-19 – had made its first flight.

The XB-19A languished at the Davis-Monthan 'boneyard' for three years before being unceremoniously cut up for scrap in 1949. The evolution of strategic air power had seen some dramatic changes in the 15 years since the advent of 'Project A'. The United States had gone from having a second-rate air force to being a nuclear power. In just a decade of that period, the US Army Air Corps, with nary a four-engined bomber, had evolved into the USAAF and taken delivery of more than 33,000 of them.

By 1949, 'Project A' and 'Project D' were just historic footnotes. The engineers who had wielded slide rules in the creation of these great aircraft had gone on to better – if not bigger – things. Only a few years after their inception, 'Project A' and 'Project D' had been superseded by events that could barely have been foreseen when they were conceived. Nevertheless, these aircraft had been the conceptual first steps in the process that led to the USAAF becoming the most powerful strategic air force in the world during World War II.

Bill Yenne

XB-19A: engine testbed

Its test work with Air Technical Service Command complete, the XB-19 was turned over to the Fisher Aircraft Division of General Motors, to be reworked as a flying testbed for a new liquid-cooled engine earmarked for a number of new types.

Redesignated XB-19A, the aircraft was equipped with four 2,600-hp (1939-kW) Allison V-3420-11 W-type, 24-cylinder turbo-supercharged engines (each comprising a pair of V-1710 12-cylinder engines mounted on a single crankcase, their crankshafts geared together to drive a single propeller shaft). New nacelles and engine mounts were required, the new installations having a noticeably 'nose high' thrust line compared to the XB-19's R-3350s. Each V-3420 drove a Curtiss Electric propeller with a diameter of 16 ft (4.88 m), though these were soon replaced with 18-ft 2-in (5.54-m) propellers to make better use of the engines' power. The inner propellers were reversible to aid braking on landing.

The modified aircraft served as a transport and undertook war bond drives around the US for the rest of the war. Among its final assignments was with the All Weather Flying Center, assisting in the development of systems and procedures to improve flight safety in hazardous weather conditions.

Left and below: Towards the end of the war the front upper turret was also removed and the aircraft stripped of its 'OD' paintwork. In these views the aircraft carries remains of markings applied while at the AWFC, including an insignia red tail with a yellow chevron marking.

Bottom: In its new guise as the XB-19A, 38-471 was initially finished in the overall olive drab/neutral grey in which it had operated since late 1941. Note that, by this time, the rear upper turret had been deleted.

US Spitfire operations

Part 2: The Mediterranean theatre

Both the 31st and 52nd Fighter Groups left the Eighth Air Force in England in late 1942, bound for North Africa and a new assignment with the Twelfth Air Force supporting the Allied invasion of North Africa. Still equipped with Spitfires, in the absence of suitable American aircraft, the two groups fought their way eastwards and across the Mediterranean to Corsica and the Italian mainland.

The invasion of North Africa, some 11 months after the Japanese attack on Pearl Harbor, was intended to wrest control of the African continent from German and Italian forces. American President Franklin D. Roosevelt and British Prime Minister Winston Churchill had quickly agreed to defeat Germany and Italy first, before turning to the Japanese. However, any clash of arms and troops between American and German ground forces would require massive and secret sea transportation to support a land invasion, and preparations for such a huge invasion took time.

Codenamed Operation Torch, the invasion began on 8 November 1942 and marked the first clash between American ground forces and German and Italian forces (although, initially, the Americans were to face resistance from Vichy French forces).

The two American Spitfire fighter groups operating in England – the 31st and 52nd – were among the major portion of US Army Air Forces units stripped from the Eighth Air Force and added to the newly-created Twelfth Air Force for the invasion. Lieutenant General Dwight D. Eisenhower, until then in charge of US forces in England, was given command of the North African invasion.

Above: The Spitfire Mk Vs of the 31st and 52nd Groups served along side such American types as the Curtiss P-40 and Bell P-39. It was not unusual for Allied aircraft in North Africa to be without unit code letters, though this practice appears to have been less common in the USAAF than in the RAF. ES264 'V' is a 52nd FG machine.

Top: A 308th FS Spitfire Mk VIII meets its replacement – a newly delivered olive drab P-51B Mustang – at Castel Volturno, March 1944.

Under reverse Lend-Lease, the Twelfth Air Force received 274 Spitfire Mk Vs for use by the 31st and 52nd Fighter Groups. This well known view shows aircraft assembled at North Front airfield on Gibraltar, awaiting issue to squadrons on the front line. Crude US markings have been applied over RAF roundels; though RAF fin flashes remain in place these were soon removed and remained a rarity on Twelfth Air Force aircraft.

When the air echelons of the two US Spitfire groups disembarked from transport ships at Gibraltar, they found the tiny airfield at North Front crowded with aircraft. The inexperienced American Spitfire groups were to fly into battle on equal terms, at least in terms of their equipment, with the combat-experienced RAF. Spitfire Mk VCs reserved for the two US groups were installed with Vokes desert filters, which not only changed the graceful shape of the Spitfire, but produced considerable drag. A Spitfire Mk VC was some 14 mph (22.5 km/h) slower than a Mk VB without the filter. Nevertheless, the Americans were eager to join the battle.

At Gibraltar, pilots were briefed by General James Doolittle's Twelfth Air Force staff for Operation Torch. German and Italian ground forces in Egypt and Libya were being pushed to the west by the British Eighth Army, and the eastward push by the Americans forced Axis forces into a closing vice in Tunisia.

Winter had begun by this time and rain often turned unprepared desert airfields and tent areas into muddy quagmires. The Spitfires' short range necessitated being near the front, and advances of American forces resulted in 15 to 20 base changes for the two groups in the 18 months the men operated Spitfires in the Mediterranean theatre.

Even with the arrival of Spitfire Mk VIIIs and IXs in 1943, tropicalised Spitfire Mk Vs remained in service with both the 31st and 52nd Fighter Groups until just before the Spitfires were replaced by Mustangs in 1944. ES306 of the 308th FS was photographed on the unprepared airstrip at Thelepte, Tunisia, in February/March 1943.

The initial task of the 31st and 52nd FGs on 8 November 1942 was to protect invasion forces landing at Meirs el Kabir, the port for Oran, Algeria. In addition, one squadron of the 52nd FG was ordered to escort General Doolittle's B-17 across the strait, to an airfield where he would set up his HQ. Tafaraoui and La Senia, major French airfields near Oran, had been liberated by US paratroopers by the time the two US fighter groups arrived.

On the day of the invasion, Royal Navy carriers sent their Hawker Sea Hurricanes to fight, and US carriers sent their Grumman Wildcats to gain air superiority over Oran's harbour. Unfortunately, total air superiority was not achieved and Vichy French fighters lay in wait. Runways at La Senia were reported to be holed by bombs or cannon shell fire, and Colonel John Hawkins, CO of the 31st FG, led his 307th, 308th and 309th Squadrons' aircraft to

New aircraft - Spitfire Mk IXs

Beginning in April 1943 the 31st FG (307th and 309th Fighter Sqns) began to receive its first Spitfire Mk IXs, introduced by the RAF in the Mediterranean the previous December to counter the threat posed by newly deployed Luftwaffe Fw 190s. The 52nd FG was issued with Mk IXs at about the same time, the new variant (powered by the new 'two-stage' Merlin 61 which offered much improved performance at altitude) serving alongside Spitfire Mk Vs in the 2nd, 4th and 5th Fighter Sqns. Pictured are EN354 *Doris June II* (part of an order for Spitfire Mk Vs converted to Mk IX standard by Rolls-Royce before delivery) of the 4th FS at La Sebala in June 1943 (above) and an unidentified aircraft of the 307th FS in Sicily. Early Mk IX deliveries were in RAF desert camouflage.

'HL-AA', a Spitfire Mk VC of the 308th FS, is pictured during a patrol over the Mediterranean in mid-1943. Interestingly, JK226 survived the war and was passed to the Greek air force in 1946.

Tafaraoui. Runways were pitted but Hawkins and some of his pilots landed. As the last Spitfires were landing, Vichy French artillery shelled the field and French Dewoitine D.520 fighters shot down one US aircraft, killing the pilot. Major Harrison Thyng, Lieutenant Carl Payne and First Lieutenant Charles C. Kenworthy, Jr, all from the 309th FS, each downed a Dewoitine D.520. Harrison Thyng became an ace, with five victories in Spitfires.

In the meantime Doolittle's B-17 transport had taken off ahead of its escort, and 12 Spitfires of the 2nd FS, led by Lieutenant Colonel Graham 'Windy' West, Deputy Commander of the 52nd FG, raced ahead to make the rendezvous. The pilots struggled through thick cloud to keep Doolittle's transport in sight. A few pilots of the 4th FS had navigation problems and landed in outlying areas, but soon returned to Tafaraoui. The 5th FS did not fly in combat that day as it was scheduled to arrive later.

The next day, the French shelled Tafaraoui. The Allies also encountered resistance at

The dusty, unprepared strip at Thelepte, Tunusia was typical of many in North Africa. The need for the cumbersome Vokes tropical filters to protect engines from excessive wear is immediately apparent.

Casablanca and Algiers, the two other Torch invasion points. When it was reported that an armoured column of French Foreign Legion forces was moving north from its base at nearby Sidi Bel Abbes towards Oran, pilots scrounged fuel abandoned by the French, used 5-US gal (19-litre) cans to load it into their Spitfires, and took off. The 31st FG and the two available squadrons of the 52nd FG flew several strafing missions and halted the column, driving it southward, then silenced French artillery around La Senia and Tafaraoui.

Ground personnel were landed on the day of the invasion but were forced to march a considerable distance inland to Tafaraoui. Men and equipment necessary to service and fuel aircraft were slow in coming. The ground echelon of the 52nd FG sailed from Scotland on the day of the invasion and was not ashore at Meirs el

Commanding Officers' aircraft – 31st Fighter Group

Below: Colonel Fred M. Dean (standing on the wing of his Spitfire Mk VC, coded 'F-MD') took command of the 31st FG one month after Opeation Torch. The group had advanced as far as Sicily when, on 15 July 1943, he was replaced by Lieutenant-Colonel Frank A. Hill (with right foot on wing).

Lieutenant-Colonel Frank A. Hill's tenure as CO of the 31st FG was comparatively brief, lasting until the conclusion of the Sicilian campaign in mid-August 1943. Here Hill is pictured with his personal Spitfire Mk IX, appropriately carrying his initials 'FA-H'. The use of initials in place of code letters was an RAF custom, accorded officers of Wing Commander rank or higher, and adopted by USAAF group COs.

Colonel Charles M. McCorkle
'Sandy' McCorkle replaced Frank Hill and oversaw the Group's transition from Spitfire to Mustang in March 1944. McCorkle finished the war with 11 aerial victories, all claimed while CO of the 31st FG – five while flying Spitfire Mk VIIIs and IXs and six on P-51B and D Mustangs. *Betty Jane* was a Spitfire Mk VIII, serial unknown.

31st Fighter Group – 'Return with Honor'

307th Fighter Squadron
Sharkmouth markings were not a common sight on Spitfire Mk Vs in any theatre; this aircraft is one of at least two examples in the 307th FS that were so-adorned. Its camouflage follows the standard desert RAF scheme, comprising dark earth and middle stone, with azure blue undersurfaces.

308th Fighter Squadron
Fargo Express was the well-known mount of Captain Leland P. Molland, CO of the 308th FS at Castel Volturno in early 1944. Molland scored 10½ kills, including 4½ while flying Spitfire Mk VIIIs. This aircraft, a late-production Mk VIII with an enlarged rudder, is often (incorrectly) depicted with the codes 'HL-X'. The standard RAF desert camouflage scheme is applied.

309th Fighter Squadron
From the end of 1943 replacement Spitfire Mk IXs were delivered in RAF temperate day fighter colours of the period, i.e. dark green/ocean grey camouflage over medium sea grey undersurfaces. Note also the red-bordered national insignia, adopted for two months during mid-1943. *Thurla Mae III* was flown by Lieutenant Robert Belmont.

Kabir until 11 November. The 2nd and 4th Squadrons of the 52nd FG were in combat right away, but the men of the 5th FS arrived by ship on 12 November.

The 31st FG transferred to the bombed French main air base at La Senia on 14 November followed by the 52nd FG three days later. Following negotiations between General Eisenhower's staff and leaders of the Vichy French in North Africa, resistance ended in Morocco and Algeria on 13 November, and by 18 November in Tunisia. Northern Tunisia, however, was soon to become an armed German camp: German forces were defeated at El Alamein in late October 1942 and Tobruk fell to British forces on 13 November, forcing the Germans to head for Tunisia. The Luftwaffe was far to the east in late November and it mounted only a few raids against American forces at Oran. Excepting losses and victories on the first day of the invasion, the 31st and 52nd Groups saw little air action and did not achieve another credited victory until 30 November. (The 57th FG had been operating in the Middle East since August, scoring victories. The 1st, 14th, 33rd and other US fighter groups took the lion's share of aerial victories from November.)

The large influx of aircraft meant any airfield used in North Africa was crowded by a great number and variety of Allied aircraft. The 31st

and 52nd Groups were never the sole occupants of airfields, and were often based apart from each other. Individual squadrons were frequently separated from their group.

The 31st FG began flying patrols and on 20 December the 307th FS moved east to Maison Blanche, Algeria, from where pilots flew transport-escort missions. Spitfires carried external 'slipper' tanks to improve range. The 309th FS remained at La Senia while the 308th FS flew to Casablanca in January 1943, where it was given the honour of flying cover for the (now famous) strategy conference between Roosevelt and Churchill and the top generals and admirals of both countries. Bad weather set in, lasting intermittently through the winter.

Severe air fighting lay farther to the east, and the 2nd FS was given the choicest assignment

of those offered to the two Spitfire groups: on 25 November it flew to a muddy field at Algiers and two days later it moved on to Bône in Algeria, just west of Tunisia, where it joined No. 81 Squadron of No. 322 Wing, RAF.

Allied harbour defence at Bône was a key aspect of maintaining the Americans' eastward push, and the Luftwaffe defended its own shore and harbour areas around Bizerte and Cape Bon in Tunisia. Bône and Cape Bon were not far apart, and pilots of the 2nd FS soon achieved success. Luftwaffe strength in North Africa was supplemented by air units based on nearby Sicily, just across the water from Bizerte. The 2nd FS engaged the Luftwaffe as it harassed defences at Bône, and the Luftwaffe fought when Allied air units raided Bizerte, finding itself in the thick of combat three weeks

This 307th FS Spitfire Mk V fell to 'friendly' AA fire near Salerno in September 1943. Misidentification was not confined to AA gunners; the caption to this USAAF photograph describes the aircraft as a "Curtiss P-40". In the background vehicles of the 817th Engineer Aviation Battalion roll off an LST and onto Italian soil.

Left: 4th FS Spitfire Mk Vs and IXs, some with RAF fin flashes, await their next assignment at an unknown airfield in North Africa. Ground crews shelter under their wings from the baking sun.

Below: This 4th FS Mk IX was finished in an unusual mottled light brown blended into a light sky blue on the underside of the aircraft. Pictured at La Sebala in Tunisia, the aircraft is on alert with its pilot, 1st Lt James W. Puffer, strapped in and a 'trolley acc' plugged in. Puffer's crew chief relaxes under the port wing. The dust filter fitted to the aircraft's carburettor intake appears to be an in-the-field modification.

after arrival. The 4th FS soon followed the 2nd FS to the east, arriving at Orleansville, Algeria, on 2 January 1943, preparing to relieve the 2nd FS at Bône.

The first victories in North Africa for the 52nd FG came when Captain James S. Coward and Captain Harold R. Warren, Jr, both of the 2nd FS, 52nd FG, downed Bf 109s on 30 November 1942. Captain Arnold E. Vinson, 2nd FS, got 2⅓ victories in a row on 2, 3 and 4 December. He became a 5⅓-victory ace before being killed in aerial combat in April 1943. Captain Norman McDonald of the 2nd FS damaged three Bf 109s in December then scored kills through the first eight months of 1943, ending with a total of 11½, four of which were achieved in the P-51D when he flew with the 325th FG. Second Lieutenant John Der Ludlow and Captain Luis T. Zendegui of the 2nd FS each downed an enemy aircraft on Christmas Day. Captains James E. Peck and Arnold Vinson, and First Lieutenant John F. Pope, of the 2nd FS each gained victories on 2 January 1943. The 2nd FS had quickly demonstrated its abilities when tasked with aerial combat.

When the 4th FS replaced the 2nd FS at Bône, Major Robert Levine of the 4th FS, later to command the 52nd FG, achieved his first of three victories on 8 January 1943.

31st Group under RAF command

Bad weather did not prevent the 31st FG from moving to Thelepte, Tunisia, in early February 1943 via Youks-les-Bains, Algeria, to be near the front line of the ground war. The 31st FG came under the command of an RAF Air Vice Marshal commanding the Northwest African Tactical Air Force (NWATAF). Thelepte's

proximity to the ground war was soon brought home very forcefully to the Americans. On 10 February 1943, Spitfires escorted P-39s and A-20s attacking Maknassy, Tunisia, well south of Cape Bon. While based at Thelepte, First Lieutenant J.D. 'Jerry' Collinsworth of the 307th FS was in a flight of 12 Spitfires escorting P-39s on a strafing mission when he downed an Fw 190 near Sidi-bou-Zid on 15 February; it was his first of six victories, all in Spitfires. Kenworthy and Thyng, veterans of the invasion day, and First Lieutenant Alvin D. Callender also scored single victories on 15 February 1943. The 31st FG had once again demonstrated its skills.

On 16 February 1943, an advancing German ground offensive came near enough to

Thelepte that the men heard cannon fire. At 02.00 on 17 February, Colonel Fred M. Dean, CO of the 31st FG, was ordered by phone to evacuate Thelepte. Trucks were loaded with men, equipment and necessary papers, and pilots flew serviceable aircraft to Youks-les-Bains before daylight. Six unserviceable Spitfires had to be destroyed. Some personnel and aircraft were hastily taken to Tebessa, Algeria, just west of the border with Tunisia. The German offensive rushed onward through the Kasserine Pass on 18 February, causing a severe setback for American ground forces. Four days after their arrival at Tebessa, squadrons split up and flew to Du Kouif or Youks as the Germans flooded toward Tebessa. Counter-attacks drove the Germans back.

Below: Its propeller smashed and wings damaged after a wheels-up landing, Spitfire Mk VB ER120 of the 5th FS receives attention from mechanics at a busy depot "somewhere in North Africa" during February 1943. A set of P-40 Warhawk wings and the hulk of a B-26 Marauder are visible in the background. This early production Mk V still carries an Operation Torch flag marking and was probably among the original Spitfires supplied to the USAAF in North Africa.

Above: Some major overhaul work on the American Spitfires was carried out by the RAF. In this view RAF maintenance unit personnel run up a recently repaired 4th FS Spitfire Mk VC at an aerodrome in North Africa. To the right is an RAF Hurricane.

Right: The 52nd FG reached the end of a long stay in Tunisia on 30 July 1943, when it moved to Corsica. Spectacular snow-covered mountains provide the backdrop to this view of a 4th FS Spitfire Mk IX at an airstrip on the island.

Below: From the time it left Sicily for Corsica the 52nd FG had dive-bombing as its primary role. Bomb rack-equipped Spitfire Mk Vs carried the bombs, while Mk IXs provided top cover. 5th FS aircraft are pictured.

On 22 February Captain Frank A. Hill of the 308th FS downed a Ju 87 with his Spitfire Mk V near Du Kouif. Hill's first victory had been over Dieppe on 19 August 1942 and his eventual total in Spitfires was seven, making him the top Spitfire ace of the 31st FG. Colonel Hawkins was reassigned in early 1943 and Colonel Dean, the next group CO, was rotated stateside in July 1943, handing command of the 31st FG to Hill. The 31st FG moved to Kalaa Djerda on 24 February, with the exception of one squadron based at Du Kouif, and then the group returned to Thelepte. Three pilots of the 4th FS downed aircraft on 24 February, including Captain

William M. Houston. Lieutenant Colonel West took command of the 52nd FG on 1 March from Colonel Dixon Allison. In early March the 52nd FG moved to Telergma, Algeria, where General Doolittle awarded decorations, and then the group moved to Thelepte later in March.

On 21 March, 31st FG pilots downed 4½ aircraft and First Lieutenant Maurice K. Langberg of the 307th FS was credited with two. First Lieutenant Carl W. Payne of the 309th FS was awarded half a kill of a Ju 87. Payne later became an ace in Spitfires. On 23 March, Captain Theodore Sweetland and Luftwaffe Major Joachin Muencheberg collided head-on

in combat near Es Sened, and both were killed. Muencheberg was known in Germany as an *expert*, with 134 confirmed victories, and Sweetland was officially credited with a single aerial victory.

On 1 April one of the few staff sergeant fighter pilots in the 52nd FG – SSgt James Edward Butler of the 2nd FS – downed an Fw 190 on 1 April. Officially credited with a total of four victories, Butler has been credited with five in some sources.

Top-scoring US Spitfire ace

The man who was to be the top US Spitfire ace, 1st Lt Sylvan Feld of the 4th FS, got his second kill, a Ju 88, on the same day as Butler's victory, and eventually Feld was credited with nine victories, all in Spitfires between 22 March and 6 June 1943. On 6 April 1943, Second Lieutenant Virgil Fields of the 307th FS downed a Bf 109. It was his first victory. Before he was shot down and killed on 6 February 1944, Fields (by then a Major) had been credited with six victories, all in Spitfires.

The 2nd FS had a big day on 3 April 1943, downing 12 enemy aircraft, mostly Ju 87s. Staff Sergeant Butler, Captain George V. Williams, 1st

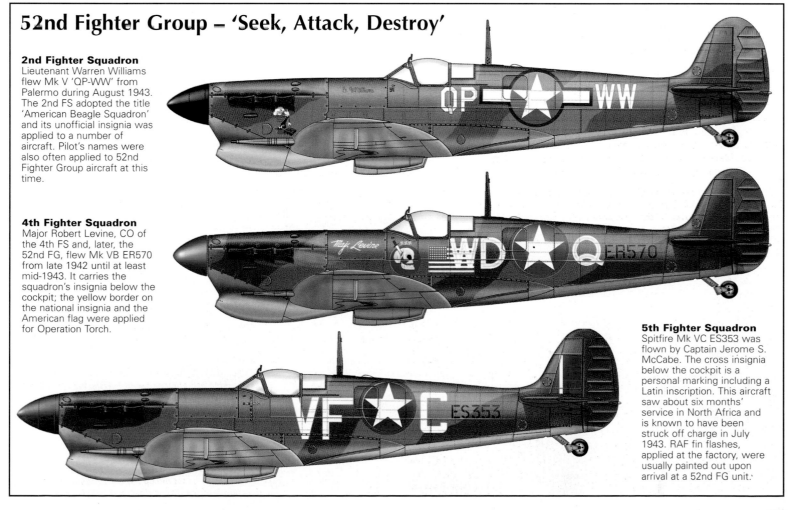

52nd Fighter Group – 'Seek, Attack, Destroy'

2nd Fighter Squadron
Lieutenant Warren Williams flew Mk V 'QP-WW' from Palermo during August 1943. The 2nd FS adopted the title 'American Beagle Squadron' and its unofficial insignia was applied to a number of aircraft. Pilot's names were also often applied to 52nd Fighter Group aircraft at this time.

4th Fighter Squadron
Major Robert Levine, CO of the 4th FS and, later, the 52nd FG, flew Mk VB ER570 from late 1942 until at least mid-1943. It carries the squadron's insignia below the cockpit; the yellow border on the national insignia and the American flag were applied for Operation Torch.

5th Fighter Squadron
Spitfire Mk VC ES353 was flown by Captain Jerome S. McCabe. The cross insignia below the cockpit is a personal marking including a Latin inscription. This aircraft saw about six months' service in North Africa and is known to have been struck off charge in July 1943. RAF fin flashes, applied at the factory, were usually painted out upon arrival at a 52nd FG unit.

Lt Jerome C. Simpson and 1st Lt John F. Pope each got two. Captain McDonald downed three Ju 87s. On 9 April Butler downed two aircraft and 1st Lt Fred Ohr got his first, a Ju 88, north of Kairouan. Ohr's next kill came on 31 May 1944 in a P-51B, and later he became an ace with a total of six victories. Ohr, whose original name was Oh, is the only Korean-American ace. He later was CO of the 2nd FS. Also on 9 April, 1st Lt Victor Cabas of the 4th FS downed 1½ Ju 88s northeast of Kairouan. Cabas was later an ace with five victories, including his first, an Fw 190 over Dieppe when he was flying with No. 403 Sqn, RCAF.

1st Lieutenant William Skinner inspects the damage caused to his 308th FS Spitfire Mk V by an 88-mm flak shell, Monte Corvino, 1 October 1943. By this time the 308th was re-equipping with Mk VIIIs

The 52nd FG moved to Sbeitla, Tunisia, on 6 April and then to La Sers, Tunisia, six days afterward. The war in North Africa was nearing its end, as Allied air superiority prevented the Luftwaffe from resupplying ground forces. The 31st FG moved closer to the front lines, going to Djilma at about the same time that the 52nd Group moved to La Sers.

The fight for Tunisia

German forces were bottled up in Tunisia, but Hitler wanted Tunisia held, and he ordered fleets of poorly-escorted, slow-flying Ju 52/3m and gigantic six-engined Me 323 transports to fly from Sicily to Tunisia at low altitude in clear weather to maintain his forces in North Africa. Those transport aircraft were sitting ducks to fast Allied fighters. Allied high command knew ahead of time of the massive effort and on 5 April 1943 it sent its fighter groups on a 'killing spree'. Under Operation Flax RAF, SAAF and AAF fighter pilots inflicted heavy losses on the Luftwaffe.

Flax lasted for a fortnight from 5 April. The interception and massive destruction of Luftwaffe aircraft attempting to supply and/or evacuate German forces in Tunisia resulted in 48 German aircraft being downed on 5 April,

USAAF Spitfire aces of the Mediterranean theatre

Nine 31st FG pilots and five from the 52nd FG 'made ace' in the MTO on Spitfires, while a number of pilots who later claimed five or more victories, on types such as the Mustang, scored at least one kill while flying Spitfires.

Above: Pictured at Termini on Sicily in August 1943, Lieutenant-Colonel Frank Hill, CO 31st FG (right), chats with Wing Commander Brian Kingcome in front of the latter's Spitfire Mk VIII. Hill had joined the 31st FG shortly before it left the US for England in 1942 and finished the war with seven confirmed kills, all achieved while flying Spitfires. Battle of Britain ace Kingcome, who led No. 244 Wing, RAF, during the invasion of Sicily, was no stranger to the 31st; he had led the 308th FS on its first combat missions over Europe in 1942.

Above: The honour of top-scoring USAAF Spitfire pilot went to Sylvan Feld, who scored nine kills as a 1st Lieutenant with the 4th FS between late March and early June 1943, flying Spitfire Mk Vs and IXs. At the end of his tour Feld transferred to the 373rd FG, based in England and equipped with P-47Ds, his second tour commencing in September 1943. On 13 August 1944 he survived being shot down by flak over France, only to be killed the following week in an Allied bombing raid.

Left: Pictured in RAF flying gear in the cockpit of his 307th FS Spitfire Mk V at RAF Biggin Hill, Lieutenant J. D. 'Jerry' Collinsworth scored six kills on Spitfire Mk Vs and IXs, all in the Mediterranean during 1943. Wounded in Sicily, Collinsworth returned to the US in September 1943.

In mid-October the 308th FS moved to Pomigliano, a former Regia Aeronautica facility with a concrete runway and permanent buildings. The airfield was frequently strafed by the Luftwaffe; here 1st Lieutenant Harold 'Dutchy' Holland (right) chats with his crew chief while on alert with his Spitfire Mk VIII.

the same number on 10 April, and 31 on 11 April. Some 76 enemy aircraft were shot down on 18 April, in what has been called 'The Palm Sunday Massacre'; in all 400 transports were downed, plus a large number of their German and Italian escorts.

However, despite the huge numbers of Allied aircraft involved, the USAAF Spitfire groups played little part in the operation. By this time Spitfire Mk VIIIs and IXs were beginning to supplement Mk VCs in the two US groups and there has been considerable speculation as to why these aircraft were not called into battle during Flax.

It is known that the USAAF was unhappy that it had been forced to equip the 31st and 52nd FGs with foreign aircraft to begin with, even if they were Spitfires. Though a proven fighting machine, the Spitfire was not ideal for some of the roles it was asked to perform, thanks largely to its lack of range. It was recognised that, as soon as suitable American aircraft were available, both groups would be re-equipped.

Thus, while other US fighter groups in the Mediterranean, equipped with US-built aircraft, were asked to perform a full range of tasks, including an air combat role, the 31st and 52nd FGs seemed to be viewed as of secondary importance as an air combat force, even though they had often been first into battle with the invasion forces and, by the time of Operation Flax, were partially equipped with new Spitfire Mk IXs.

North Africa finally fell to the Allies on 13 May 1943; on 23 May the 52nd FG was transferred to the RAF's Mediterranean Allied Coastal Air Force (MACAF). On that day, the 52nd FG moved to La Sebala near the Cape Bon peninsula in Tunisia, where the group remained until 30 July, past the time when Pantelleria was taken and long after the invasion of Sicily. The

Above: Well known as a Korean War ace, Harrison Thyng 'cut his teeth' on the Spitfire, joining the 309th FS as its CO prior to the unit's move to England with the 31st FG. His first confirmed kill (a Vichy French Dewoitine D.520) came during the Torch landings and four more followed over the next six months. Spitfire Mk V Mary and James *was his personal aircraft.*

Above: Major Fred Ohr scored five of his six kills on Mustangs, after claiming an all-important first victory while flying a Spitfire. The only American ace of Korean ancestry, Ohr (original surname 'Oh') scored his first kill – a Junkers Ju 88 – on 9 April 1943 and went on to command the 2nd FS during 1944. His second kill came over a year after the first; by then the 52nd FG had converted to Mustangs.

Left: CO of the 307th FS from October 1943 until his death the following February, Major Virgil Fields had joined the 31st FG from the 388th FG earlier in the year. His first kill (a Bf 109) came on 6 April 1943 and was followed by five more, the last just eight days before he fell to enemy aircraft over Anzio on 6 February 1944.

Right: Little is known of the USAAF career of 1st Lieutenant Richard F. Hurd, other than that he scored six victories flying Spitfire Mk VIIIs with the 308th FS during 1943/44. Here he is pictured with his aircraft at Castel Volturno, Italy during January 1944, at which point he had just a single kill to his name. Five more followed between 20 February and 21 March.

Spitfire HF.Mk VIIIs of the 308th Fighter Squadron

While the other squadrons in the group received Spitfire Mk IXs, the 308th was re-equipped entirely with Spitfire Mk VIIIs, beginning in August 1943. Among those delivered to the 31st Fighter Group were a number of HF.Mk VIIIs, equipped with Merlin 70 engines and extended wingtips for high-altitude operation. This example is pictured at Pomigliano, Italy, in October 1943.

52nd FG remained far behind the front lines. US forces worked closely with the British, and the 52nd FG was subordinated to the RAF. It ceased to be a fighter group in all but name, flying reconnaissance and convoy protection patrols and harbour patrol for almost one year. US Spitfire pilots grew discouraged at being kept from the aerial front in the war. The 52nd FG was not employed again as an air superiority fighter group until it was transferred in early 1944 to the newly-created US 15th Air Force after receiving P-51 Mustangs.

Pantelleria

The brief campaign to take the isolated and poorly defended island of Pantelleria began on 14 May and ended on 11 June. On 9 June, the 308th FS was credited with eight victories. Captain Thomas B. Fleming and First Lieutenant Walter J. Overend downed two aircraft each. On the next day, the 31st FG got 12. First Lieutenant Robert O. Rahn of the 309th

FS downed one. Rahn became a well-known test pilot with the Douglas Aircraft Company after the war. First Lieutenant Dale E. Shafer, Jr, of the same squadron downed an MC.202 over Pantelleria. Shafer's war total was seven, four in Spitfires and three in Mustangs, including a twin-jet Arado Ar 234B in April 1945 when he was with the 503rd FS. First Lieutenant John White also scored two on 10 June, an Fw 190 and a Bf 109 over Pantelleria. 31st FG pilots did well on 11 June over or near Pantelleria. First Lieutenant Charles R. Fischette of the 307th FS shot down two Bf 109s 10 miles (16 km) north-west of the island, making him an ace in Spitfires.

Disaster struck the 52nd FG at La Sebala on 22 June. A small grass fire on the airstrip spread to one of a number of abandoned German aircraft and turned into a major blaze. Alerted to the conflagration, Colonel West, First Lieutenant Howard H. Brians and Sergeant Bernard Schreiber ran to attempt to put the fire

out. However, the flames set off a heavy-calibre shell in the wrecked Luftwaffe aircraft, killing Brians and Schreiber, and injuring West's legs so badly they had to be amputated. Colonel James S. Coward became Group CO.

On 30 June the 31st FG was moved to the island of Gozo, to the north of Malta and only 60 miles (97 km) from Sicily, which put that group at the forefront of air action as the day of the invasion of Sicily approached – 10 July.

On 11 July, pilots of all three squadrons downed seven aircraft. Captain Carl Payne of the 309th FS damaged a Do 217 with a Spitfire Mk VB on one sortie and downed an Fw 190 with a Spitfire Mk VIII later that same day. On 8 August, Second Lieutenant Richard F. Hurd of the 308th FS shot down his first aircraft, an Fw 190. He scored all of his six victories in a Spitfire Mk VIII.

Sicily falls

On 30 July the 52nd FG's long stay at La Sebala ended when the 2nd FS flew into Bocca di Falco, near Palermo, Sicily. Ground crew and the other squadrons followed. The group was needed for dive-bombing missions in Italy.

At about the same time the 31st FG flew into Ponte Olivo near Gela, Sicily, and by 2 August was at Termini, a few miles from the Italian coastline. It was here, three weeks later, that the group witnessed the arrival of an Italian delegation that signed the armistice at Termini on 17 August.

On 31 August Italy signed an armistice with the Allies. Although some Italian forces now joined the Allies in the war against Germany, German forces remained in the north of the country and on the islands of Corsica and Sardinia. The latter was abandoned and Corsica reinforced, but the Allies launched a ground campaign and, by 4 October, Corsica was in Allied hands. It became an Allied staging area and forward island air base.

In the meantime the 31st FG had moved forward to Milazzo, Sicily, to continue combat operations over Italy. Colonel Charles McCorkle had by now taken command of the 31st and on 30 September, in a Spitfire Mk IX, downed an Me 210 – the first of five victories in Spitfires.

In November, Colonel Marvin McNickle took command of the 52nd FG, and the 2nd and 5th Squadrons flew into Borgo, Corsica, and the 4th FS moved to Calvi, Corsica. Ground personnel moved by sea to Corsica on 1 December.

The first of many Spitfire dive-bombing missions was flown by the 52nd FG on 28 December at San Vincenzo, Italy, led by Major Robert Levine. These continued to be part of the 52nd's repertoire until the group received Mustangs. On 23 January 1944, one such mission encountered Luftwaffe twin-engined bombers heading for Allied ships; the Spitfire pilots jettisoned their bombs and waded into the enemy. Second Lieutenant Mike Encinias and First Lieutenant James W. Bickford of the 2nd FS downed two aircraft each, and two other pilots got one apiece. Lieutenant Colonel Robert Levine replaced McNickle as Group CO at the end of February 1944.

Air actions over Italy began before the invasion of Salerno on 9 September 1943. On 11 August, pilots of the 307th FS scored well. Captain Royal N. Baker, 308th FS, downed an Fw 190 with his Spitfire Mk VIII. Baker had three victories in Spitfires, a half kill in a P-47

Above: Mk VIIIs of the 308th FS, equipped with 45-Imp gal (205-litre) slipper tanks, taxi prior to take-off from Castel Volturno in early 1944. Behind the aircraft coded 'HL-G' is a machine with an RAF roundel and no unit codes – presumably a newly delivered replacement aircraft.

Right: Lady Ellen III was Spitfire Mk IX MH894 of the 309th FS, finished in RAF temperate day fighter colours. It later served with No. 326 (Free French) Sqn, RAF.

10 March 1944 saw the first P-51B Mustangs for the 31st FG ferried to Castel Volturno, Italy, from Algiers; on 1 April the Group joined the 15th Air Force. These views show aircraft from each of the Group's squadrons at the time of the Mustangs' arrival. The 309th FS aircraft (above) include newer Mk IXs delivered in late 1943 in RAF temperate camouflage, though the squadron commander's aircraft ('WZ-VV', with the red wing stripe, nearest the camera) is an earlier machine in a faded desert scheme. Mk VIIIs (as flown by the 308th FS, right) continued to be delivered in RAF desert camouflage until the end; this view shows a newly arrived P-51B being run-up in the background. The 307th FS (below) also possessed a mixture of both 'temperate' and 'desert' Mk IXs.

and, later, 13 kills in F-86s in the Korean War. (He went on to fly 140 missions in Vietnam during 1968/69, retiring as a Lieutenant General in 1975.)

No US Spitfire pilot scored for an entire month after Baker's victory, and then single victories were recorded on various days. Not until 7 December 1943 was there a flurry of kills on a single day, when five pilots of the 309th FS were credited with six aircraft. Among them was First Lieutenant James O. Tyler of the 4th FS, who claimed an Me 210 in his Spitfire Mk IX. Tyler went on to score six in P-51Bs and Cs, for a final total of eight.

On 20 January 1944, the 31st FG downed four. Second Lieutenant Leland Molland, 308th FS, got his first, a Bf 109, in his Spitfire VIII. Molland eventually had four kills in Spitfires and six in Mustangs. The next day, Molland and three others in his squadron scored one each.

The landings at Anzio took place on 22 January 1944. On 28 January, Second Lieutenant Frank J. Haberle, 307th FS, got three, and others did well. Molland downed two on 22 February and others were credited. On 18 March, pilots of the 31st FG scored six. Second Lieutenant N. H. Youngblood Ricks of the 308th FS downed one.

Mustangs

In early March 1944, men of the 31st FG were told they would be re-equipping with Mustangs. On 10 March, Mustangs were ferried from Algiers to Castel Volturno, Italy, where the group was, by then, based. The group then

moved to San Severo. The final mass flight of 31st FG Spitfires took place on 29 March. From 1 April, the 31st FG was transferred from the XII Air Support Command to the US Fifteenth Air Force. On 16 April 1944, 52 P-51Bs and Cs escorted B-17s and B-24s on a 400-mile trip (644-km) to Romania and back. No Spitfire mission had gone as far.

The 52nd FG was transferred from MACAF to the 15th AF. At Aghione, Corsica, where it had received and trained on Mustangs, it flew its first Mustang bomber escort mission on 10 May 1944. The group moved to Madna, north of Foggia, Italy, on 13 May 1944.

The importance and success of the 52nd FG changed almost overnight when it received Mustangs and became part of the strategic 15th AF, flying bomber escort missions over Germany. In two months – May and June 1944 – the group shot down more aircraft than it had while flying Spitfires for 20 months. It set a

record in June 1944 for the MTO, being credited with 102 kills, and earned two Unit Citations for extremely successful escort missions on 9 June and 31 August 1944. Spitfire credits were 164⅓, and for Mustangs, 257, for a total of 421⅓.

The 31st FG was officially credited with 570½ victories, of which approximately 192 were scored in Spitfires.

Thus the USAAF's association with the Spitfire in a combat role came to an end. For whatever reason the exploits and successes of the 31st and 52nd Fighter Groups over those 18 months from late 1942 received scant coverage. Most histories of the US Air Force compiled since virtually ignore the use of Spitfires in World War II, even though these Groups, and others based in the UK, operated Spitfires exclusively for most of the war and made a major contribution in the theatres in which they served.

Paul Ludwig

The 31st FG kept a number of their beloved Spitfires as 'hacks'. Clipped-wing, late production Spitfire Mk VIII JF470 was retained by the 308th FS well into 1945. It was finished in an unusual overall grey with azure blue undersurfaces.

Picture acknowledgments

Front cover: Lockheed Martin, Saab, Aleksey Mikheyev. **4:** Lockheed Martin, Eurofighter (two). **5:** Yaso Niwa, Timm Ziegenthaler. **6:** NASA/AFFTC, Chris Ryan. **7:** AFFTC (two), via Shlomo Aloni. **8:** Alexander Mladenov, US Navy. **9:** US Navy, David Donald, Daniel J. March. **10:** US Navy, 81 Wing via Nigel Pittaway. **11:** Chris Knott/API (two). **12:** Dick Lohuis, USAF, US Navy. **13:** USAF (three). **14:** Shlomo Aloni, Antoine Roels, Dirk Lamarque. **15:** Timm Ziegenthaler, USAF (two). **16:** Richard Collens (two), via Richard Collens. **17:** Richard Collens, Bob Fischer. **18:** Bob Fischer (three), Rudolf/Bogotá via Bob Fischer. **19:** David Donald (two). **20-21:** Chris Knott and Tim Spearman/API. **22-23:** Aleksey Mikheyev. **24:** Kamov (three), Aleksey Mikheyev. **25-30:** Aleksey Mikheyev. **31:** Aleksey Mikheyev (two), Sergey Sergeyev, Peter R. March. **32-34:** Aleksey Mikheyev. **35:** Aleksey Mikheyev (three), Phazotron. **36-37:** Aleksey Mikheyev. **38-41:** Peter Steinemann. **42:** Peter Steinemann (four), Daniel J. March. **43:** Peter Steinemann (three). **44-45:** AFFTC via Terry Panopalis (TP). **46:** Lockheed Martin via TP, AFFTC via TP. **47:** AFFTC via TP, Lockheed Martin via TP. **48:** AFFTC (two), Lockheed Martin via TP. **49:** Lockheed Martin, Ted Carlson/Fotodynamics. **50:** AFFTC, Lockheed Martin (two). **51:** AFFTC (two), Lockheed Martin. **52:** Lockheed Martin (two), AFFTC. **53:** Lockheed Martin via TP, Boeing (four). **54:** Pratt & Whitney (five). **55:** Peter R. March, AFFTC via TP, AFFTC. **56:** AFFTC via TP, AFFTC, Lockheed Martin (two). **57:** Lockheed Martin via TP, AFFTC. **58:** AFFTC via TP, Lockheed Martin via TP, Lockheed Martin. **59:** AFFTC, Lockheed Martin. **60:** Lockheed Martin (three), AFFTC via TP. **61:** Lockheed Martin via TP, AFFTC. **62:** AFFTC via TP, Lockheed Martin, Ted Carlson/ Fotodynamics. **63:** AFFTC via TP, Lockheed Martin via TP. **64:** AFFTC via TP, Boeing, AFFTC, Lockheed Martin via TP. **65:** AFFTC via TP, Lockheed Martin, Lockheed Martin via TP. **66:** AFFTC (three). **67:** Pratt & Whitney. **68:** Lockheed Martin. **71:** Lockheed Martin via TP. **72:** Lockheed Martin via TP, Boeing (two), Lockheed Martin. **73:** AFFTC via TP, Ted Carlson/Fotodynamics, USAF. **74-79:** Chris Knott and Tim Spearman/API. **81-91:** Simon Watson/Wingman Aviation. **94:** USAF via Larry Davis (LD). **95:** USAF via LD, Paul Swendrowski via LD, William Fairbrother via LD. **96:** USAF Museum via LD (three), USAF via LD (three). **97:** Fred Hoffman via LD, Norm Taylor via LD, William Schwehm via LD. **98:** USAF via LD (three), Mike Campbell via LD, USAF Museum via LD, Mick Roth via LD (two). **99:** Don Jay via LD, E-Systems via LD, USAF via LD (two). **100:** USAF via LD (three), via LD (three). **101:** USAF via LD, Kurt Minert via LD, Don Jay via LD. **102:** Norm Taylor via LD, USAF via LD, William Fairbrother via LD (two). **103:** USAF via LD (three), via LD. **104:** USAF via LD (two), William Fairbrother via LD. **105:** via LD, Frank Eaton via LD (six). **106:** USAF via LD (two), William Fairbrother via LD. **107:** USAF via LD, Mike Campbell via LD, Frank Eaton via LD, via LD (two). **108:** Dornier, Aerospace. **109:** Bundesarchiv via Dr Alfred Price. **110-111:** Bundesarchiv via Dr Alfred Price (four). **112:** via Dr Alfred Price (four). **114:** Aerospace, Dornier, via Dr Alfred Price. **115:** Dornier, via Dr Alfred Price (three). **116:** Aerospace, via Dr Alfred Price (three). **117:** via Dr Alfred Price (two), Aerospace, Dornier, Bundesarchiv via Dr Alfred Price. **118:** Dornier (three), via Dr Alfred Price (two), USAF via Dr Alfred Price. **119:** Aerospace, Dornier four), via Dr Alfred Price. **120:** via Dr Alfred Price, Dornier, Aerospace (two). **121:** Dornier, IWM via Dr Alfred Price. **122:** Dornier (two), Aerospace. **124-125:** Saab. **126:** Peter Liander. **127:** Torstein Landström via Bo Widfeldt (BW) (two), Saab via BW (two). **128:** Saab via BW (two), Bo Dahlin via BW. **129-130:** Saab via BW. **131:** Saab via BW (two), Bo Bjernekull via BW. **132:** Saab via BW (three). **133:** Saab via BW (three), Air Historic Research via BW. **134:** Hans Bladh via BW (two), Flygvapnet via BW, Saab via BW. **135:** Hans Bladh via BW, Sven Stridsberg via BW, Air Historic Research via BW. **136:** Anders Nylén, Saab via BW, Sven Stridsberg via BW. **137:** Ingemar Thuresson via BW, Ericsson via BW, Peter Liander. **138:** Volvo Flygmotor via BW (two), Saab via BW. **139:** Ericsson via BW (two), Flygvapnet via BW, Saab via BW. **143:** Saab, Aerospace (two), Saab via BW (two). **144:** Saab via BW (three), Peter Liander. **145:** Saab via BW, Ingemar Thuresson via BW (two), Lars Soldéns via BW. **146:** via BW (all). **147:** Peter Liander, Jan Jørgensen. **148:** Peter Liander, Saab via BW, Air Historic Research via BW. **150-151:** Jyrki Laukkanen via BW. **152:** Saab via BW (three). **153:** Flyvevåbnet via BW, Saab via BW. **154:** Flyvevåbnet via BW (four). **155:** Flyvevåbnet via BW (three), Saab via BW. **156:** Anders Nylén, Bundesministerium für Landesverteidigung via BW (two), **157:** Sölve Fasth via BW, Lars Soldéns via BW, Anders Nylén. **158:** Luigino Caliaro, Jan Jørgensen (two), Anders Nylén. **159:** Werner Münzenmaier, Anders Nylén. **160:** Anders Nylén, Van-Son Hayn/F10 via BW. **161:** Saab via BW, Lars Lundin via BW (two), Saab, Anders Nylén, 2./F21 via BW. **162:** Jan Jørgensen (five), Peter R. March. **163:** Jan Jørgensen (two), Stefan Jonger, Jim Winchester, Peter R. March (two). **164:** Douglas, USAF. **165-168:** via Bill Yenne. **169:** via Bill Yenne, Douglas. **170:** via Bill Yenne. **171:** via Bill Yenne (four), USAF. **172:** William Skinner via Paul Ludwig (PL), via PL, via Dr Alfred Price. **173:** Frank Hill via PL, via PL, William Skinner via PL. **174:** Hagins via PL, IWM via PL, Frank Hill via PL (two). **175:** USAAF via Dr Alfred Price. **176:** via PL (two), USAAF via Dr Alfred Price, IWM via PL, via Dr Alfred Price. **177:** via PL, Frank Sherman via PL. **178:** William Skinner via PL, Frank Hill via PL, via PL (two). **179:** William Skinner via PL (two), via PL, John Fawcett via PL, USAAF via Dr Alfred Price. **180:** William Skinner via PL, John Fawcett via PL (two). **181:** John Fawcett via PL, Haings via PL, William Skinner via PL, Arthur Bleiler via PL.